CRYSTALS
AND CRYSTAL STRUCTURES

CRYSTALS
AND CRYSTAL STRUCTURES

Second Edition

RICHARD J. D. TILLEY

Emeritus Professor,
Cardiff University, UK

WILEY

This edition first published 2020

"John Wiley & Sons Inc. (1e, 2006)".

Registered Offices
John Wiley & Sons, Inc., 111 River Street, Hoboken, NJ 07030, USA
John Wiley & Sons Ltd, The Atrium, Southern Gate, Chichester, West Sussex, PO19 8SQ, UK

Editorial Office
John Wiley & Sons Ltd, The Atrium, Southern Gate, Chichester, West Sussex, PO19 8SQ, UK

For details of our global editorial offices, customer services, and more information about Wiley products visit us at www.wiley.com.

Wiley also publishes its books in a variety of electronic formats and by print-on-demand. Some content that appears in standard print versions of this book may not be available in other formats.

Library of Congress Cataloging-in-Publication Data
Names: Tilley, R. J. D., author.
Title: Crystals and crystal structures / Richard J.D. Tilley, Emeritus Professor, University of Cardiff.
Description: Second edition. | Hoboken, NJ : Wiley, 2020. | Includes bibliographical references and index.
Identifiers: LCCN 2020000435 (print) | LCCN 2020000436 (ebook) | ISBN 9781119548386 (paperback) | ISBN 9781119548614 (adobe pdf) | ISBN 9781119548591 (epub)
Subjects: LCSH: Crystals. | Crystallography.
Classification: LCC QD905.2 .T56 2020 (print) | LCC QD905.2 (ebook) | DDC 548–dc23
LC record available at https://lccn.loc.gov/2020000435
LC ebook record available at https://lccn.loc.gov/2020000436

Cover Design: Wiley
Cover Image: Hexagonal quartz Courtesy Richard Tilley
5.4 High Quartz crystal Courtesy of Richard Tilley

Set in 9/13pt Ubuntu by SPi Global, Pondicherry, India

10 9 8 7 6 5 4 3 2 1

Contents

Preface

Crystallography – the study of crystals, their structures and properties – plays an important role in a wide range of disciplines, including biology, chemistry, materials science and technology, mineralogy and physics, as well as engineering. The scientific breakthroughs in the first half of the twentieth century, leading to applications as diverse as nuclear power and semiconductor technology, were built, to a considerable extent, upon detailed understanding of metallic and inorganic crystal structures. In the latter half of the twentieth century, molecular biology, founded upon the determination of molecular crystal structures, has led to a deep-rooted change in medicine, with the structures of biologically important molecules such as insulin, the epoch-making determination of the crystal structure of DNA, and recent studies of protein structures being pivotal.

With these applications in mind, this book is designed as an introductory text for students and others who need to understand crystals and crystallography without necessarily becoming crystallographers. The aim is to explain how crystal structures are described for anyone coming to this area of study for the first time. At the end of it, a student should be able to read scientific papers and articles describing a crystal structure, or use crystallographic databases, with complete confidence and understanding. In addition, the book contains an introduction to areas of crystallography, such as modulated structures, quasicrystals and protein crystallography, that are currently the subject of active research.

The book is organised into nine chapters. The first of these provides an introduction to the subject. Chapters 2, 3 and 4 are concerned with fundamental crystallographic concepts: symmetry, lattices and lattice geometry, while the newly-added Chapter 5 outlines the close relationship between crystal symmetry and physical properties. Chapter 6 is focused upon how to build or display a crystal structure, given the information normally available in databases. Chapter 7 outlines the relationship between diffraction and structure, including information on X-ray, electron and neutron diffraction. Chapter 8 describes the principal ways in which structures are depicted in order to reveal chemical and physical relationships and biochemical reactivity. The final chapter, Chapter 9, is concerned with defects in crystals and recent developments in crystallography, such as the recognition of incommensurately modulated crystals and quasicrystals.

Compared with the first edition, all chapters have been rewritten and Chapter 5 is a new addition. Many figures have been redrawn and clarified and additional problems have been added in most chapters. However, the overall format of the book is carried over from the first edition. Each chapter is prefaced by three 'Introductory Questions'. These are questions that have been asked by students in the past, and to some extent provide a focus for the student at the beginning of each chapter. The answers are given at the end of the chapter. Within each chapter, new crystallographic concepts, invariably linked to a definition of the idea, are given in bold type when encountered for the first time. All chapters are provided with a set of problems and exercises, designed to reinforce the concepts introduced. These are of two types. A multiple choice 'Quick Quiz' is meant to be tackled rapidly, and aims to reveal gaps in the reader's grasp of the preceding material. The 'Calculations and Questions' sections are more traditional, and contain numerical exercises that reinforce concepts that are often described in terms of mathematical relationships or equations. A revised and updated bibliography provides students with sources for further exploration of the subject.

It is a pleasure to acknowledge the considerable assistance received in the preparation of this edition. The staff of the Trevithick Library, Cardiff University, were of great help in obtaining access to both historical literature and the rapidly expanding current scientific literature in the areas dealt with here. In addition, the staff at John Wiley, notably Emma Strickland, gave encouragement and assistance at all times during the production of this book. Last, but by no means least, I thank my wife Anne, without whose continual support this volume would not have been possible.

Richard J. D. Tilley
February 2020

Chapter 1
Crystals and Crystal Structures

What is a crystal system?
What are unit cells?
What information is needed to specify a crystal structure?

Crystals are homogeneous solids that possess a long-range three-dimensional array of ordered atoms. That is, the arrangement of the atoms in one small volume of a crystal is identical (excepting localised mistakes or defects that can arise during crystal growth or that are inserted deliberately) to that in any other similar but remote part of the crystal. Crystallography is the study of crystals and describes the ways in which the component atoms are arranged in crystals and how the long-range order is achieved. Many chemical (including biochemical) and physical properties of solids depend upon crystal structure, and knowledge of crystallography is essential if the properties of materials are to be understood and exploited.

1.1 CRYSTAL FAMILIES AND CRYSTAL SYSTEMS

Crystallography first developed as an observational science: an adjunct to the study of minerals. Minerals were (and still are) described by their morphology or **habit**, the shape that a mineral specimen may exhibit, which may vary from an amorphous mass to a well-formed gemstone. Indeed, the regular and beautiful shapes of naturally occurring crystals attracted attention from the earliest times, and the relationship between crystal shape and the disposition of crystal faces of a well-formed crystal provides an obvious means of classification. For example, some crystals resemble cubes or octahedra, whilst others are brick-like or form prisms with a hexagonal cross-section (Figure 1.1).

The external shape of a well-formed crystal reflects the internal order of the solid,

Crystals and Crystal Structures, Second Edition. Richard J. D. Tilley.

(a)

(b)

Figure 1.1 (a) Quartz (SiO_2) crystals, showing hexagonal morphology; (b) pyrite (FeS_2, fool's gold) crystals showing cubic and octahedral morphology.

especially the presence of internal symmetry. Symmetry will be developed later (especially in Chapters 3 and 4), but for the moment we can note that the most important symmetries displayed by crystals, and used in their classification, are the mirror plane, across which two parts of the crystal are related by reflection, and rotation axes. There are four different rotation axes: the diad or two-fold, in which successive rotations by (360/2)° leave the crystal unchanged; the triad or three-fold, in which successive rotations by (360/3)° leave the crystal unchanged; the tetrad or four-fold, in which successive rotations by (360/4)° leave the crystal unchanged; and the hexad or six-fold, in which successive rotations by (360/6)° leave the crystal unchanged. Careful measurement of mineral specimens using these symmetry criteria have allowed crystals to be classified in terms of six **crystal families** – anorthic, monoclinic, orthorhombic, tetragonal, hexagonal, and isometric or cubic – later expanded slightly by crystallographers into seven **crystal systems** (Table 1.1).

The crystal systems are sets of reference axes, which have a direction as well as a magnitude, and hence are vectors.[1] The allocation of a crystal to a particular system is made on the basis of the internal symmetries that are inferred from the crystal habit, which includes the apparent external symmetry. For instance, a crystal that resembled a hexagonal prism would be allocated to the hexagonal crystal system. It is common sense to allocate one reference axis to lie parallel to the hexagonal prism length and the other two axes to be normal to two chosen faces of the hexagonal prism. The axis lying along the prism length is really the defining axis for this designation because it is the unique hexad axis about which the crystal could be rotated by successive 60° turns to reproduce the same shape. Similarly, if a crystal has the form of a brick with a square cross-section, it is assigned to the tetragonal system, and the axes are positioned parallel to the three edges of the crystal. The defining axis this time is a tetrad perpendicular to the square cross-section, and this time the crystal can be rotated by

[1] Vectors are set in **bold** typeface throughout this book.

Table 1.1 The seven crystal systems

Crystal family	Crystal system	Required symmetry[a]	Axial relationships
Anorthic	Triclinic	None	$a \neq b \neq c$, $\alpha \neq 90°$, $\beta \neq 90°$, $\gamma \neq 90°$
Monoclinic	Monoclinic	1 diad or 1 mirror plane	$a \neq b \neq c$, $\alpha = 90°$, $\beta \neq 90°$, $\gamma = 90°$
Orthorhombic	Orthorhombic	3 diads or 1 diad + 2 mirror planes	$a \neq b \neq c$, $\alpha = 90°$, $\beta = 90°$, $\gamma = 90°$
Tetragonal	Tetragonal	1 tetrad	$a = b \neq c$, $\alpha = 90°$, $\beta = 90°$, $\gamma = 90°$
Hexagonal	Trigonal	1 triad	$a = b = c$, $\alpha = \beta = \gamma$ (rhombohedral axes); or $a' = b' \neq c$, $\alpha' = 90°$, $\beta' = 90°$, $\gamma' = 120°$ (hexagonal axes)
	Hexagonal	1 hexad	$a = b \neq c$, $\alpha = 90°$, $\beta = 90°$, $\gamma = 120°$
Isometric	Cubic	3 tetrads	$a = b = c$, $\alpha = 90°$, $\beta = 90°$, $\gamma = 90°$

[a]Symmetry nomenclature is expanded in Chapters 3 and 4.

successive 90° turns to regenerate the same shape. Thus the allocation of the reference axes and the placement of a crystal into a crystal family depends upon the external symmetry of the crystal.

The three reference axes for each family, allocated with respect to the crystal symmetry, are labelled **a**, **b** and **c**, and the angles between the positive directions of the axes are α, β, and γ, where α lies between +**b** and +**c**, β lies between +**a** and +**c**, and γ lies between +**a** and +**b** (Figure 1.2). The angles are chosen to be greater or equal to 90° except for the trigonal crystal system, as described below. In the figures, the **a**-axis is represented as projecting out of the plane of the page, towards the reader, the **b**-axis points to the right and the **c**-axis points towards the top of the page. This arrangement is a **right-handed coordinate system.** Measurements on mineral specimens could give absolute values for the inter-axial angles, but only relative axial lengths could be derived. These relative lengths are written *a*, *b* and *c*.

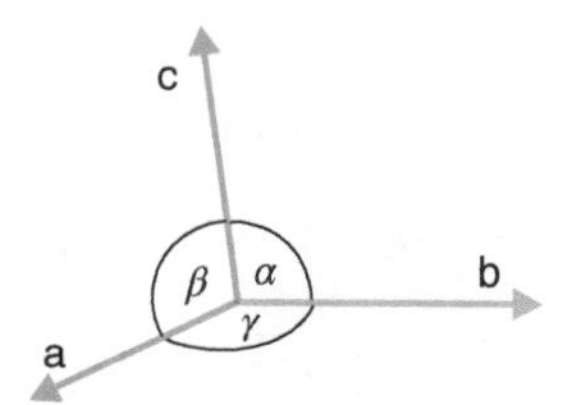

Figure 1.2 Reference axes used to characterise the seven crystal systems.

The most symmetric of the crystal systems is the cubic or isometric system, in which the three tetrad axes are arranged at 90° to each other and the axial lengths are identical. These form the familiar Cartesian axes. The tetragonal system is similar, with mutually perpendicular axes, but two of these, usually designated **a** (= **b**), are of equal length, whilst the third, the tetrad axis, usually designated **c**, is longer or shorter than the other two. The orthorhombic system has three mutually perpendicular axes of different lengths parallel to the three diads. The monoclinic system is also defined by three unequal axes. Two of these, conventionally chosen as **a** and **c**, are at an oblique angle, β, whilst the third **c**, normally parallel to the diad axis or mirror normal, is perpendicular to the plane containing **a** and **b**. The least symmetrical crystal system is the triclinic, which has three unequal axes at oblique angles.

The hexagonal crystal system has two axes of equal length, designated **a** (= **b**), at an angle, γ, of 120°. The **c**-axis lies perpendicular to the plane containing **a** and **b** and lies parallel to the hexad axis. The trigonal system has three axes of equal length, each enclosing equal angles α (= β = γ), forming a rhombohedron. The axes are called rhombohedral axes and the triad axis is allocated to the body diagonal of the rhombohedron. Crystals described in terms of rhombohedral axes are often more conveniently described in terms of a hexagonal set of axes. In this case, the hexagonal **c**-axis is parallel to the rhombohedral body diagonal, which is parallel to the triad axis (Figure 1.3). The relationship between the two sets of axes is given by the *vector* equations:

$$\mathbf{a}_R = \tfrac{2}{3}\,\mathbf{a}_H + \tfrac{1}{3}\,\mathbf{b}_H + \tfrac{1}{3}\,\mathbf{c}_H$$

$$\mathbf{b}_R = -\tfrac{1}{3}\,\mathbf{a}_H + \tfrac{1}{3}\,\mathbf{b}_H + \tfrac{1}{3}\,\mathbf{c}_H$$

$$\mathbf{c}_R = -\tfrac{1}{3}\,\mathbf{a}_H - \tfrac{2}{3}\,\mathbf{b}_H + \tfrac{1}{3}\,\mathbf{c}_H$$

$$\mathbf{a}_H = \mathbf{a}_R - \mathbf{b}_R$$

$$\mathbf{b}_H = \mathbf{b}_R - \mathbf{c}_R$$

$$\mathbf{c}_H = \mathbf{a}_R + \mathbf{b}_R + \mathbf{c}_R$$

where the subscripts R and H stand for rhombohedral and hexagonal respectively. (Note that in these equations the vectors **a**, **b** and **c** are added vectorially, not arithmetically [see Appendix A].) The arithmetical relationships between the axial lengths are given by:

$$a_H = 2a_R \sin\frac{\alpha}{2} \qquad c_H = a_R\sqrt{3 + 6\cos\alpha}$$

$$a_R = \tfrac{1}{3}\sqrt{3a_H{}^2 + c_H{}^2} \qquad \sin\frac{\alpha}{2} = \frac{3a_H}{2\sqrt{3a_H{}^2 + c_H{}^2}}$$

where the subscripts R and H stand for rhombohedral and hexagonal respectively.

1.2 UNIT CELLS AND MILLER INDICES

Observations of the cleavage of crystals, that is, the way in which they can be fractured along certain directions in such a manner that the two resultant fragments have perfect faces, suggested that all crystals might be built up by a stacking of infinitesimally small, geometrically regular elementary volumes, each unique to the crystal under consideration. These elementary

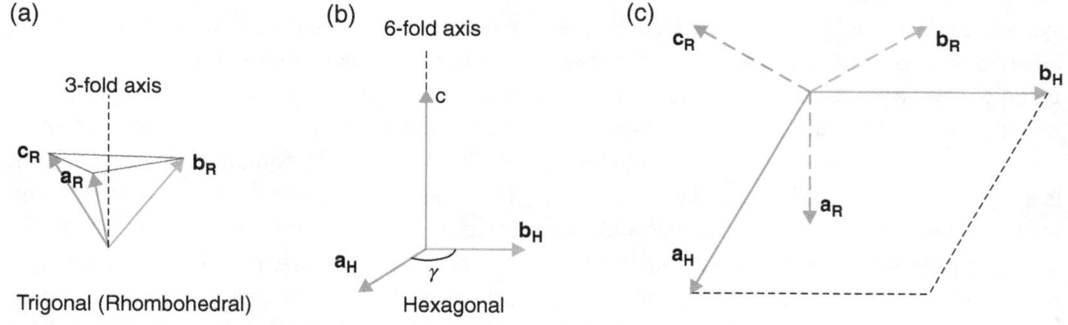

Figure 1.3 Rhombohedral (trigonal) and hexagonal axes: (a, b) axes in equivalent orientations with the trigonal three-fold axis parallel to the hexagonal six-fold axis (**c**-) axis; (c) superposition of both sets of axes, projected down the hexagonal **c**-axis (the rhombohedral three-fold axis).

volumes, the edges of which could be considered to be parallel to the axial vectors **a**, **b** and **c**, of the seven crystal systems, eventually came to be termed **morphological unit cells**. The relative axial lengths, *a*, *b*, and *c* were taken as equal to the relative lengths of the unit cell sides. The values *a*, *b*, *c*, α, β and γ are termed the **morphological unit cell parameters**. The unit cell type was then given the same name as the corresponding crystal system: triclinic, monoclinic, orthorhombic, tetragonal, trigonal, hexagonal or cubic. The absolute lengths of the axes, also written *a*, *b* and *c*, determined by diffraction techniques, yield the **structural unit cell** of the material. Unit cell parameters now refer only to these structural values, but the morphological names (triclinic, monoclinic, etc.) are still used. In these terms, the vectors that define the edges of the structural unit cell, **a**, **b**, **c**, are called the **basis vectors** of the direct lattice, whilst *a*, *b*, *c* are the lengths of the basis vectors or lengths of the structural unit cell edges, and are called the **lattice parameters**.

A central concept in crystallography is that the whole of a crystal can be built up by stacking identical copies of the unit cell in exactly the same orientation. That is to say, a crystal is characterised by both **translational** and **orientational** long-range order. The unit cells are displaced repeatedly in three dimensions (**translational** long-range order), without any rotation or reflection (**orientational** long-range order). This leads to severe restrictions upon the shape (strictly speaking the symmetry) of a unit cell. The fact that some unit cell shapes are not allowed is easily demonstrated in two dimensions, as it is apparent that regular pentagons (a plane figure with five equal sides and five equal internal angles), for example, cannot pack together without leaving gaps (Figure 1.4).

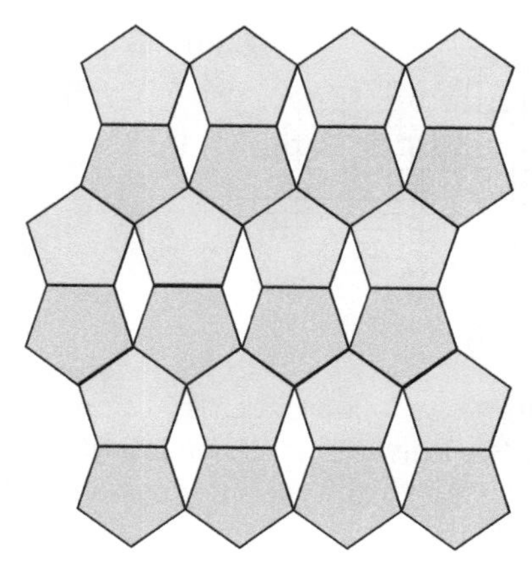

Figure 1.4 Irrespective of how they are arranged, regular pentagons cannot fill a plane completely; spaces always appear between some of the pentagons.

Not only can unit cells be stacked by translation alone to yield the internal structure of the crystal, but, depending on the rate at which the unit cells are stacked in different directions (i.e. the rate at which the crystal grows in three dimensions), different facets of the crystal become emphasised, whilst others are suppressed, so producing a variety of external shapes or habits (Figure 1.5), thus explaining the observation that a single mineral can occur in differing crystal morphologies.

The faces of a crystal, irrespective of the overall shape of the crystal, can always be labelled with respect to the crystal axes. Each face is given a set of three integers, (*h k l*), called **Miller indices**. These are such that the crystal face in question makes intercepts on the three axes of *a*/*h*, *b*/*k*, and *c*/*l*. A crystal face that intersects the axes in exactly the axial ratios is given importance as the **parametral plane**, with indices (1 1 1). Miller indices are used to label any plane, internal or external, in a crystal, as described in Chapter 2, and the nomenclature is not just confined to the external faces of a crystal.

Figure 1.5 (a) schematic depiction of a crystal built of rectangular (orthorhombic) unit cells. The unit cells must be imagined to be much smaller than depicted, thus producing smooth facets. (b) Varying crystal habits derived by differing rates of directional crystal growth.

1.3 THE DETERMINATION OF CRYSTAL STRUCTURES

The descriptions of crystals above are mainly the result of the application of optical techniques. However, the absolute arrangement of the atoms in a crystal cannot be determined in this way. This limitation was overcome in the early years of the twentieth century, when it was discovered that X-rays were scattered, or **diffracted**, by crystals in a way that could be interpreted to yield the absolute arrangement of the atoms in a crystal, the **crystal structure**. X-ray diffraction remains the most widespread technique used for structure determination, but diffraction of electrons and neutrons is also of great importance, as these reveal features that are complementary to those observed with X-rays.

The physics of diffraction by crystals has been worked out in detail. It is found that the incident radiation is scattered in a characteristic way, called a **diffraction pattern**. The positions and intensities of the diffracted beams are a function of the arrangements of the atoms in space and some other atomic properties, such as the atomic number of the atoms. Thus, if the positions and the intensities of the diffracted beams are recorded, it is possible to deduce the arrangement of the atoms in the crystal and their chemical nature. The determination of crystal structures by use of the diffraction of radiation is outlined in Chapter 7.

1.4 THE DESCRIPTION OF CRYSTAL STRUCTURES

Any crystal structure is described in terms of its unit cell. Although unit cells are correctly defined in terms of the symmetries present, for the moment we will simply assume that the unit cell is an experimentally determined small volume of the crystal, which, if translated (repeated) in three dimensions without rotation or reflection, will build up the entire structure. The minimum amount of information needed to specify a crystal structure is the unit cell type (i.e. cubic, tetragonal, etc.), the unit cell parameters, and the positions of all of the atoms in the unit cell. The atomic contents of the unit cell are a simple multiple, *Z*, of the composition of the material. The value of *Z* is equal to the number of **formula units** of the solid in the unit cell. Atom positions are expressed in terms of three coordinates, *x*, *y* and *z*. These are taken as **fractions** of *a*, *b* and *c*, the unit cell sides, for example ½, ½, ¼. The *x*, *y* and

z coordinates are plotted with respect to the unit cell axes, not to a Cartesian set of axes. The position of an atom can also be expressed as a vector, **r**:

$$\mathbf{r} = x\mathbf{a} + y\mathbf{b} + z\mathbf{c}$$

where **a**, **b** and **c** are the basis vectors (unit cell axes). To summarise:

a, b, c are the basis vectors of the direct lattice
a, b, c are the lengths of the basis vectors or lengths of the unit cell edges
x, y, z are the coordinates of a point as fractions of *a, b, c*.

In addition:

X, Y, Z refer to Cartesian axes when required.

In figures, the conventional origin is placed at the left rear corner of the unit cell. The **a**-axis is represented as projecting out of the plane of the page, towards the reader, the **b**-axis points to the right and the **c**-axis points towards the top of the page (Figure 1.6). In projections, the origin is set at the upper left corner of the unit cell projection. A frequently encountered projection is that perpendicular to the **c**-axis. In this case, the **a**-axis is drawn pointing down, (from top to bottom of the page), and the **b**-axis pointing to the right. In projections the x and y coordinates can be determined from the figure. The z position is usually given on the figure as a fraction.

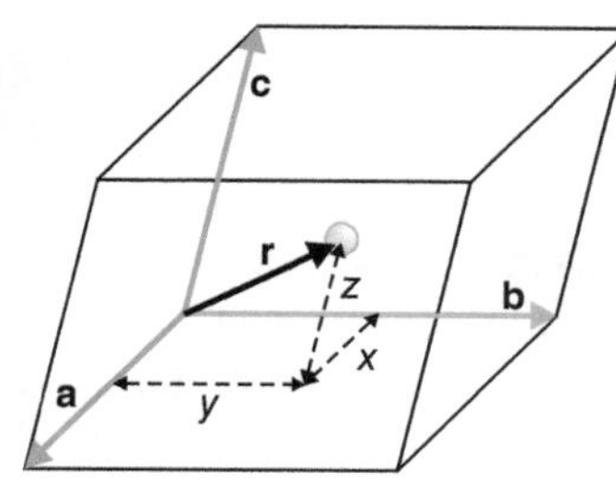

Figure 1.6 The position of an atom, x, y and z, in a unit cell.

An atom at a cell corner is given the coordinates (0, 0, 0). An atom at the centre of the face of a unit cell is given the coordinates (½, ½, 0) if it lies between the **a**- and **b**-axes, (½, 0, ½) if between the **a**- and **c**-axes, and (0, ½, ½) if between the **b**- and **c**-axes. An atom at the centre of a unit cell would have a position specified as (½, ½, ½), irrespective of the type of unit cell. Atoms at the centres of the cell edges are specified at positions (½, 0, 0), (0, ½, 0), or (0, 0, ½), for atoms on the **a**-, **b**- and **c**-axis, (Figure 1.7). Stacking of the unit cells to build a structure will ensure that an atom at the unit cell origin will appear at every corner, and atoms on unit cell edges or faces will appear on all of the cell edges and faces.

A vast number of structures have been determined, and it is very convenient to group those with topologically identical structures

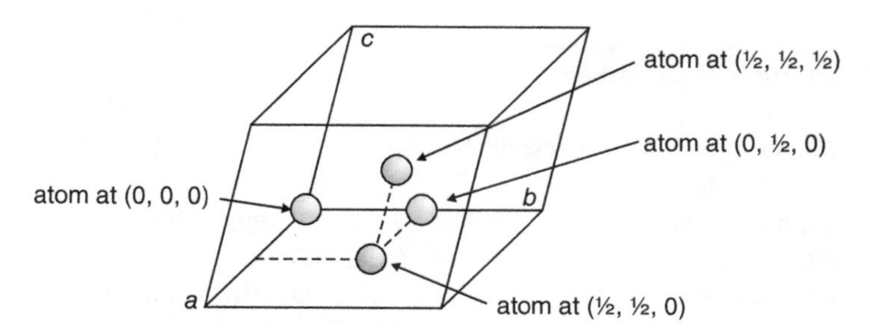

Figure 1.7 Atoms at positions 0, 0, 0; 0, ½ 0; ½, ½, 0; and ½, ½, ½ in a unit cell.

together. On going from one member of the group to another, the atoms in the unit cell differ, reflecting a change in chemical composition, and the atomic coordinates and unit cell dimensions change slightly, reflecting any differences in atomic size, but relative atom positions are identical or very similar. Frequently, the group name is taken from the name of a mineral, as mineral crystals were the first solids used for structure determination. Thus, all crystals with a structure similar to that of sodium chloride, NaCl (the mineral halite), are said to adopt the *halite* structure. These materials then all have a general formula MX, where M is a metal atom and X a non-metal atom, for example, MgO. Similarly, crystals with a structure similar to the rutile form of titanium dioxide, TiO_2, are grouped with the *rutile* structure. These all have a general formula MX_2, for example FeF_2. As a final example, compounds with a similar structure to the mineral fluorite, (sometimes called fluorspar), CaF_2, are said to adopt the *fluorite* structure. These also have a general formula MX_2, an example being UO_2. Examples of these three structures follow. Crystallographic details of a number of simple inorganic structures are given in Appendix B. Some mineral names of common structures are found in Table 1.2 and Appendix B.

A system of nomenclature that is useful for describing relatively simple structures is that originally set out in 1920, in Volume 1 of the German publication *Strukturbericht*. It assigned a letter code to each structure: A for materials with only one atom type, B for two different atoms, and so on. Each new structure was characterised further by the allocation of a numeral, so that the crystal structures of elements were given symbols A1, A2, A3, and so on. Simple binary compounds were given symbols B1, B2, and so on, and binary compounds with more complex structures C1, C2, D1, D2, and so on. As the number of crystal structures and the complexity displayed increased, the system became unworkable. Nevertheless, the terminology is still used for the description of simple structures. Some *Strukturbericht* symbols are given in Table 1.2.

1.5 CRYSTAL STRUCTURES: METALS

In these simple structures, only one atom type is present, and all the atom positions are listed. (More complex crystal structures will contain many atoms in the unit cell, and those of proteins contain thousands. To make the list of atom positions manageable use is made of the

Table 1.2 *Strukturbericht* symbols and names for simple structure types

Symbol and name	Examples	Symbol and name	Examples
A1, cubic close-packed, copper	Cu, Ag, Au, Al	A2, body-centred cubic, iron	Fe, Mo, W, Na
A3, hexagonal close-packed, magnesium	Mg, Be, Zn, Cd	A4, diamond	C, Si, Ge, Sn
B1, halite, rock salt	NaCl, KCl, NiO, MgO	B2, caesium chloride	CsCl, CsBr, AgMg, BaCd
B3, zincblende	ZnS, ZnSe, BeS, CdS	B4, wurtzite	ZnS, ZnO, BeO, CdS, GaN
C1, fluorite	CaF_2, BaF_2, UO_2, ThO_2	C4, rutile	TiO_2, SnO_2, MgF_2, FeF_2

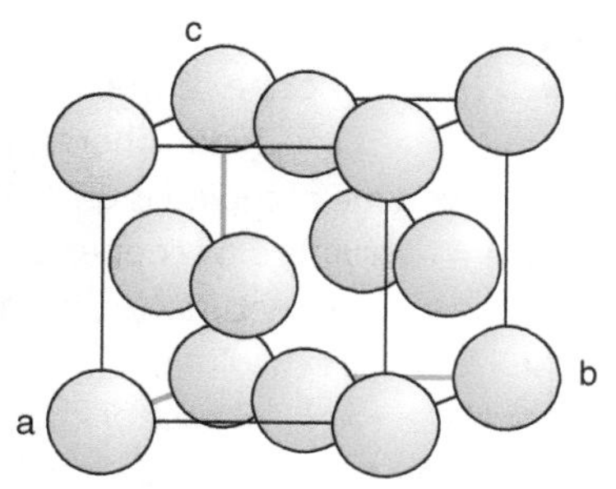

Figure 1.8 The cubic unit cell of the A1, copper, structure.

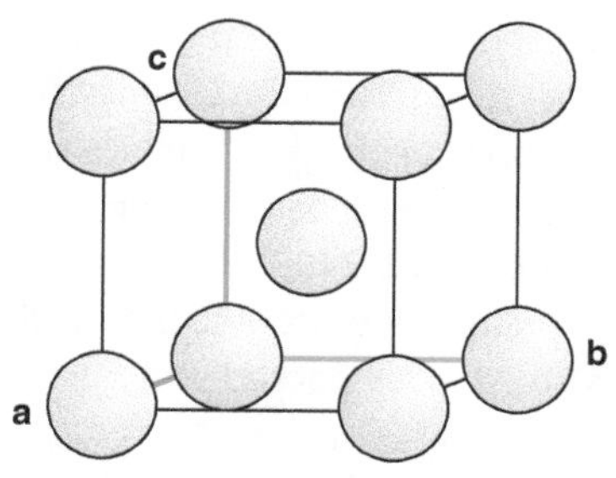

Figure 1.9 The cubic unit cell of the A2, tungsten, structure.

underlying symmetry elements present in the unit cell, which allows a minimum number of atoms to be specified such that application of the symmetry operators generates the whole cell contents, described in more detail in Chapter 7.)

1.5.1 The Cubic Close-packed (A1) Structure of Copper

A number of elemental metals crystallise with the cubic A1 structure, also called the copper structure.

Unit cell: cubic
Lattice parameter[2] for copper: $a = 0.360$ nm
$Z = 4$ Cu
Atom positions: 0, 0, 0; ½, ½, 0; 0,½, ½; ½, 0, ½

There are four copper atoms in the unit cell (Figure 1.8). Besides a number of metals, the noble gases, Ne(s), Ar(s), Kr(s) and Xe(s), also adopt this structure in the solid state. This structure is often called the face-centred cubic (*fcc*) structure or the cubic close-packed (*ccp*) structure, but the *Strukturbericht* symbol, A1, is the most compact notation. Each atom has 12 nearest neighbours, and if the atoms are supposed to be hard touching spheres, the fraction of the volume occupied is 0.7405. More information on this structure is given in Chapter 7.

[2] Lattice parameters and interatomic distances in crystal structures are usually reported in Ångström units, Å in crystallographic literature. 1 Å is equal to 10^{-10} m, that is, 10 Å = 1 nm. In this book, the SI unit of length, nm, will be used most often, but Å will be used on occasion, to conform to crystallographic practice.

1.5.2 The Body-Centred Cubic (A2) Structure of Tungsten

A second common structure adopted by metallic elements is that of the cubic structure of tungsten.

Unit cell: cubic
Lattice parameter for tungsten: $a = 0.316$ nm
$Z = 2$ W
Atom positions: 0, 0, 0; ½, ½, ½

There are two tungsten atoms in the body-centred unit cell (Figure 1.9). This structure is often called the body-centred cubic (*bcc*) structure, but the *Strukturbericht* symbol, A2, is a more compact designation. In this structure, each atom has eight nearest neighbours and six next-nearest neighbours at only 15% greater distance. If the atoms are supposed to be hard touching spheres, the fraction of the volume occupied is 0.6802. This is less than either the A1 structure above or the A3 structure that follows, both of which have a volume fraction of occupied space of 0.7405. The A2 structure is often the high-temperature structure of a metal that has an A1 close-packed structure at lower temperatures.

1.5.3 The Hexagonal (A3) Structure of Magnesium

The third common structure adopted by elemental metals is the hexagonal magnesium structure.

Unit cell: hexagonal
Lattice parameters for magnesium: a = 0.321 nm, c = 0.521 nm
Z = 2 Mg
Atom positions: 0, 0, 0; ⅓, ⅔, ½

(The atoms can also be placed at ⅔, ⅓, ¼; ⅓, ⅔, ¾, by changing the unit cell origin. This is preferred for some purposes.) There are two magnesium atoms in the unit cell. The structure is shown in perspective (Figure 1.10a) and projected down the **c**-axis (Figure 1.10b). This structure is often referred to as the hexagonal close-packed (*hcp*) structure. If the atoms are supposed to be hard touching spheres, the fraction of the volume occupied is 0.7405, equal to that in the A1 structure of copper, and the ratio of the lattice parameters, c/a, is equal to $\sqrt{8}/\sqrt{3}$, = 1.633. The ideal volume, V, of the unit cell, equal to the area of the base of the unit cell multiplied by the height of the unit cell, is:

$$V = \frac{\sqrt{3}}{2} a^2 c \approx 0.8660 a^2 c$$

More information on this structure, and the relationship between the A1 and A3 structures, is given in Chapter 8.

1.6 CRYSTAL STRUCTURES: BINARY COMPOUNDS

1.6.1 The *Halite* (Rock Salt, NaCl) Structure

The general formula of crystals with the *halite* structure is MX. The mineral halite, which names the group, is sodium chloride, NaCl, and the structure is often called the rock salt or NaCl structure.

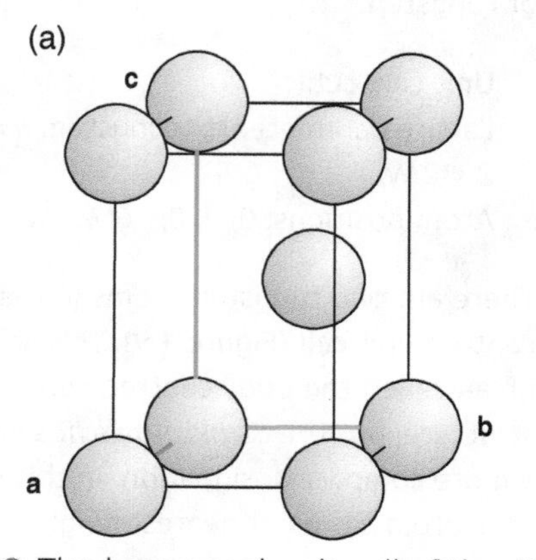

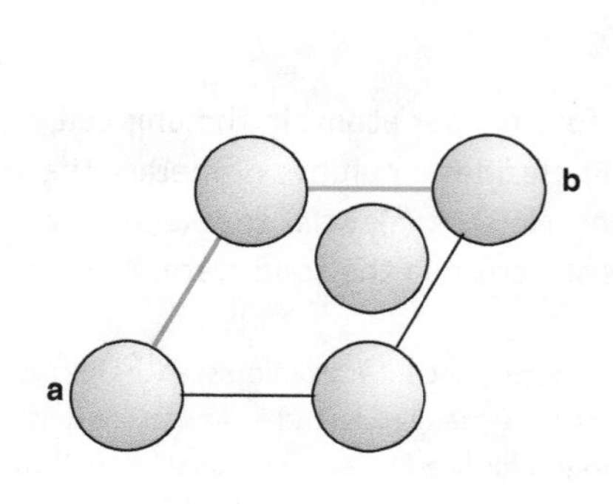

Figure 1.10 The hexagonal unit cell of the A3, magnesium, structure: (a) perspective view; (b) projection down the **c**-axis.

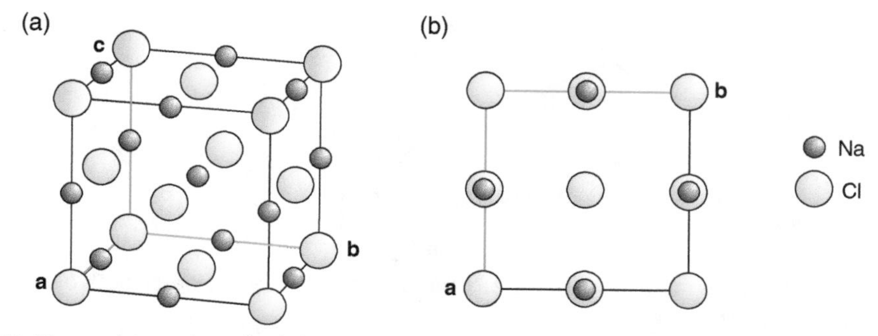

Figure 1.11 The cubic unit cell of the B1 (halite) structure of NaCl: (a) perspective view; (b) projection down the **c**-axis.

Unit cell: cubic
Lattice parameter for halite:
a (= *b* = *c*) = 0.5500 nm
Z = 4 {NaCl}
Atom positions: Na: ½, 0, 0; 0, 0, ½; 0, ½, 0; ½, ½, ½
Cl: 0, 0, 0; ½, ½, 0; ½, 0, ½; 0, ½, ½

There are four sodium and four chlorine atoms in the unit cell. Each atom is surrounded by six atoms of the opposite type at the corners of a regular octahedron. A perspective view of the structure is shown in Figure 1.11a, and a projection, down the **c**-axis, in Figure 1.11b.

This structure is adopted by many oxides, sulfides, halides and nitrides with a formula MX.

1.6.2 The *Rutile* Structure

The general formula of crystals with the *rutile* structure is MX_2. The mineral rutile, which names the group, is one of the structures adopted by titanium dioxide, TiO_2. (The other commonly encountered form of TiO_2 is called anatase. Other structures for TiO_2 are also known.)

Unit cell: tetragonal
Lattice parameters for rutile: *a* (= *b*) = 0.4594 nm, *c* = 0.2959 nm
Z = 2 {TiO_2}
Atom positions: Ti: 0, 0, 0; ½, ½, ½
O: 3/10, 3/10, 0; 4/5, 1/5, ½; 7/10, 7/10, 0; 1/5, 4/5, ½

There are two titanium and four oxygen atoms in the unit cell. In this structure, each titanium atom is surrounded by six oxygen atoms at the corners of a regular octahedron. A perspective view of the rutile structure is shown in Figure 1.12a, and a projection down the **c**-axis in Figure 1.12b.

This structure is relatively common and adopted by a number of oxides and fluorides of medium-sized cations.

1.6.3 The *Fluorite* Structure

The general formula of crystals with the *fluorite* structure is MX_2. The mineral fluorite, calcium fluoride, CaF_2, which names the group, is sometimes also called fluorspar.

Unit cell: cubic
Lattice parameter(s) for fluorite:
a (= *b* = *c*) = 0.5463 nm
Z = 4 {CaF_2}
Atom positions: Ca: 0, 0, 0; ½, ½, 0; 0, ½, ½; ½, 0, ½
F: ¼, ¾, ¼; ¼, ¾, ¾; ¼, ¼, ¼; ¼, ¼, ¾; ¾ ¼ ¼; ¾ ¼ ¾; ¾ ¾ ¼; ¾ ¾ ¾

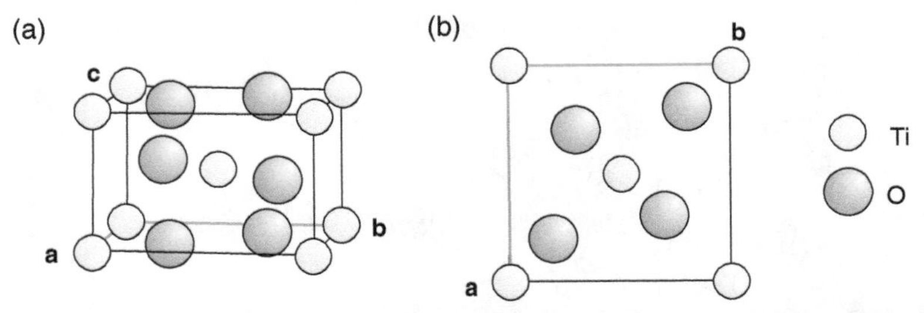

Figure 1.12 The tetragonal unit cell of the rutile structure TiO_2: (a) perspective view; (b) projection down the **c**-axis.

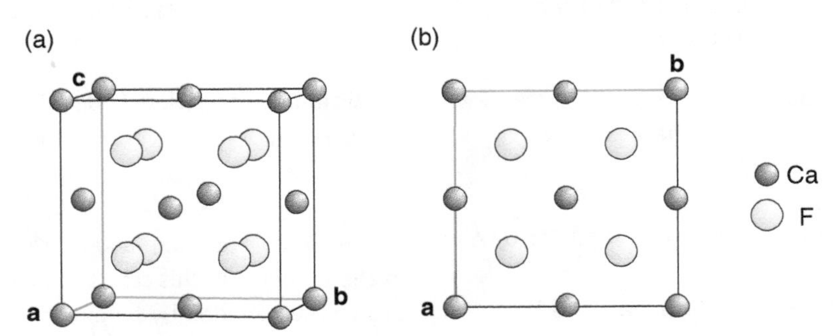

Figure 1.13 The cubic unit cell of the fluorite structure CaF_2: (a) perspective view; (b) projection down the **c**-axis.

There are four calcium and eight fluorine atoms in the unit cell. Each calcium atom is surrounded by eight fluorine atoms at the corners of a cube. Each fluorine atom is surrounded by four calcium atoms at the vertices of a tetrahedron (see also Chapter 7). A perspective view of the structure is shown in Figure 1.13a, and a projection of the structure down the **c**-axis in Figure 1.13b.

This structure is adopted by a number of oxides and halides of large divalent cations.

1.7 THE CUBIC *PEROVSKITE* STRUCTURE

The general formula of crystals with the *perovskite* structure is ABX_3, and all are structurally related to the mineral perovskite, $CaTiO_3$. The simplest (ideal) structure is that adopted by strontium titanate, $SrTiO_3$.

Unit cell: cubic
Lattice parameter(s) for strontium titanate: a $(= b = c) = 0.3905$ nm
$Z = 1$ $\{SrTiO_3\}$
Atom positions: Sr: 0, 0, 0
Ti: ½, ½, ½
O: ½, ½, 0; ½, 0, ½; 0, ½, ½

There is one Ti atom at the cell centre and one Sr atom at the cell corners. The O atoms lie at the centres of the cell faces. Each Ti is surrounded by six O atoms arranged to give a regular octahedral coordination polyhedron, often stressed when describing perovskites in general (Figure 1.14).

The perovskite ABX_3 structure or slightly distorted variants of it are adopted by many compounds containing a large (A) and a

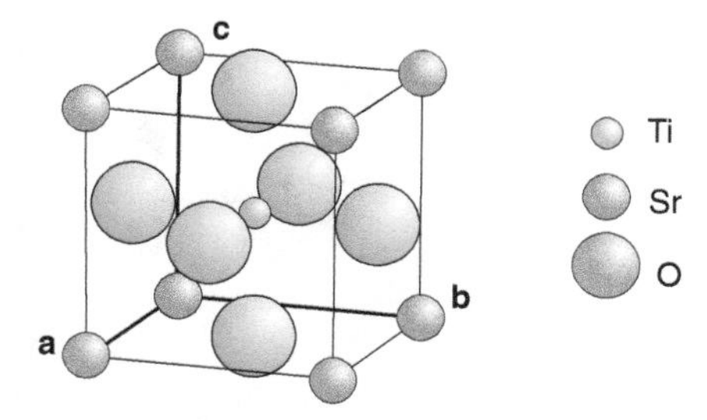

Figure 1.14 The cubic unit cell of the perovskite structure $SrTiO_3$.

medium-sized (B) cation when combined with anions (X) including oxygen, halogens and nitrogen.

1.8 THE STRUCTURE OF UREA

The structures of molecular crystals tend to have a different significance to those of the inorganic and mineral structures described above. Frequently, the information of most importance is the molecular geometry, and how the molecules are arranged in the crystallographic unit cell is often of secondary interest. To introduce the changed emphasis when dealing with molecular materials, the crystal structure of the organic compound urea is described. Urea is a very simple molecule, with the formula CH_4N_2O. The unit cell is small and of high symmetry. It was one of the earliest organic structures to be investigated using the methods of X-ray crystallography, and in these initial investigations the data was not precise enough to locate the hydrogen atoms. (The location of hydrogen atoms in a structure remains a problem to present times; see also Chapters 7 and 8.) The crystallographic data for urea are as follows:[3]

Unit cell: tetragonal
Lattice parameters: a (= b) = 0.5589 nm, c = 0.46947 nm
Z = 2 {CH_4N_2O}
Atom positions: C1: 0, 0.5000, 0.3283
C2: 0.5000, 0, 0.6717
O1: 0, 0.5000, 0.5963
O2: 0.5000, 0, 0.4037
N1: 0.1447, 0.6447, 0.1784
N2: 0.8553, 0.3553, 0.1784
N3: 0.6447, 0.8553, 0.8216
N4: 0.3553, 0.1447, 0.8216
H1: 0.2552, 0.7552, 0.2845
H2: 0.1428, 0.6428, 0.9661
H3 0.8448, 0.2448, 0.2845
H4: 0.8572, 0.3572, 0.9661
H5: 0.7552, 0.7448, 0.7155
H6: 0.2448, 0.2552, 0.7155
H7: 0.6429, 0.8571, 0.0339
H8: 0.3571, 0.1428, 0.0339

Notice that atoms of the same chemical type are numbered sequentially. This is because each different atom occupies a site with a different symmetry to the others (see Chapter 6). The atoms in a unit cell (including hydrogen) are shown in Figure 1.15a. This turns out to be not very helpful, and an organic chemist would have difficulty in recognising it as urea. This is because the molecules lie along the unit cell sides, so that a whole molecule is not displayed in the unit cell, but only molecular fragments. (The unit cell is chosen in this way because of symmetry constraints, described in the following chapters.) The chemical structural formula for urea is drawn in Figure 1.15b, and this is compared with a molecule of urea viewed front on (Figure 1.15c), and edge on (Figure 1.15d) extracted from the crystallographic data. The crystal structure is redrawn in Figure 1.15e and f with the atoms linked to form molecules. This latter depiction now agrees with chemical intuition, and

[3] Data adapted from: Zavodnik, V., Stash, A., Tsirelson, V., et al. (1999). *Acta Crystallographica* **B55**: p45.

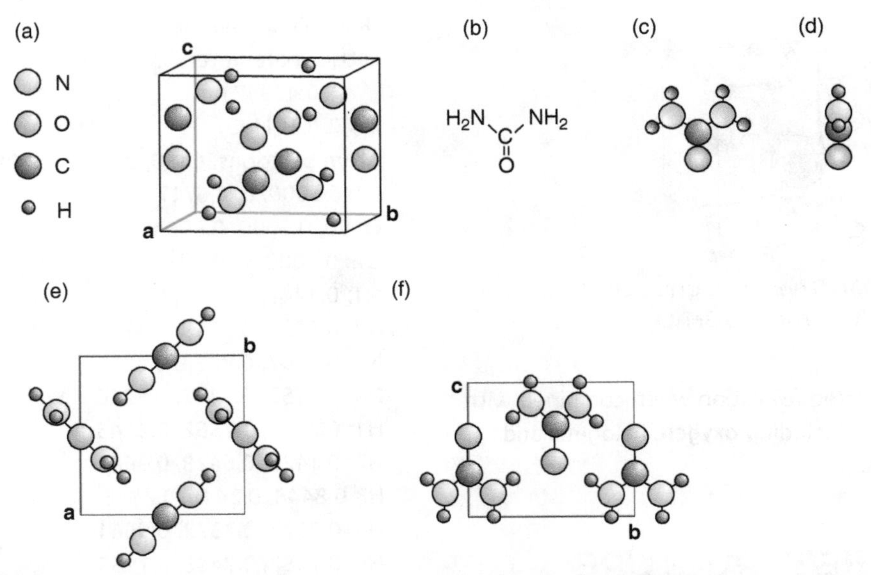

Figure 1.15 The crystal structure of urea: (a) perspective view of the tetragonal unit cell of urea; (b) structural formula of urea; (c) a 'ball and stick' representation of urea 'face on', as in (b); (d) 'sideways on'; (e) projection of the structure cell down the **c**-axis; (f) projection of the structure down the **a**-axis.

shows how the molecules are arranged in space.

Note that the list of atoms in the unit cell is becoming lengthy, albeit that this is an extremely simple structure. The means used by crystallographers to reduce these lists to manageable proportions, by using the symmetry of the crystal, are explained in Chapter 6.

1.9 THE DENSITY OF A CRYSTAL

The theoretical density of a crystal can be found by calculating the mass of all the atoms in the unit cell. The mass of an atom, m_A, is its molar mass (grams mol^{-1}) divided by the Avogadro constant, N_A (6.02214×10^{23} mol^{-1}).

$$m_A = \text{molar mass}/N_A \text{ (grams)}$$
$$= \text{molar mass}/(1000 \times N_A) \text{ (kilograms)}$$

The total mass of all of the atoms in the unit cell is then

$$n_1\, m_1 + n_2\, m_2 + n_3\, m_3 \ldots/(1000 \times N_A)$$

where n_1 is the number of atoms of type 1, with a molar mass of m_1, and so on. This is written in a more compact form as

$$\sum_{i=1}^{q} n_i m_i/(1000 N_A)$$

where there are q different atom types in the unit cell. The density, ρ, is simply the total mass is divided by the unit cell volume, V:

$$\rho = \left\{ \sum_{i=1}^{q} n_i m_i/(1000 N_A) \right\} / V$$

For example, the theoretical density of halite is calculated in the following way. First count the number of different atom types in the unit cell.

To count the number of atoms in a unit cell, use the information:

an atom within the cell counts as 1
an atom in a face counts as 1/2
an atom on an edge counts as 1/4
an atom on a corner counts as 1/8

A quick method to count the number of atoms in a unit cell is to displace the unit cell outline to remove all atoms from corners, edges, and faces. The atoms remaining, which represent the unit cell contents, are all within the boundary of the unit cell and count as 1.

The unit cell of the halite structure contains 4 sodium (Na) and 4 chlorine (Cl) atoms. The mass of the unit cell, m, is then given by:

$$m = [(4 \times 22.99) + (4 \times 35.453)]/1000 \times N_A$$
$$= 3.882 \times 10^{-25}\text{ kg}$$

where 22.99 g mol^{-1} is the molar mass of sodium, 35.453 g mol^{-1} is the molar mass of chlorine, and N_A is the Avogadro constant, 6.02214×10^{23} mol^{-1}.

The volume, V, of the cubic unit cell is given by a^3, thus:

$$V = (0.5500 \times 10^{-9})^3\text{ m}^3$$
$$= 1.66375 \times 10^{-28}\text{m}^3$$

The density, ρ, is given by the mass m divided by the volume, V:

$$\rho = 3.882 \times 10^{-25}\text{kg}/1.66375 \times 10^{-28}\text{m}^3$$
$$= 2333\ \text{kg m}^{-3}$$

The measured density is 2165 kg m^{-3}. The theoretical density is almost always slightly greater than the measured density because real crystals contain defects that act so as to reduce the total mass per unit volume.

ANSWERS TO INTRODUCTORY QUESTIONS

What is a crystal system?

A crystal system is a set of reference axes, used to define the geometry of crystals and crystal structures, that were originally derived using the apparent external symmetry of mineral crystals. There are seven crystal systems: cubic, tetragonal, orthorhombic, monoclinic, triclinic, hexagonal, and trigonal. As the crystal systems are sets of reference axes, they have a direction as well as a magnitude, and hence are vectors. They must be specified by length and interaxial angles.

The three reference axes are labelled **a**, **b** and **c**, and the angles between the positive direction of the axes as α, β and γ, where α lies between +**b** and +**c**, β lies between +**a** and +**c**, and γ lies between +**a** and +**b**. The angles are chosen to be greater or equal to 90° except for the trigonal system. The axes are chosen by reference to the symmetry of the crystal under consideration, and some or all of the axes are chosen to be parallel to conspicuous symmetry elements displayed by the crystal. In figures, the **a**-axis is represented as projecting out of the plane of the page, towards the reader, the **b**-axis points to the right and the **c**-axis points towards the top of the page.

What are unit cells?

All crystals can be built by the regular stacking of a small volume of material called the unit cell. A central concept in crystallography is that the whole of a crystal can be built up by stacking identical copies of the unit cell in exactly the same orientation. That is to say, a crystal is characterised by both translational and orientational long-range order. The unit cells are displaced repeatedly in three dimensions (translational long-range order), without any rotation or reflection (orientational long-range order). The edges of the unit cell are generally taken to be parallel to the axial vectors **a**, **b** and **c**, of the seven crystal systems. The lengths of the unit cell sides are written *a*, *b* and *c*, and the angles between the unit cell edges are written α, β and γ. The collected values *a*, *b*, *c*, α, β and γ for a crystal structure are termed the unit cell parameters. Note that there is no unique unit cell for any crystal structure, and alternatives can usually be specified if this is helpful.

What information is needed to specify a crystal structure?

The minimum amount of information needed to specify a crystal structure is the unit cell type (i.e. cubic, tetragonal, etc.), the unit cell (lattice) parameters and the positions of all of the atoms in the unit cell. The atomic contents of the unit cell are a simple multiple, *Z*, of the composition of the material. The value of *Z* is equal to the number of formula units of the solid in the unit cell. Atom positions are expressed in terms of three coordinates, *x*, *y* and *z*. These are taken as fractions of *a*, *b* and *c*, the unit cell sides, for example, ½, ½, ¼.

PROBLEMS AND EXERCISES

Quick Quiz

1. The number of crystal systems is:
   a. 5
   b. 6
   c. 7
2. The angle between the **a**- and **c**-axes in a unit cell is labelled:
   a. α
   b. β
   c. γ
3. A tetragonal unit cell is defined by:
   a. $a = b = c$, $\alpha = \beta = \gamma = 90°$
   b. $a = b \neq c$, $\alpha = \beta = \gamma = 90°$
   c. $a \neq b \neq c$, $\alpha = \beta = \gamma = 90°$
4. A crystal is built by the stacking of unit cells with:
   a. orientational and translational long-range order
   b. orientational long-range order
   c. translational long-range order
5. Miller indices are used to label
   a. crystal shapes
   b. crystal faces
   c. crystal sizes
6. Crystal structures are often determined by the scattering of:
   a. light
   b. microwaves
   c. X-rays
7. In crystallography the letter *Z* specifies:
   a. the number of atoms in a unit cell
   b. the number of formula units in a unit cell
   c. the number of molecules in a unit cell
8. The position of an atom at the corner of a monoclinic unit cell is specified as:
   a. 1, 0, 0
   b. 1, 1, 1
   c. 0, 0, 0
9. The number of atoms in the unit cell of the halite structure is:
   a. 2
   b. 4
   c. 8
10. When determining the number of atoms in a unit cell, an atom in a face counts as:
    a. ½
    b. ¼
    c. ⅛

Calculations and Questions

1.1. The rhombohedral unit cell of arsenic, As, has unit cell parameters $a_R = 0.412$ nm, $\alpha = 54.17°$. Use graphical vector addition (Appendix A) to determine the equivalent hexagonal lattice parameter a_H. Check your answer arithmetically and calculate a value for the hexagonal lattice parameter c_H.

1.2. Cassiterite, tin dioxide, SnO_2, adopts the *rutile* structure, with a tetragonal unit cell, lattice parameters, a = 0.4738 nm, c = 0.3188 nm, with Sn atoms at: 0, 0, 0; ½, ½, ½; and O atoms at: $^3/_{10}$, $^3/_{10}$, 0; ⅘, ⅕, ½; $^7/_{10}$, $^7/_{10}$, 0; ⅕, ⅘, ½. Taking one corner of the unit cell as origin, determine the atom positions in nm and calculate the unit cell volume in nm^3. Draw a projection of the structure down the **c**-axis, with a scale of 1 cm = 0.1 nm.

1.3. The structure of $SrTiO_3$, is cubic, with a lattice parameter a = 0.3905 nm. The atoms are at the positions: Sr, ½, ½, ½; Ti, 0, 0, 0; 0, ½, 0. Calculate the bond lengths (interatomic distances) from Ti to the surrounding O atoms. What is the shape of the O coordination polyhedron around the Ti?

1.4. (a) Ferrous fluoride, FeF_2, adopts the tetragonal *rutile* structure, with lattice parameters a = 0.4697 nm, c = 0.3309 nm. The molar masses are Fe, 55.847 g mol^{-1}, F, 18.998 g mol^{-1}. Calculate the density of this compound. (b) Barium fluoride, BaF_2, adopts the cubic *fluorite* structure, with lattice parameter a = 0.6200 nm. The molar masses are Ba, 137.327 g mol^{-1}, F, 18.998 g mol^{-1}. Calculate the density of this compound.

1.5. Strontium chloride, $SrCl_2$, adopts the *fluorite* structure, and has a density of 3052 kg m^{-3}. The molar masses of the atoms are Sr, 87.62 g mol^{-1}, Cl, 35.45 g mol^{-1}. Estimate the lattice parameter, a, of this compound.

1.6. Molybdenum, Mo, adopts the A2 (tungsten) structure. The density of the metal is 10 222 kg mol^{-1} and the cubic lattice parameter is a = 0.3147 nm. Estimate the molar mass of molybdenum.

1.7. A metal difluoride, MF_2, adopts the tetragonal *rutile* structure, with lattice parameters a = 0.4621 nm, c = 0.3052 nm and density 3148 kg m^{-3}. The molar mass of fluorine, F, is 18.998 g mol^{-1}. Estimate the molar mass of the metal and hence attempt to identify it.

1.8. The density of anthracene, $C_{14}H_{10}$, is 1250 kg m^{-3} and the unit cell volume is 475.35 × 10^{-30} m^3. Determine the number of anthracene molecules, Z, which occur in a unit cell. The molar masses are: C, 12.011 g mol^{-1}, H, 1.008 g mol^{-1}.

1.9. Assuming CaF_2, fluorite, is ionic, and the F^- ions just touch, calculate the ionic radius of F^-. The lattice parameter of (cubic) CaF_2 is 0.5463 nm.

1.10 Estimate the C=O bond length in urea. The structure is tetragonal with lattice parameters a = 0.5589 nm, c = 0.46947 nm.

Chapter 2
Lattices, Planes and Directions

How does a crystal lattice differ from a crystal structure?
What is a primitive unit cell?
What are Miller-Bravais indices used for?

The development of the idea of a lattice was amongst the earliest mathematical explorations in crystallography. Crystal structures and crystal lattices are different, although these terms are frequently (and incorrectly) used as synonyms. A crystal **structure** is built of **atoms**. A crystal **lattice** is an infinite pattern of **points**, each of which must have the same surroundings in the same orientation. A lattice is a mathematical concept.

All crystal structures can be built up from a lattice by placing an atom or a group of atoms at each lattice point. The crystal structure of a simple metal and that of a complex protein may both be described in terms of the same lattice, but whereas the number of atoms allocated to each lattice point is often just one for a simple metallic crystal, it may easily be 1000 for a protein crystal. The group of atoms associated with each lattice point is called the **motif** (or sometimes the **basis**).

2.1 TWO-DIMENSIONAL LATTICES

In two dimensions, if any lattice point is chosen as the origin, the position of any other lattice point is defined by the vector $\mathbf{P}(uv)$:

$$\mathbf{P}(uv) = u\mathbf{a} + v\mathbf{b} \tag{2.1}$$

where the vectors **a** and **b** define a parallelogram and u and v are integers. The parallelogram is the **unit cell** of the lattice, with sides of length a and b. The coordinates of the lattice points are indexed as u, v (Figure 2.1). Standard crystallographic terminology writes negative values as $\overline{u}$ and $\overline{v}$ (pronounced u bar and v bar). To agree with the convention for crystal systems given in Table 1.1, it is usual to label the angle between the lattice vectors as γ. The **lattice parameters** are the lengths of

Crystals and Crystal Structures, Second Edition. Richard J. D. Tilley.

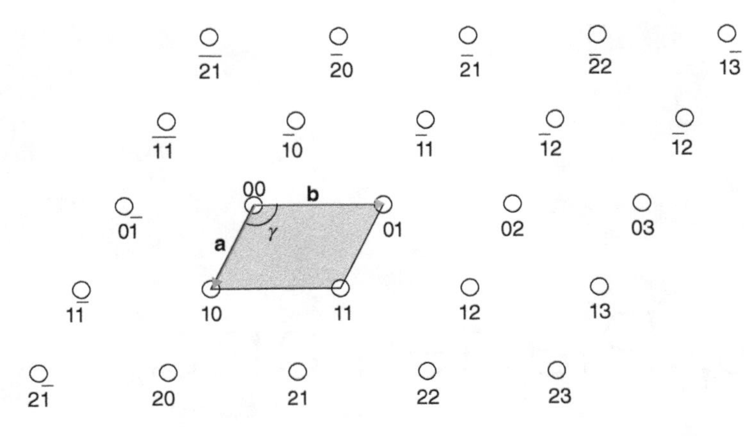

Figure 2.1 Part of an infinite lattice: the numbers are the indices, u, v, of each lattice point. The unit cell is shaded. Note that the lattice points are exaggerated in size and do not represent atoms.

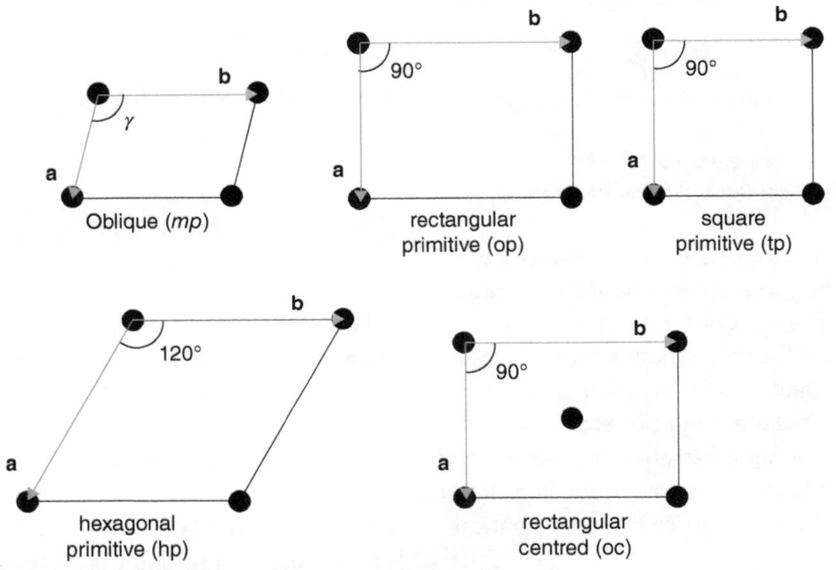

Figure 2.2 The unit cells of the five plane lattices: (a) oblique (*mp*); (b) rectangular (*op*); (c) square (*tp*); (d) hexagonal (*hp*); (e) rectangular centred (*oc*).

the axial vectors and the angle between them, a, b, and γ. The choice of the vectors **a** and **b**, which are called the **basis vectors** of the lattice, is completely arbitrary, and any number of unit cells can be constructed. However, for crystallographic purposes it is most convenient to choose as small a unit cell as possible and one that reveals the symmetry of the lattice.

Despite the multiplicity of possible unit cells, only five unique two-dimensional or **plane lattices** are possible (Figure 2.2). Those unit cells that contain only one lattice point are called **primitive** cells, and are labelled *p*. They are normally drawn with a lattice point at each cell corner, but it is easy to see that the unit cell contains just one lattice point by mentally displacing the unit cell outline slightly. There are four primitive plane lattices: oblique (*mp*), rectangular (*op*), square (*tp*) and hexagonal (*hp*).

The fifth plane lattice contains one lattice point at each corner and one in the unit cell centre. Such unit cells are called **centred** cells, and are labelled *c*. In this particular case, as the lattice is rectangular it is designated *oc*. The unit cell contains two lattice points, as can be verified by displacing the unit cell outline slightly. It is easily seen that this lattice could also be drawn as a primitive lattice (Figure 2.3). This latter lattice, known as a rhombic (*rp*) lattice, has the two basis vectors of equal length, and an interaxial angle, γ, different from 90°. In terms of the basis vectors of the centred *oc* cell, **a** and **b**, the basis vectors of the rhombic *rp* cell, **a**′ and **b**′ are:

$$\mathbf{a}' = \tfrac{1}{2}(\mathbf{a} + \mathbf{b})$$

$$\mathbf{b}' = \tfrac{1}{2}(-\mathbf{a} + \mathbf{b})$$

Note that this is a vector equation and the terms are to be added vectorially (see Appendix A).

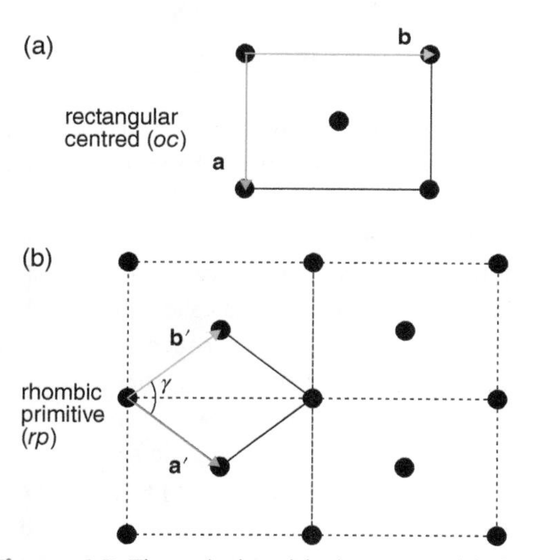

Figure 2.3 The relationship between (a) the rectangular (*oc*) lattice and (b) the equivalent rhombic primitive (*rp*) designation of the same lattice.

The five plane lattices are summarised in Table 2.1.

Although the lattice points in any primitive cell can be indexed in accordance with Eq. (2.1), using integer values of *u* and *v*, this is not possible with the *oc* lattice. For example, taking the basis vectors as the unit cell sides, the coordinates of the two lattice points in the unit cell are 0, 0 and ½, ½. For crystallographic purposes it is better to choose a basis that reflects the symmetry of the lattice rather than stick to the rigid definition given by Eq. (2.1). Because of this, lattices in crystallography are defined in terms of **conventional crystallographic bases** (and hence **conventional crystallographic unit cells**). In this formalism, the definition of Eq. (2.1) is relaxed, so that the coefficients of each vector, *u*, *v*, must be either **integral** or **rational**. (A rational number is a number that can be written as *a*/*b* where both *a* and *b* are integers.) The *oc* lattice makes use of this definition, as the rectangular nature of the lattice is of prime importance. Remember, though, that the *oc* and *rp* definitions are simply alternative descriptions of the same array of points. The lattice remains unique. It is simply necessary to use the description that makes things easiest.

Each lattice has a characteristic symmetry. The *mp* lattice has a diad running through each lattice point, as do the *oc* and *op* lattices. The hexagonal *hp* lattice is characterised by a hexad running through each lattice point and the square lattice, *tp*, has a tetrad through each lattice point and the unit cell centre. The combination of these symmetry elements with the translation of the lattice can produce other symmetry elements. For example, in the *tp* lattice the tetrads plus translation generate diads halfway along each unit cell edge and mirrors which run between all the tetrads and between all the diads. However, it is sufficient

Table 2.1 The five plane lattices

Crystal system (lattice type)	Lattice symbol	Lattice parameters
Oblique	*mp*	$a, b, \gamma \neq 90°$
Rectangular primitive	*op*	$a, b, \gamma = 90°$
Rectangular centred	*oc*	$a, b, \gamma = 90°$
	rp	$a' (= b'), \gamma \neq 90°$
Square	*tp*	$a (= b), \gamma = 90°$
Hexagonal	*hp*	$a (= b), \gamma = 120°$

just to specify the tetrads and the lattice type to account for the symmetries present.

It is an axiomatic principle of crystallography that a lattice cannot take on the symmetry of a regular pentagon. It is easy to see this. Suppose that a lattice fragment is drawn in which a lattice point is surrounded by five others arranged at the vertices of a regular pentagon (Figure 2.4a). Now each point in a lattice must have the exactly the same surroundings as any other lattice point. Displacement of the fragment by a lattice vector (Figure 2.4b and c) will show that some points are seen to be closer than others, which means that the construction is not a lattice. However, the fragment can form part of a pattern by placing the fragment itself upon each point of a lattice, in this case an *op* lattice (Figure 2.4d).

2.2 UNIT CELLS

The unit cells described above are conventional crystallographic unit cells. However, the method of unit cell construction described is not unique. Other shapes can be found that will fill the space and reproduce the lattice. Although these are not often used in crystallography, they are encountered in other areas of science. The commonest of these is the **Wigner-Seitz** cell.

This cell is constructed by drawing a line from each lattice point to its nearest neighbours (Figure 2.5a). A second set of lines is then drawn normal to the first, through the mid-points (Figure 2.5b). The polygon so defined (Figure 2.5c) is called the **Dirichlet region** or the Wigner-Seitz cell. Because of the method of construction, a Wigner-Seitz cell will always be primitive. Three-dimensional equivalents are described in Section 2.5.

2.3 THE RECIPROCAL LATTICE IN TWO DIMENSIONS

Many of the physical properties of crystals, as well as the geometry of the three-dimensional patterns of radiation diffracted by crystals (see Chapter 7), are most easily described by using the **reciprocal lattice**. The two-dimensional (plane) lattices, sometimes called the **direct lattices**, are said to occupy **real space**, and the reciprocal lattice occupies **reciprocal space**. The concept of the reciprocal lattice is straightforward. (Remember, the reciprocal lattice is simply another lattice.) It is defined in terms of two basis vectors labelled $\mathbf{a}^*$ and $\mathbf{b}^*$. The direction of these vectors is perpendicular to the end faces of the direct lattice unit cell. The lengths of the basis vectors of the

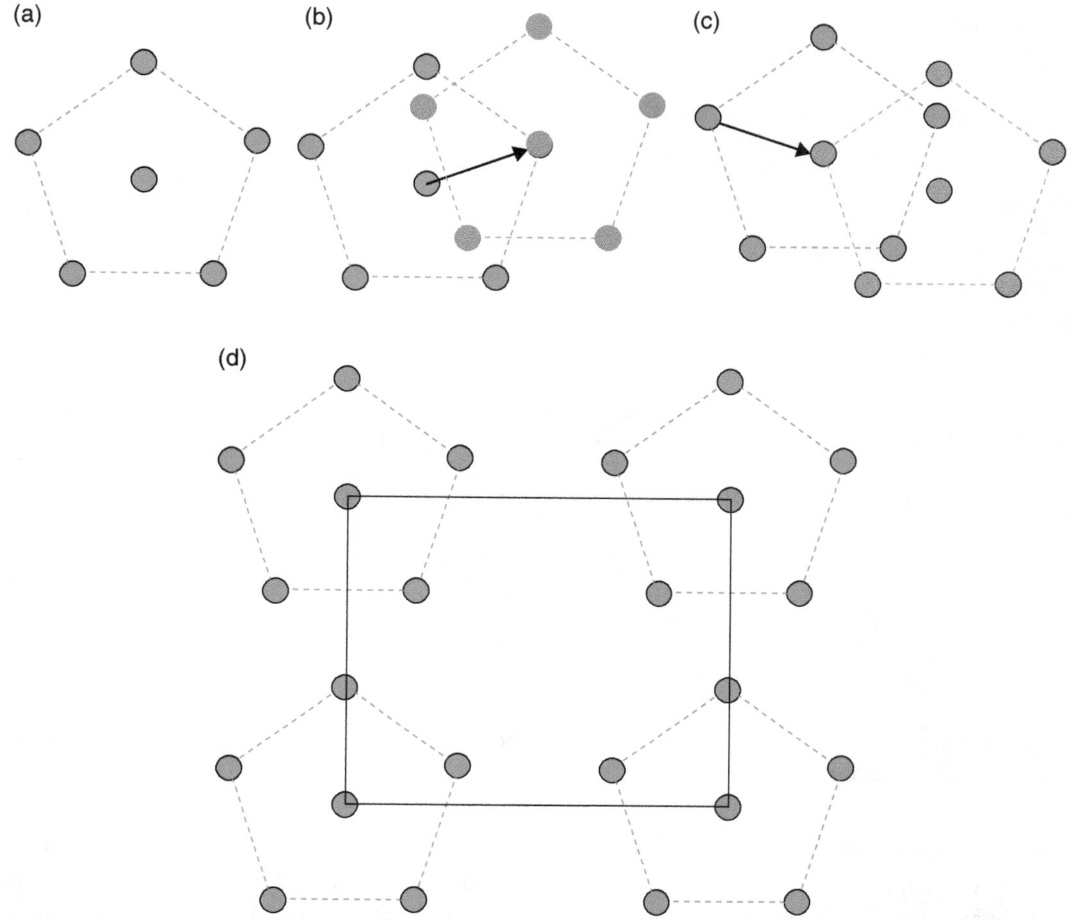

Figure 2.4 (a) A fragment of a 'lattice' with pentagonal symmetry; (b, c) displacement of the fragment by a lattice vector does not extend the lattice; (d) fragments with pentagonal symmetry can be part of a pattern by placing each on a lattice point.

reciprocal lattice are the inverse of the perpendicular distance from the lattice origin to the end faces of the direct lattice unit cell. For the square and rectangular plane lattices, this is simply the inverse of the lattice spacing:

$$a^* = 1/a, \quad b^* = 1/b$$

For the oblique and hexagonal plane lattices, these are given by:

$$a^* = 1/d_{10}, \quad b^* = 1/d_{01}$$

where the perpendicular distances between the rows of lattice points are labelled d_{10} and d_{01}.

The construction of a reciprocal plane lattice is simple, as illustrated for the oblique plane (*mp*) lattice. Draw the lattice and mark the unit cell (Figure 2.6a). Draw lines perpendicular to the two sides of the unit cell. These lines give the axial directions of the reciprocal lattice basis vectors (Figure 2.6b). Determine the perpendicular distances from the origin of the direct lattice to the end faces of the unit cell, d_{10}

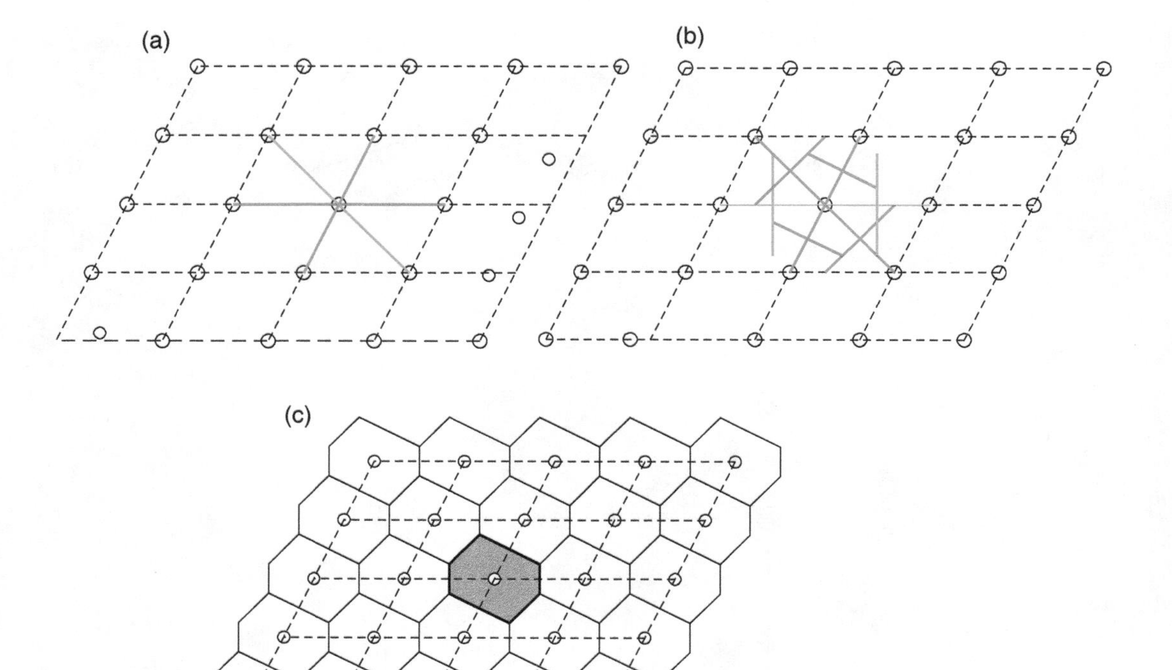

Figure 2.5 The construction of a Wigner-Seitz cell or Dirichlet region: the polygon formed (c, shaded) is the cell.

and d_{01} (Figure 2.6c). The inverse of these distances, $1/d_{10}$ and $1/d_{01}$, are the reciprocal lattice axial lengths, a^* and b^*.

$$1/d_{10} = a^*$$

$$1/d_{01} = b^*$$

Mark the lattice points at the appropriate reciprocal distances, and complete the lattice (Figure 2.6d). Note that in this case, as in all real and reciprocal lattice pairs, the vector joining the origin of the reciprocal lattice to a lattice point *hk* is perpendicular to the (*hk*) planes[1] in the real lattice and of length $1/d_{hk}$ (Figure 2.6e).

[1] For the definition of *h* and *k*, see Section 2.7.

It will be seen that the angle between the reciprocal axes, $\mathbf{a}^*$ and $\mathbf{b}^*$, is $(180 - \gamma) = \gamma^*$, when the angle between the direct axes, $\mathbf{a}$ and $\mathbf{b}$, is γ. It is thus simple to construct the reciprocal lattice by drawing $\mathbf{a}^*$ and $\mathbf{b}^*$ at an angle of $(180 - \gamma)$ and marking out the lattice with the appropriate spacing a^* and b^*.

Sometimes it can be advantageous to construct the reciprocal lattice of the centred rectangular *oc* lattice using the primitive unit cell. In this way it will be found that the primitive reciprocal lattice so formed can also be described as a centred rectangular lattice. This is a general feature of reciprocal lattices. Each direct lattice generates a reciprocal lattice of the same type, that is, *mp* → *mp*, *oc* → *oc*, and so on. In addition, the reciprocal lattice of a reciprocal lattice is the direct lattice.

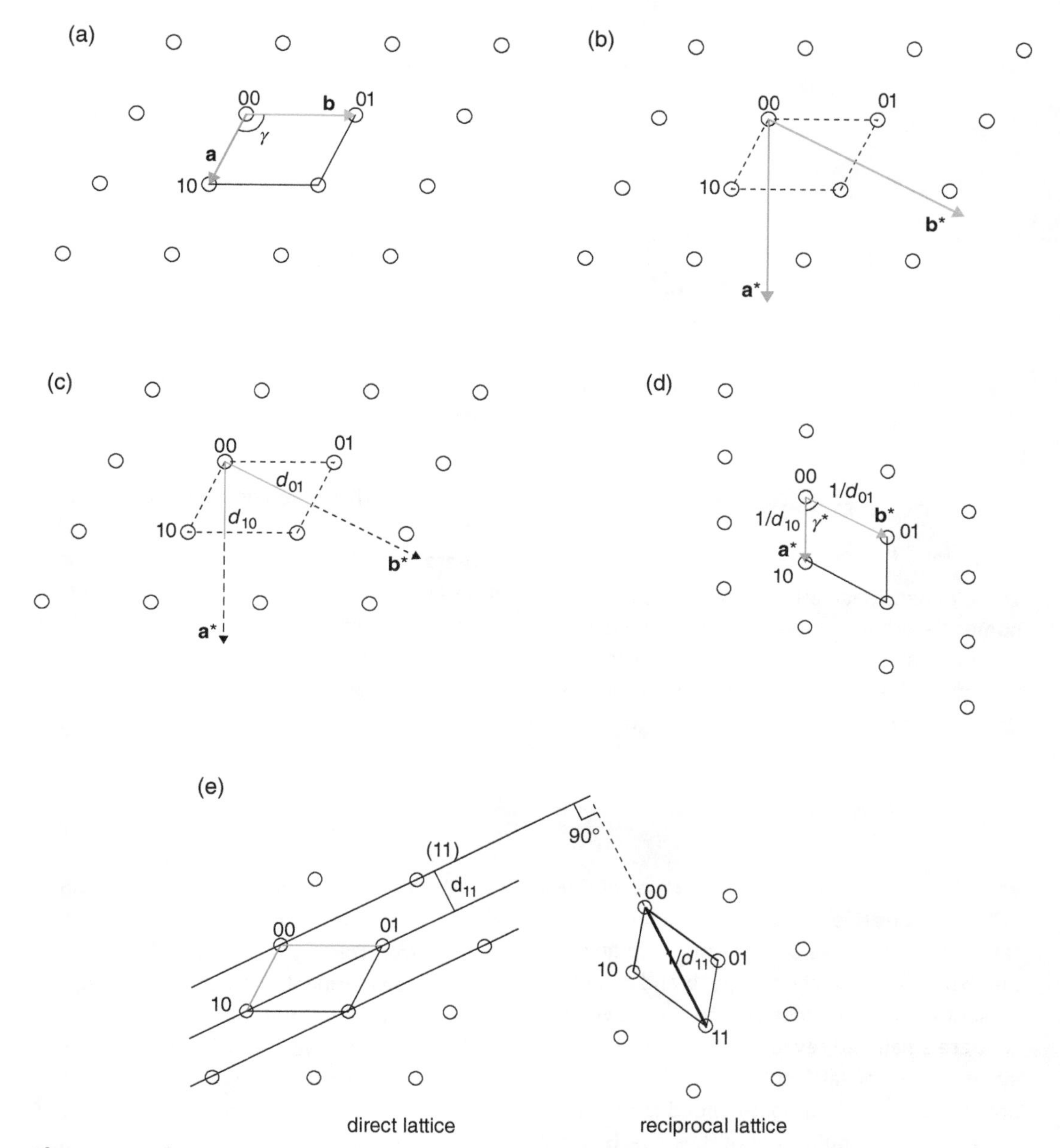

Figure 2.6 The construction of a reciprocal lattice (see text for details).

A construction in reciprocal space identical to that used to delineate the Wigner-Seitz cell in direct space gives a cell known as the first **Brillouin zone** (Figure 2.7). The zone is constructed by drawing the perpendicular bisectors of the lines connecting the origin, 00, to the nearest neighbouring lattice points, in an identical fashion to that used to obtain the Wigner-Seitz cell in real space. The first Brillouin zone of a lattice is thus a primitive cell.

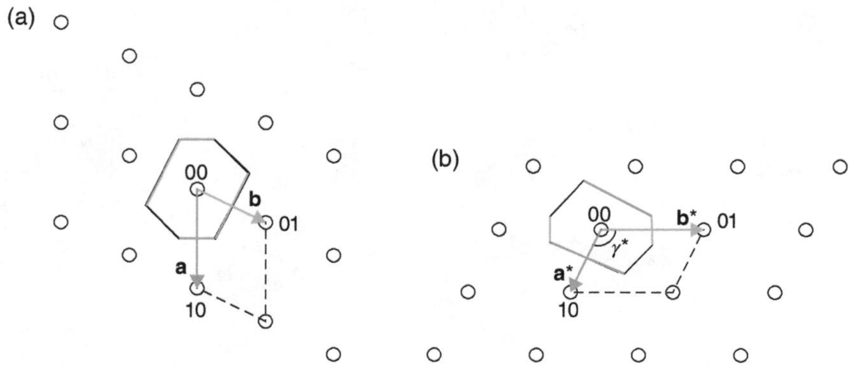

Figure 2.7 The first Brillouin zone of a reciprocal lattice: (a) the real lattice and Wigner-Seitz cell; (b) the reciprocal lattice and first Brillouin zone.

2.4 THREE-DIMENSIONAL LATTICES

Three-dimensional lattices use the same nomenclature as the two-dimensional lattices described above. If any lattice point is chosen as the origin, the position of any other lattice point is defined by the vector **P**(*uvw*):

$$\mathbf{P}(uvw) = u\mathbf{a} + v\mathbf{b} + w\mathbf{c}$$

where **a**, **b** and **c** are the basis vectors, and *u*, *v* and *w* are positive or negative integers or rational numbers. As before, there are any number of ways of choosing **a**, **b** and **c**, and crystallographic convention is to choose vectors that are small and reveal the underlying symmetry of the lattice. The three basis vectors are chosen to form a right-handed coordinate system, that is, **a** points out of the page, **b** points to the right and **c** is vertical. The parallelepiped formed by the three basis vectors **a**, **b** and **c**, defines the unit cell of the lattice, with edges of length *a*, *b* and *c*. The numerical values of the unit cell edges and the angles between them are collectively called the lattice parameters. It follows from the above description that the unit cell is not unique and is chosen for convenience and to reveal the underlying symmetry of the crystal.

There are only 14 possible three-dimensional lattices, called **Bravais** lattices (Figure 2.8). Bravais lattices are sometimes called direct lattices. Each Bravais lattice has a characteristic symmetry.

(i) The **a**, **b** and **c** basis vectors for a cubic lattice are parallel to the three tetrad (fourfold) symmetry axes. The conventional unit cell has $a_1 = a_2 = a_3$, $\alpha = \beta = \gamma = 90°$.
(ii) The basis vector **c** for the tetragonal lattice is taken along the unique tetrad (fourfold) symmetry axis. The conventional unit cell has $a_1 = a_2$, c, $\alpha = \beta = \gamma = 90°$.
(iii) The basis vectors **a**, **b** and **c** for an orthorhombic crystal lie along three mutually perpendicular diad (twofold) symmetry axes. The conventional unit cell has a, b, c, $\alpha = \beta = \gamma = 90°$.
(iv) The basis vector **c** for the hexagonal lattice lies parallel to the unique hexad (sixfold) symmetry axis. The conventional unit cell has $a_1 = a_2$ (= b), c, $\alpha = \beta = 90°$, $\gamma = 120°$. The lattice points in a primitive hexagonal lattice are stacked vertically above each other

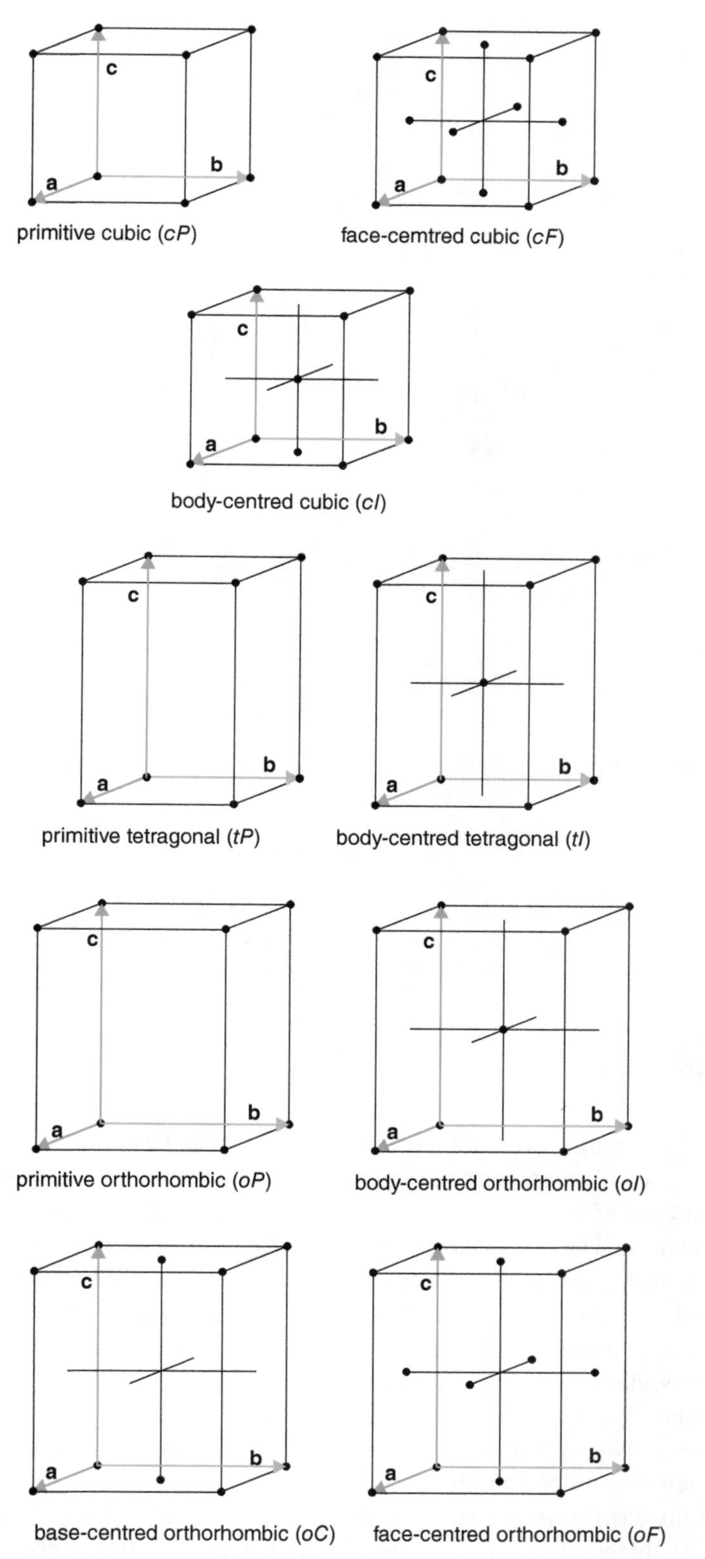

Figure 2.8 The 14 Bravais lattices. Note that the lattice points are exaggerated in size and are not atoms. The monoclinic lattices have been drawn with the **b**-axis vertical, to emphasise that it is normal to the plane containing the **a**- and **c**-axes.

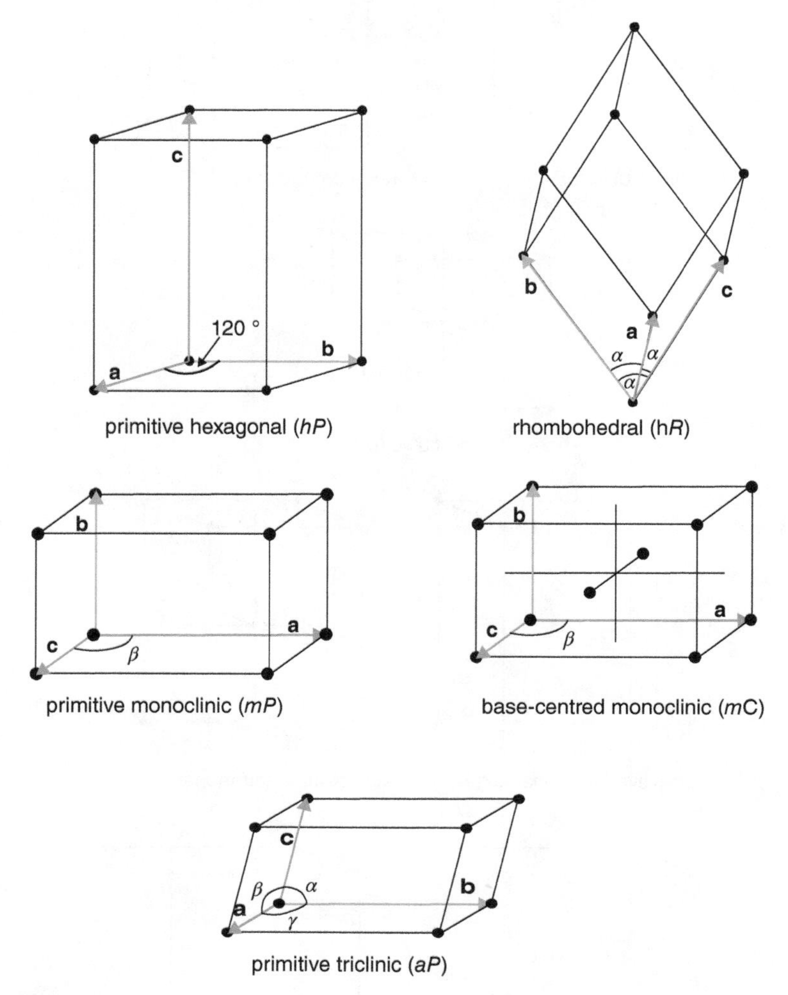

Figure 2.8 (Continued)

in the **c**-direction, represented as ... AAA ... where A refers to the position of each layer of lattice points.

(v) A rhombohedral lattice is characterised by a triad (threefold) symmetry axis. The rhombohedral axes, **a**, **b** and **c** are the shortest non-coplanar lattice vectors symmetrically equivalent with respect to the threefold axis. The unit cell has $a_1 = a_2 = a_3$, $\alpha = \beta = \gamma$. This lattice is often described in terms of a hexagonal lattice, in which the hexagonal **c**-axis lies along the threefold symmetry axis, with **a** and **b** normal to it and enclosing an angle of 120°, chosen as for the hexagonal system.

(vi) The unique symmetry direction in monoclinic lattices is a diad (twofold) symmetry axis, conventionally labelled **b**. The conventional unit cell has a, b, c, $\alpha = \gamma = 90°$, β.

(vii) A triclinic cell is chosen as primitive. The conventional unit cell has a, b, c, α, β, γ.

The smallest unit cell possible for any of the lattices, the one that contains just one lattice

Table 2.2 Bravais lattices

Crystal system	Lattice symbol	Required symmetry	Lattice parameters
Triclinic	*aP*	None	$a, b, c, \alpha, \beta, \gamma$
Monoclinic primitive	*mP*	1 diad	a, b, c, β ($\alpha = \gamma = 90°$)
Monoclinic centred	*mC*		
Orthorhombic primitive	*oP*	3 diads	$a, b, c, \alpha = \beta = \gamma = 90°$
Orthorhombic C-face-centred	*oC*		
Orthorhombic body-centred	*oI*		
Orthorhombic face-centred	*oF*		
Tetragonal primitive	*tP*	1 tetrad	a (= b), $c, \alpha = \beta = \gamma = 90°$
Tetragonal body-centred	*tI*		
Trigonal	*hR*	1 triad	$a = b = c, \alpha = \beta = \gamma$ (primitive cell); a (= b), $c, \alpha = \beta = 90°, \gamma = 120°$ (hexagonal cell)
Hexagonal primitive	*hP*	1 hexad	a (= b), $c, \alpha = \beta = 90°, \gamma = 120°$
Cubic primitive	*cP*	3 tetrads	a (= b = c), $\alpha = \beta = \gamma = 90°$
Cubic body-centred	*cI*		
Cubic face-centred	*cF*		

point, is the primitive unit cell. A primitive unit cell, usually drawn with a lattice point at each corner, is labelled *P*. All other lattice unit cells contain more than one lattice point. A unit cell with a lattice point at each corner and one at the centre of the unit cell (thus containing two lattice points in total) is called a **body-centred** unit cell, and labelled *I*. A unit cell with a lattice point in the middle of each face, thus containing four lattice points, is called a **face-centred** unit cell, and labelled *F*. A unit cell that has just one of the faces of the unit cell centred, thus containing two lattice points, is labelled **A-face-centred**, *A*, if the centred faces cut the **a**-axis, **B-face-centred**, *B*, if the centred faces cut the **b**-axis and **C-face-centred**, *C*, if the centred faces cut the **c**-axis. The 14 Bravais lattices are summarised in Table 2.2. Note that all of the non-primitive Bravais lattices can be described in terms of alternative primitive unit cells.

As in two dimensions, a lattice with a three-dimensional unit cell derived from regular pentagons, such as an icosahedron, cannot be constructed.

2.5 RHOMBOHEDRAL, HEXAGONAL AND CUBIC LATTICES

The rhombohedral lattice is often described in terms of a hexagonal lattice, in which **c** lies along the threefold symmetry axis, with **a** and **c** chosen as for the hexagonal system. When viewed along the threefold symmetry axis, each layer of lattice points sits in the centres of the triangles of lattice points in the layer below.

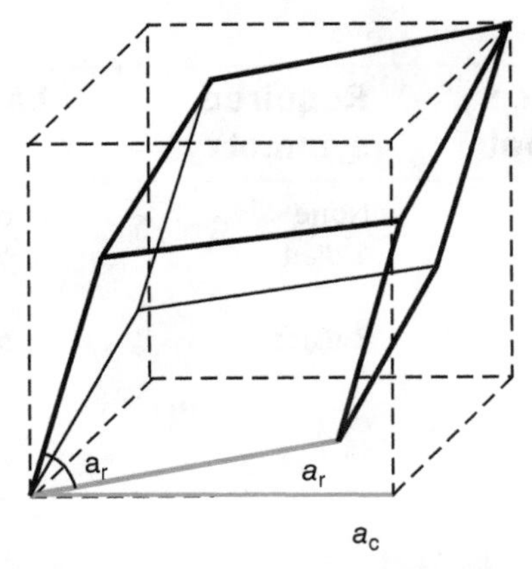

Figure 2.9 The relationship between a cubic lattice and a rhombohedral lattice.

These positions repeat after three layers, and the stacking can be represented as ... ABCABC .... This nomenclature is also used for the stacking of spheres (Section 8.2).

The rhombohedral primitive lattice can also, for special values of the rhombohedral angle α, be described in terms of a cubic lattice. If a cubic unit cell is compressed or stretched along one of the cube body diagonals it becomes rhombohedral (Figure 2.9). The compression or stretching direction becomes the unique threefold axis that characterises the system. In the case that the rhombohedral angle α_r is exactly equal to 90°, the unit cell is cubic where the lattice parameter of the cubic cell, a_c is given by:

$$a_c = a_r$$

where a_r is the rhombohedral unit cell parameter. When the cubic cell diagonal is stretched until the rhombohedral angle α is exactly equal to 60°, the points of the primitive rhombohedral lattice can be indexed in terms of a face-centred cubic lattice, where the lattice parameter of the cubic cell is given by:

$$a_c = (\sqrt{2})\, a_r$$

If the cubic unit cell diagonal is compressed until the rhombohedral angle α is equal to 109.47°, the points of the primitive rhombohedral lattice can be indexed in terms of a body-centred cubic lattice, and a body-centred unit cell can be chosen with lattice parameter:

$$a_c = 2\, a_r/\sqrt{3}$$

2.6 ALTERNATIVE UNIT CELLS

As outlined above, a number of alternative unit cells can be described for any lattice. The most widely used is the Wigner-Seitz cell, constructed in three dimensions in an analogous way to that described in Section 2.2. The Wigner-Seitz cell of a body-centred cubic Bravais *I* lattice is drawn in Figure 2.10a and b. It has the form of a truncated octahedron, centred upon any lattice point, with the square faces lying on the cube faces of the Bravais lattice unit cell. The Wigner-Seitz cell of a face-centred cubic Bravais *F* lattice is drawn in Figure 2.10c and d. The polyhedron is a regular rhombic dodecahedron. Note that it is displaced by ½ **a** with respect to the crystallographic cell, and is centred upon the lattice point marked * in Figure 2.10c. The lattice points labelled A, B, C help to make the relationship between the two cells clearer.

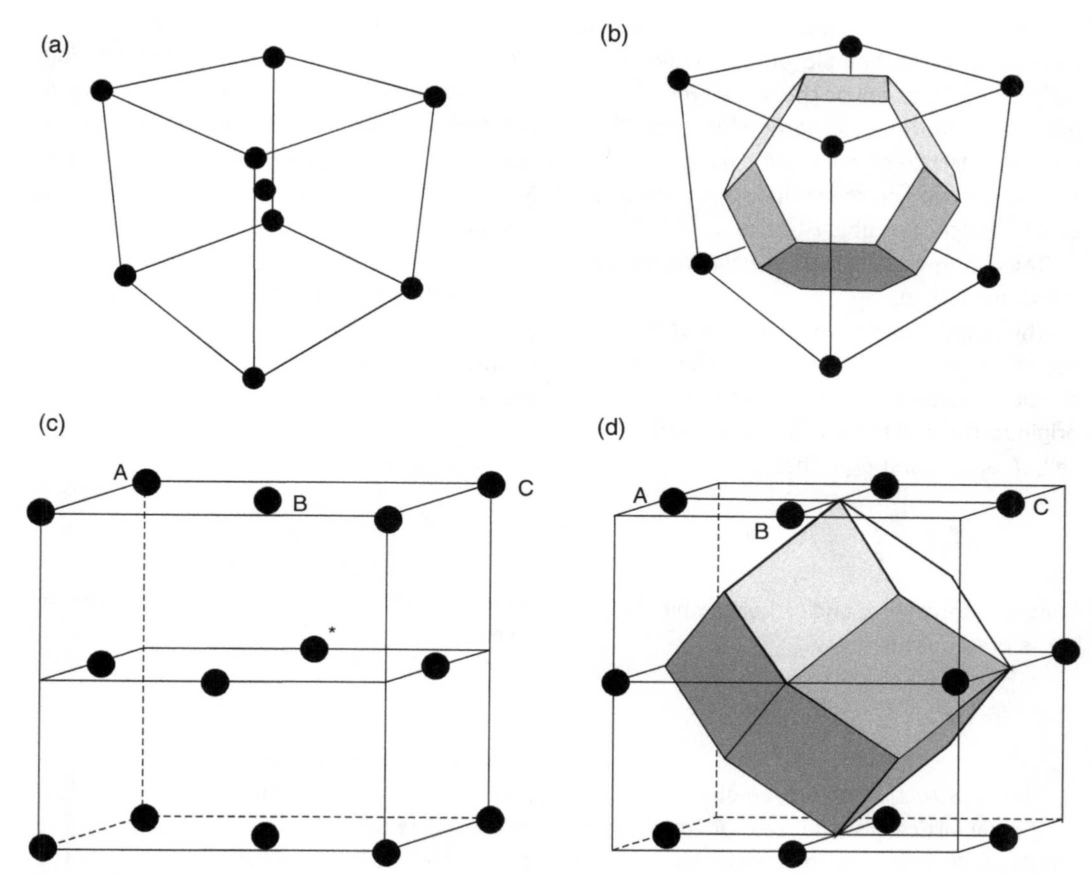

Figure 2.10 Wigner-Seitz cells: (a) the body-centred cubic lattice; (b) the Wigner-Seitz cell of (a); (c) the face-centred cubic lattice; (d) the Wigner-Seitz cell of (c). The face-centred cubic lattice point marked * forms the central lattice point in the Wigner-Seitz cell.

Recall that a Wigner-Seitz cell is always primitive. Other unit cells are sometimes of use when crystal structures are discussed (see Chapter 8).

2.7 THE RECIPROCAL LATTICE IN THREE DIMENSIONS

As with the two-dimensional lattices, the three-dimensional direct (Bravais) lattices are said to occupy real space, and can be used to construct three-dimensional reciprocal lattices occupying reciprocal space. Each reciprocal lattice is defined in terms of three basis vectors $\mathbf{a}^*$, $\mathbf{b}^*$ and $\mathbf{c}^*$. The direction of these vectors is perpendicular to the end faces of the direct lattice unit cell. This means that a direct axis will be perpendicular to a reciprocal axis if they have different labels, that is, $\mathbf{a}$ is perpendicular to $\mathbf{b}^*$ and $\mathbf{c}^*$. The reciprocal lattice axes are parallel to the direct lattice axes for cubic, tetragonal and orthorhombic primitive direct lattices, that is, $\mathbf{a}$ is parallel to $\mathbf{a}^*$, $\mathbf{b}$ is parallel to $\mathbf{b}^*$ and $\mathbf{c}$ is parallel to $\mathbf{c}^*$.

A direct lattice of a particular type, (triclinic, monoclinic, orthorhombic, etc.), will give a reciprocal lattice cell of the same type (triclinic, monoclinic, orthorhombic, etc.). The reciprocal lattice of the cubic *F* direct lattice is a cubic *I* lattice and the reciprocal lattice of the cubic *I* lattice is a cubic *F* lattice.

The reciprocal lattice of a reciprocal lattice is the direct lattice.

The lengths of the basis vectors of the reciprocal lattice, a^*, b^* and c^*, are the inverse of the perpendicular distance from the lattice origin to the end faces of the direct lattice unit cell, d_{100}, d_{010} and d_{001}, that is:

$$a^* = 1/d_{100}, b^* = 1/d_{010}, c^* = 1/d_{001}$$

For cubic, tetragonal and orthorhombic crystals, these are equivalent to:

$$a^* = 1/a, b^* = 1/b, c^* = 1/c$$

The construction of a three-dimensional reciprocal lattice is similar to that for a plane lattice: as an example, the construction of the reciprocal lattice of a *P* monoclinic lattice is described. The direct lattice has a lozenge-shaped unit cell, with the **b**-axis normal to the **a**- and **c**-axes (Figure 2.11a). To construct the sheet containing the **a*** and **c*** axes, draw the direct lattice unit cell projected down **b**, and draw normals to the end faces of the unit cell (Figure 2.11b). These give the directions of the reciprocal lattice **a*** and **c*** axes. The reciprocals of the perpendicular distances from the origin to the faces of the unit cell give the axial lengths (Figure 2.11c). These allow the reciprocal lattice plane to be drawn (Figure 2.11d).

The **b**-axis is normal to **a** and **c**, and so the **b*** axis is parallel to **b** and normal to the section containing **a*** and **c***, as drawn. The length of the **b*** axis is equal to $1/b$. The reciprocal lattice layer containing the 010 point is then identical to Figure 2.11d, and is stacked a distance of b^* vertically below it, whilst the layer containing the $0\bar{1}0$ point is vertically above it (Figure 2.11e). Other layers then follow in the same way.

For some purposes, it is convenient to multiply the length of the reciprocal axes by a constant. Thus, physics texts frequently use a reciprocal lattice spacing 2π times that given above, that is:

$$a^* = 2\pi/d_{100}, b^* = 2\pi/d_{010}, c^* = 2\pi/d_{001}$$

Crystallographers often use a reciprocal lattice scale multiplied by λ, the wavelength of the radiation used to obtain a diffraction pattern, so that:

$$a^* = \lambda/d_{100}, b^* = \lambda/d_{010}, c^* = \lambda/d_{001}$$

As with two-dimensional lattices, the procedure required to construct the (primitive) Wigner-Seitz cell in the direct lattice yields a cell called the first Brillouin zone when applied to the reciprocal lattice. The lattice that is reciprocal to a real space face-centred cubic *F* lattice is a body-centred cubic *I* lattice. The (primitive) Wigner-Seitz cell of the cubic body-centred *I* lattice (Figure 2.10a), a truncated octahedron (Figure 2.10b), is therefore identical in shape to the (primitive) first Brillouin zone of a face-centred cubic *F* lattice. In the same way, the lattice that is reciprocal to a real space body-centred cubic *I* lattice is a face-centred cubic *F* lattice. The Wigner-Seitz cell of the real space face-centred cubic *F* lattice (Figure 2.10c), a regular rhombic dodecahedron (Figure 2.10d), is identical in shape to the first Brillouin zone of a body-centred cubic *I* lattice (Table 2.3).

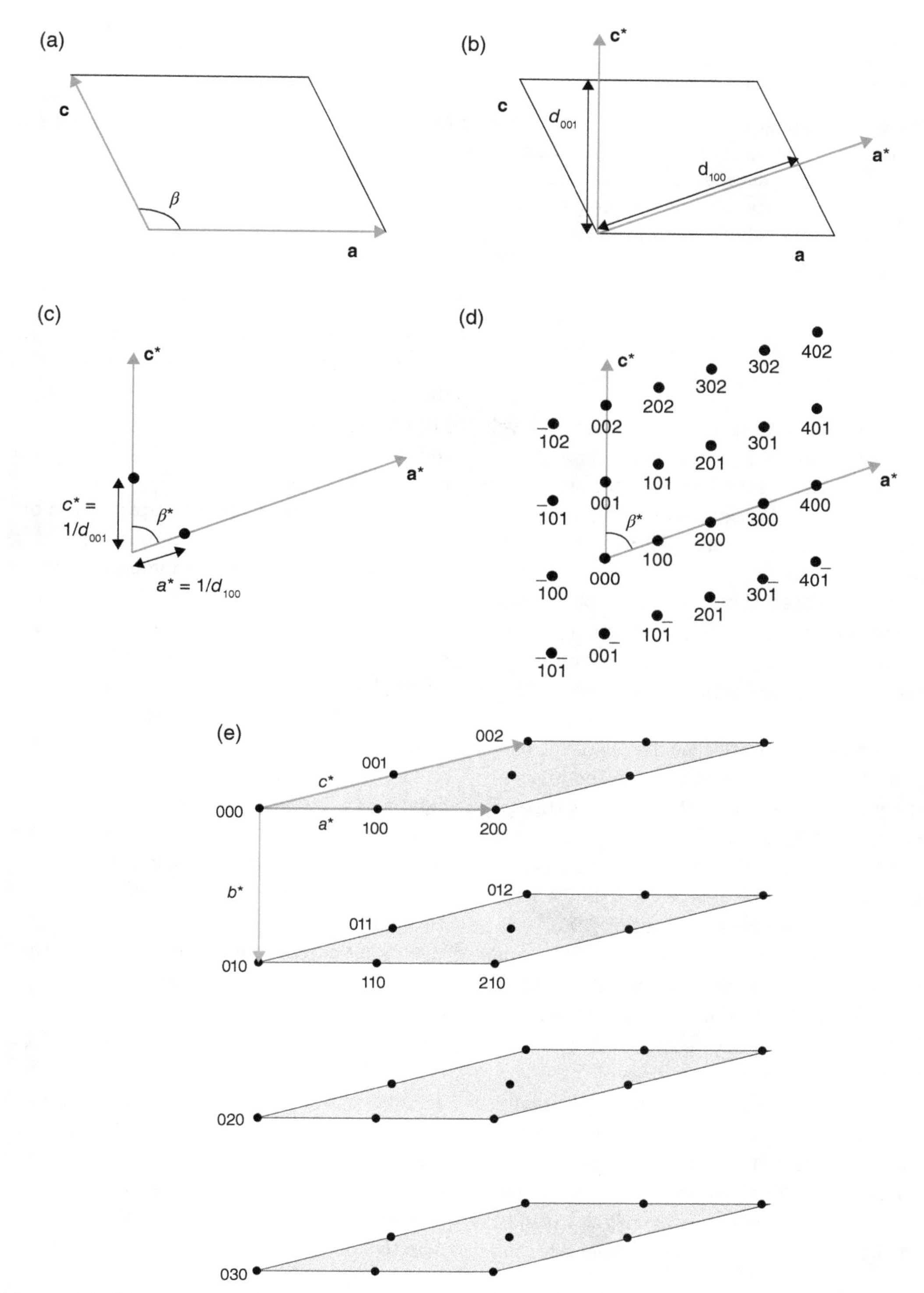

Figure 2.11 The construction of a monoclinic reciprocal lattice; see text for details.

Table 2.3 Reciprocal and real space cells

Lattice	Direct lattice	Reciprocal lattice
Plane	Oblique *mp*: Wigner-Seitz cell Figure 2.7a	Oblique *mp*: 1st Brillouin zone Figure 2.7b
Cubic 3d	*F* lattice: Wigner-Seitz cell is a rhombic dodecahedron, Figure 2.10d	*I* lattice: 1st Brillouin zone is a truncated octahedron, Figure 2.10b
	I lattice: Wigner-Seitz cell is a truncated octahedron, Figure 2.10b	*F* lattice: 1st Brillouin zone is a rhombic dodecahedron, Figure 2.10d

2.8 LATTICE PLANES AND MILLER INDICES

As described in Chapter 1, the facets of a well-formed crystal or internal planes through a crystal structure are specified in terms of **Miller indices**, *h*, *k* and *l*, written in round brackets, (*hkl*). The same terminology is used to specify planes in a lattice.

Miller indices, (*hkl*), represent not just one plane, but the set of all identical parallel lattice planes. The values of *h*, *k* and *l* are the reciprocals of the fractions of a unit cell edge, *a*, *b* and *c* respectively, intersected by an appropriate plane. This means that a set of planes that lie parallel to a unit cell edge is given the index 0 (zero) regardless of the lattice geometry. Thus a set of planes that pass across the ends of the unit cells, cutting the **a**-axis at a position 1 *a*, and parallel to the **b**- and **c**-axes of the unit cell, has Miller indices of (100) (Figure 2.12a and b). The same principles apply to the other planes shown. The set of planes that lies parallel to the **a**- and **c**-axes, and intersecting the end of each unit cell at a position 1 *b*, have Miller indices of (010) (Figure 2.12c and d). The set of planes that lies parallel to the **a**- and **b**-axes, and intersecting the end of each unit cell at a position 1 *c*, have Miller indices (001) (Figure 2.12e and f). Planes cutting both the **a**-axis and **b**-axis at 1 *a* and 1 *b* will be (110) planes (Figure 2.12g and h), and planes cutting the **a**-, **b**- and **c**-axes at 1 *a*, 1 *b* and 1 *c* will be (111) (Figure 2.12i). Remember that the Miller indices refer to a family of planes, not just one. For example, Figure 2.13 shows part of the set of (122) planes, which cut the unit cell edges at 1 *a*, ½ *b* and ½ *c*.

The Miller indices for lattice planes can be determined using a simple method. For convenience, take an orthorhombic lattice, and imagine a set of planes parallel to the **c**-axis, intersecting parallel sheets of lattice points as drawn in Figure 2.14. The Miller indices of this set of planes are determined by travelling along the axes in turn and counting the number of spaces between planes encountered on passing from one lattice point to the next. Thus, in Figure 2.14, three spaces are crossed on going from one lattice point to another along the **a**-direction, so that *h* is 3. In travelling along the **b**-direction, two spaces are encountered in going from one lattice point to another, so that the value of *k* is 2. As the planes are supposed to be parallel to the **c**-axis, the value of *l* is 0. The planes have Miller indices (320). For non-zero values of *l*, repeat the sketch so as to include **a**- and **c**- or **b**- and **c**-, and repeat the procedure.

The planes described in Figures 2.12–2.14 are crossed in travelling along positive directions of the axes. Some planes may intersect one of the axes in a positive direction and the other in a negative direction. Negative intersections are

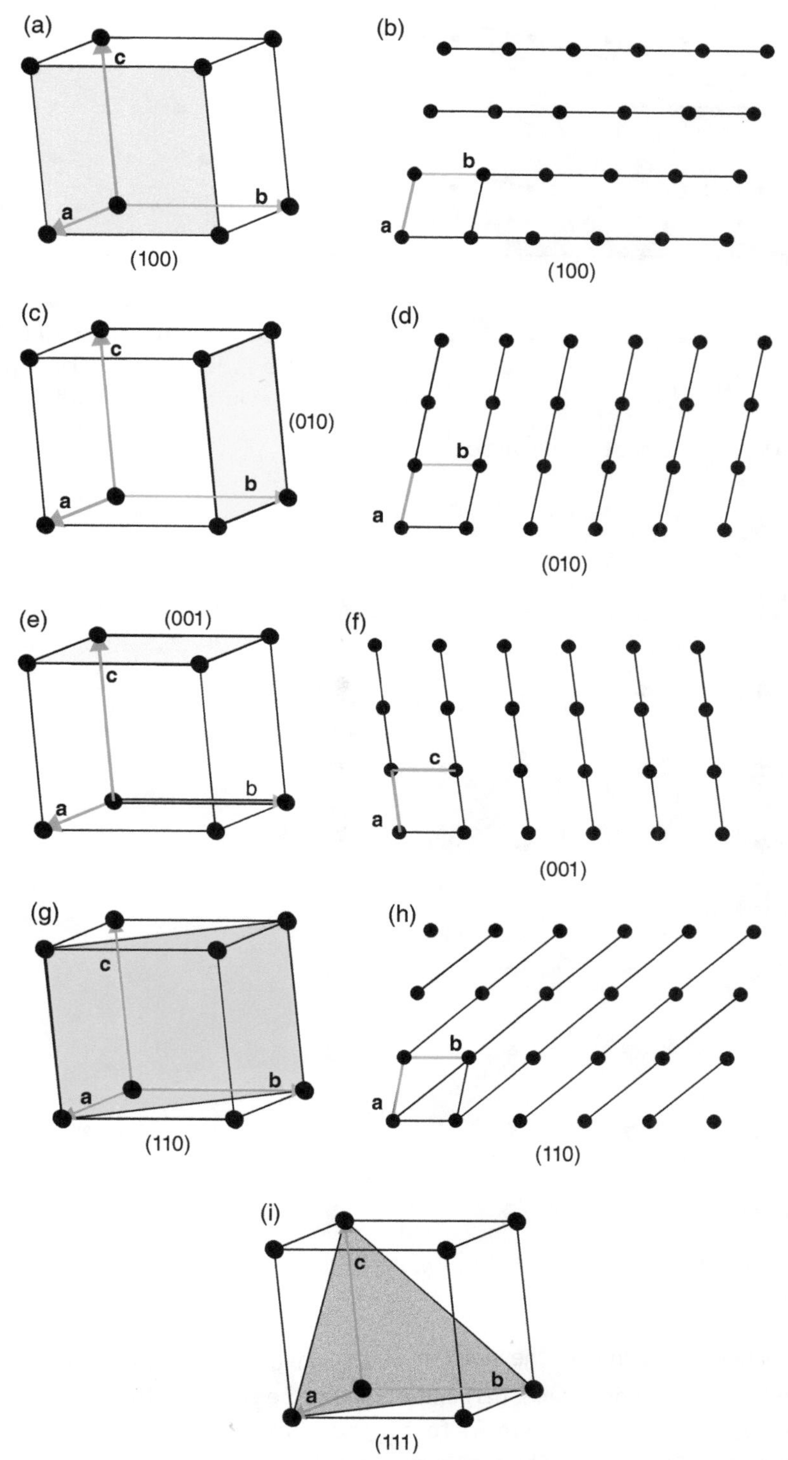

Figure 2.12 Miller indices of lattice planes: (a, b) (100); (c, d) (010); (e, f) (001), (g, h) (110); (i) (111).

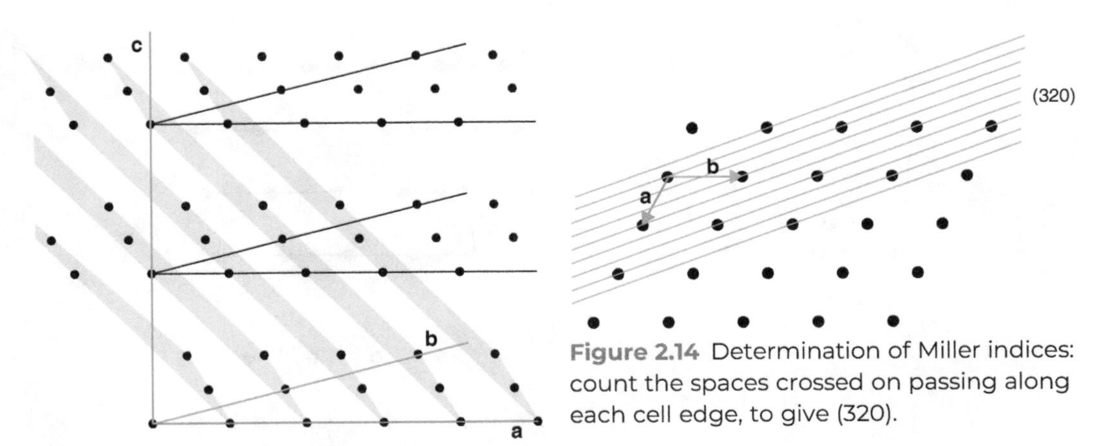

Figure 2.13 Part of the set of (122) lattice planes.

Figure 2.14 Determination of Miller indices: count the spaces crossed on passing along each cell edge, to give (320).

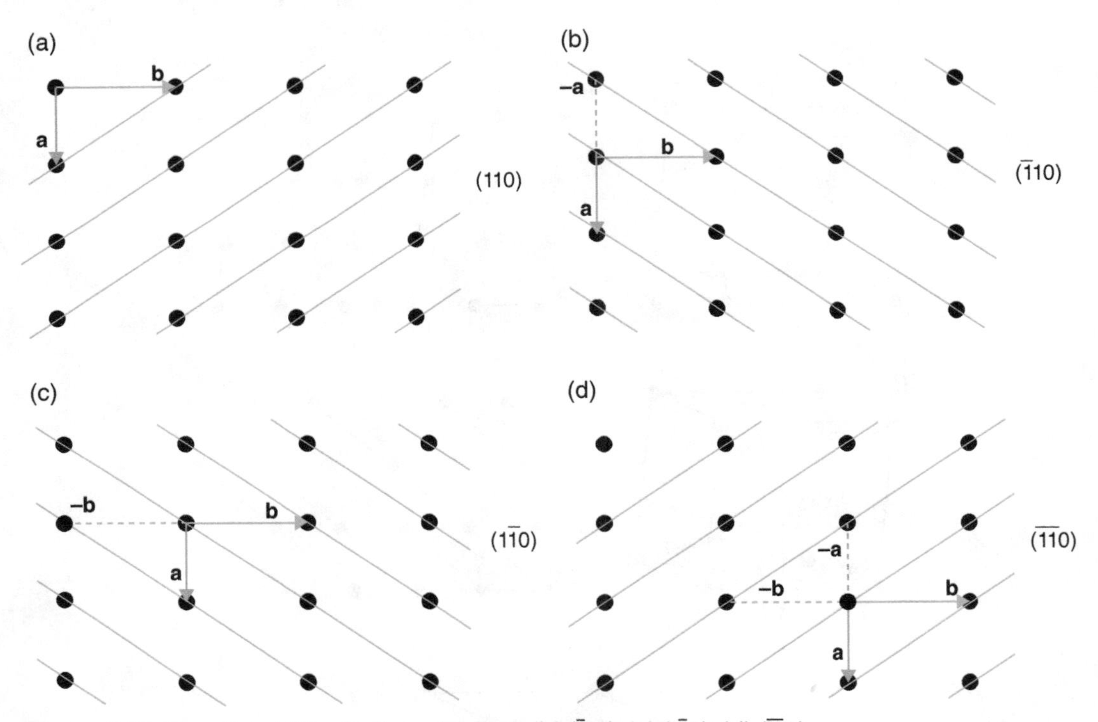

Figure 2.15 Negative Miller indices: (a) (110); (b) ($\bar{1}$10); (c) (1$\bar{1}$0); (d) ($\bar{1}\bar{1}$0).

written with a negative sign over the index, $\bar{h}$ (pronounced h bar), $\bar{k}$ (pronounced k bar) and $\bar{l}$ (pronounced l bar). For example, there are four planes related to (110), three of which involve travelling in a negative axial direction in order to count the spaces between the planes encountered. The planes shown in Figure 2.15a cut the **a**-axis at +1*a* and the **b**-axis at +1*b*, so that the planes have Miller indices (110). The planes shown in Figure 2.15b cut the **a**-axis at $-1a$ and the **b**-axis at $+1b$, so that the planes have Miller indices $(\bar{1}10)$. The planes shown in Figure 2.15c cut the **a**-axis at $+1a$ and the **b**-axis at $-1b$, so that the planes have Miller indices $(1\bar{1}0)$. Finally, the planes shown in Figure 2.15d cut the **a**-axis at $-1a$ and the **b**-axis at $-1b$, so the planes have Miller indices $(\bar{1}\bar{1}0)$. Note that the $(\bar{1}\bar{1}0)$ plane is identical to the (110) plane, as the position of the axes is arbitrary, allowing them to be placed anywhere on the diagram. Similarly, planes with Miller indices $(1\bar{1}0)$ are identical to $(\bar{1}10)$ planes.

In the three-dimensional direct and reciprocal lattice pairs, the vector joining the origin of the reciprocal lattice to a lattice point *hkl* is perpendicular to the (*hkl*) planes in the real lattice and of length $1/d_{hkl}$. An alternative method of constructing the reciprocal lattice is thus to draw the normals to the relevant (*hkl*) planes on the direct lattice and plot the reciprocal lattice points along lines parallel to the normal lines at separations of $1/d_{hkl}$. This method is often advantageous when constructing the reciprocal lattices for all of the face- or body-centred direct lattices. The same is true for the planar direct and reciprocal lattice pairs. The vector joining the origin of the reciprocal lattice to a lattice point *hk* is perpendicular to the (*hk*) planes in the real lattice and of length $1/d_{hk}$.

In lattices of high symmetry, there are always some sets of (*hkl*) planes that are identical from the point of view of symmetry. For example, in a cubic lattice, the (100), (010) and (001) planes are identical in every way. Similarly, in a tetragonal lattice, (110) and $(\bar{1}10)$ planes are identical. Curly brackets, {hkl}, designate these related planes. Thus, in the cubic system, the symbol {100} represents the set of planes (100), $(\bar{1}00)$, (010), $(0\bar{1}0)$, (001), and $(00\bar{1})$, the symbol {110} represents the set of planes (110), (101), (011), $(\bar{1}10)$, $(\bar{1}01)$ and $(0\bar{1}1)$, and the symbol {111} represents the set (111), $(11\bar{1})$, $(1\bar{1}1)$ and $(\bar{1}11)$.

2.9 HEXAGONAL LATTICES AND MILLER-BRAVAIS INDICES

The Miller indices of planes in hexagonal lattices can be ambiguous. For example, three sets of planes lying parallel to the **c**-axis, which is imagined to be perpendicular to the plane of the diagram, are shown in Figure 2.16. These planes have Miller indices A: (110), B: $(1\bar{2}0)$ and C: $(\bar{2}10)$. Although these Miller indices seem to refer to different types of plane, clearly the three planes are identical, and all are equivalent to the planes through the 'long' diagonal of the unit cell. In order to eliminate this confusion, four indices, (*hkil*), are often used to specify planes in a hexagonal crystal. These are called **Miller-Bravais indices** and are only used in the hexagonal system. The index *i* is given by:

$$h + k + i = 0, \text{or } i = -(h + k)$$

In reality this third index is not needed, as it is simply derived from the known values of *h* and *k*. However, it does help to bring out the relationship between the planes. Using four indices, the planes are A: $(11\bar{2}0)$, B: $(1\bar{2}10)$ and C: $(\bar{2}110)$. Because it is a redundant index, the value of *i* is sometimes replaced by a dot, to give indices (*hk.l*). This nomenclature emphasises that the hexagonal system is under discussion without actually including a value for *i*.

2.10 MILLER INDICES AND PLANES IN CRYSTALS

In most lattices, and all primitive lattices, there are no planes parallel to (100) with a smaller spacing than the (100) planes, because lattice planes must pass through sheets of lattice

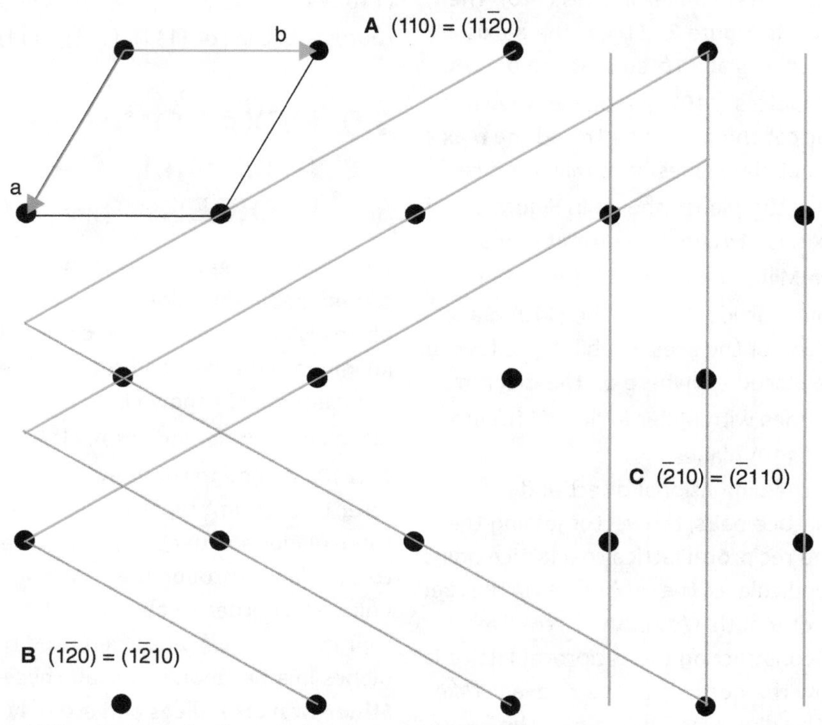

Figure 2.16 Miller-Bravais indices in hexagonal lattices.

points or nodes (see Figure 2.12). The same can be said of (010), (001), (110) and other planes. However, such smaller spacing planes can be described in crystals, and are significant. These can be characterised in the same way as lattice planes. For example, the fluorite structure, described in Chapter 1, has alternating planes of Ca and F atoms running perpendicular to the **a**-, **b**- and **c**-axes (see Figure 1.13). Taking the (100) planes as an example, these lie through the end faces of each unit cell and are perpendicular to the **b**- and **c**-axes (Figure 2.17a). They contain only Ca atoms. A similar plane, also containing only Ca atoms, runs through the middle of the cell. Using the construction described earlier (Figure 2.14), these can be indexed as (200), as two interplanar spaces are crossed in moving one lattice parameter distance (Figure 2.17b). Thus all the set of (200) planes contain only Ca atoms. The set of parallel planes with half the spacing of the (200) set will be indexed as (400), as four spaces are crossed in moving one lattice parameter distance. However, these planes are not all identical, as some contain only F atoms and others only Ca atoms (Figure 2.17c). The Miller indices of (110) (Figure 2.17d) and (220) (Figure 2.17e) planes do not show this change and the atomic composition of both sets of planes is the same.

Any general set of planes parallel to the **b**- and **c**-axes, and so only cutting the **a** cell edge, is written (*h*00). Any general set of planes parallel to the **a**- and **c**-axes, and so only cutting the **b** cell-edge has indices (0*k*0), and any general plane parallel to the **a**- and **b**-axes, and so cutting the **c** cell-edge, has indices (00*l*). Planes

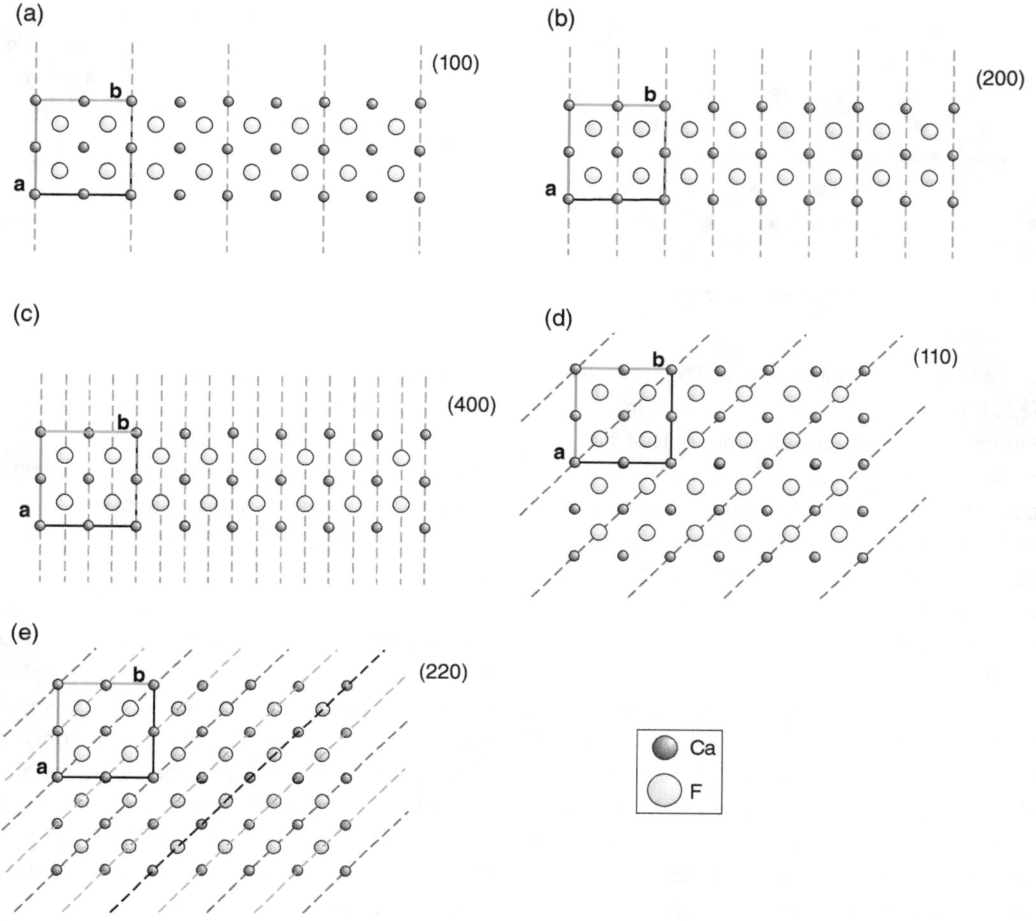

Figure 2.17 Planes in a crystal of fluorite, CaF_2: (a) (100); (b) (200); (c) (400); (d) (110); (e) (220).

that cut two edges and parallel to a third are described by indices (*hk*0), (0*kl*), or (*h*0*l*). Planes that are at an angle to all three axes have indices (*hkl*). Negative intersections and symmetrically equivalent planes are defined using the same terminology as described in Section 2.8.

2.11 DIRECTIONS

The response of a crystal to an external stimulus such as a tensile stress, electric field, and so on, is frequently dependent upon the direction of the applied stimulus. It is therefore important to be able to specify directions in both lattices and crystals in an unambiguous fashion.

The three indices *u*, *v* and *w* define the coordinates of a point within the lattice. The index *u* gives the coordinates in terms of the lattice repeat *a* along the **a**-axis, the index *v* gives the coordinates in terms of the lattice repeat *b* along the **b**-axis, and the index *w* gives the coordinates in terms of the lattice repeat *c* along the **c**-axis. Directions are written generally as [*uvw*] and are enclosed in square brackets and the direction [*uvw*] is simply the vector pointing from the origin to the lattice

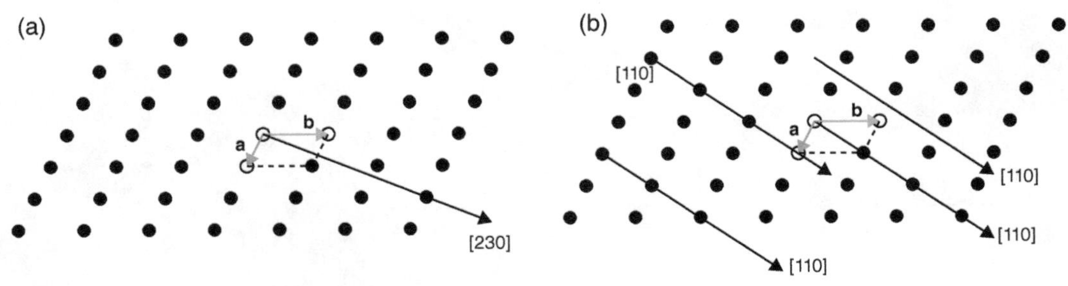

Figure 2.18 Directions in a lattice: (a) [230]; (b) [110].

point with coordinates *u, v, w*. The direction [230], with *u* = 2 and *v* = 3, is drawn in Figure 2.18a. Remember, though, that any parallel direction shares the symbol [*uvw*], because the origin of the coordinate system is not fixed and can always be moved to the starting point of the vector (Figure 2.18b). (A North wind is always a North wind, regardless of where you stand.)

The direction [100] is parallel to the **a**-axis, the direction [010] is parallel to the **b**-axis, and [001] is parallel to the **c**-axis. The direction [111] lies along the body diagonal of the unit cell. Negative values of *u, v* and *w* are written $\overline{u}$ (pronounced u bar), $\overline{v}$ (pronounced v bar) and $\overline{w}$ (pronounced w bar). Further examples of directions in a lattice are illustrated in Figure 2.19. Because directions are vectors, [*uvw*] is not identical to $[\overline{u}\overline{v}\overline{w}]$, in the same way that the direction 'North' is not the same as the direction 'South'.

Directions in a crystal are specified in the same way. In these instances the integers *u, v, w* are applied to the unit cell vectors, **a**-, **b**- and **c**-. As with Miller indices, it is sometimes convenient to group together all directions that are identical by virtue of the symmetry of the structure. These are represented by the notation <*uvw*>. In a cubic crystal the symbol <100> represents the six directions [100], $[\bar{1}00]$, [010], $[0\bar{1}0]$, [001] and $[00\bar{1}]$.

A **zone** is a set of planes, all of which are parallel to a single direction. A group of planes that intersect along a common line therefore forms a zone. The direction that is parallel to the planes, which is the same as the line of intersection, is called the **zone axis**. The zone axis [*uvw*] is perpendicular to the plane (*uvw*) in cubic crystals but *not* in crystals of other symmetry.

It is sometimes important to specify a vector with a definite length. In such cases the vector, **R**, is written by specifying the end coordinates, *u, v, w*, with respect to an origin at 0, 0, 0. Should the vector be greater or less than the specified length, it is prefixed by the appropriate scalar multiplier (see Appendix A), in accordance with normal vector arithmetic (Figure 2.20a). Crystals often contain planar boundaries which separate two parts of a crystal that are not in perfect register. Vectors are used to define the displacement of one part with respect to the other. For example, a fault involving the displacement of one part of a crystal of fluorite, CaF_2, by ¼ of the unit cell edge, i.e. **R** = ¼ [010] with respect to the other part,[2] is drawn in (Figure 2.20b).

As with Miller indices, directions in hexagonal crystals are sometimes specified in

[2] In cubic crystals, a vector such as that describing the boundary drawn in Figure 2.20b is frequently denoted as ¼ **a** [010]. Similarly, a vector 2/3 $[\bar{3}\bar{1}0]$ may be written as 2/3 **a** $[\bar{3}\bar{1}0]$. This notation is confusing, and is best avoided.

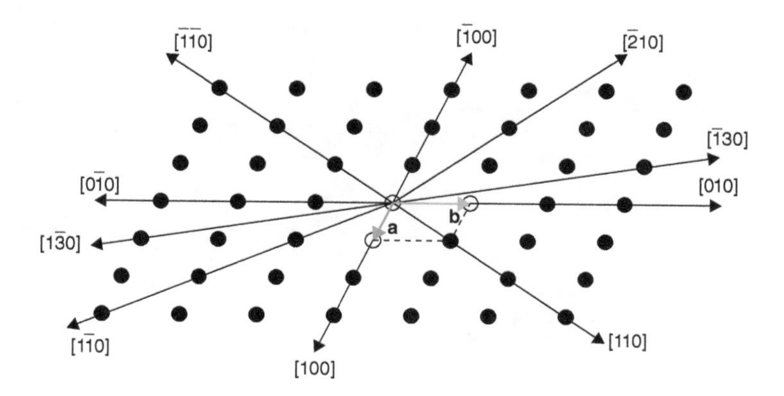

Figure 2.19 Directions in a lattice.

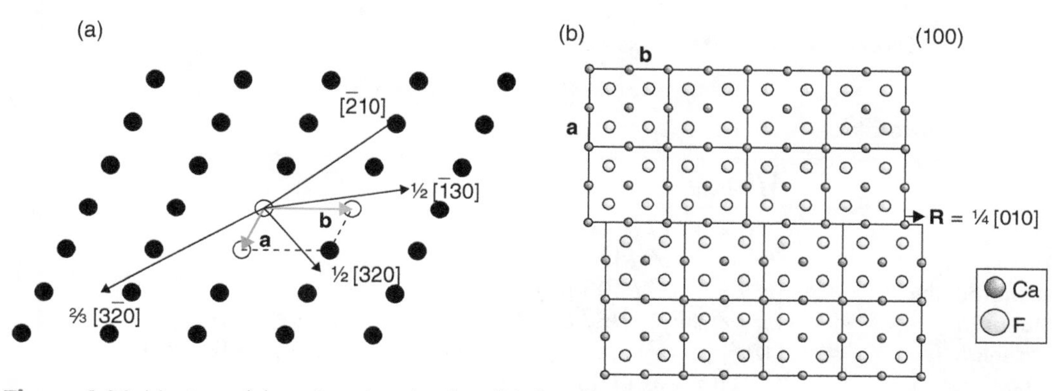

Figure 2.20 Vectors: (a) vectors in a lattice; (b) the displacement of part of a crystal of fluorite, CaF_2, by a vector, **R** = ¼ [010] with respect to the other part.

terms of a four-index system, $[u'v'tw']$, called **Weber** indices. The conversion of a three-index set to a four-index set is given by the following rules.

$$[uvw] \rightarrow [u'v'tw']$$
$$u' = n(2u - v)/3$$
$$v' = n(2v - u)/3$$
$$t = -(u' + v')$$
$$w' = nw$$

In these equations, n is a factor *sometimes* needed to make the new indices into smallest integers. Thus the direction [001] always transforms to [0001]. The three equivalent directions in the basal (0001) plane of a hexagonal lattice (Figure 2.21) are obtained by using the above transformations. The correspondence is:

$$[100] = [2\bar{1}\bar{1}0]$$
$$[010] = [\bar{1}2\bar{1}0]$$
$$[\bar{1}\bar{1}0] = [\bar{1}\bar{1}20]$$

The relationship between directions and planes depends upon the symmetry of the crystal. In cubic crystals (and *only* cubic crystals), the direction [hkl] is normal to the plane (hkl).

2.12 LATTICE GEOMETRY

The most important metrical properties of lattices and crystals for everyday crystallography are given in Tables 2.4–2.6.

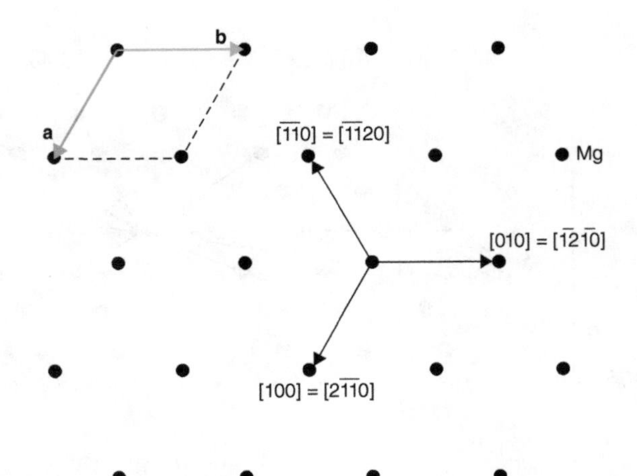

Figure 2.21 Directions in a hexagonal lattice.

Table 2.4 Interplanar spacing, d_{hkl}

System	$1/(d_{hkl})^2$
Cubic	$[h^2 + k^2 + l^2] / a^2$
Tetragonal	$[(h^2 + k^2) / a^2] + [l^2 / c^2]$
Orthorhombic	$[h^2 / a^2] + [k^2 / b^2] + [l^2 / c^2]$
Monoclinic	$[h^2 / a^2\sin^2\beta] + [k^2 / b^2] + [l^2 / c^2\sin^2\beta] - [(2hl\cos\beta) / (ac\sin^2\beta)]$
Triclinic[a]	$[1/V^2]\ \{[S_{11}h^2] + [S_{22}k^2] + [S_{33}\ l^2] + [2S_{12}hk] + [2S_{23}kl] + [2S_{13}hl]\}$
Hexagonal	$[4/3]\ [(h^2 + hk + k^2) / a^2] + [l^2 / c^2)]$
Rhombohedral	$\{[(h^2 + k^2 + l^2)\ \sin^2\alpha + 2(hk + kl + hl)(\cos^2\alpha - \cos\alpha] / [a^2(1 - 3\cos^2\alpha + 2\cos^3\alpha)]\}$

[a] V = unit cell volume.
$S_{11} = b^2c^2\sin^2\alpha$; $S_{22} = a^2c^2\sin^2\beta$; $S_{33} = a^2b^2\sin^2\gamma$.
$S_{12} = abc^2\ (\cos\alpha\ \cos\beta - \cos\gamma)$; $S_{23} = a^2bc\ (\cos\beta\ \cos\gamma - \cos\alpha)$.
$S_{13} = ab^2c\ (\cos\gamma\ \cos\alpha - \cos\beta)$.

Table 2.5 Unit cell volume, V

System	V
Cubic	a^3
Tetragonal	a^2c
Orthorhombic	abc
Monoclinic	$abc\sin\beta$
Triclinic	$abc\sqrt{(1 - \cos^2\alpha - \cos^2\beta - \cos^2\gamma + 2\cos\alpha\ \cos\beta\ \cos\gamma)}$
Hexagonal	$[\sqrt{(3)}/2]\ [a^2c] \approx 0.866\ a^2c$
Rhombohedral	$a^3\ \sqrt{(1 - 3\cos^2\alpha + 2\cos^3\alpha)}$

These are expressed most compactly using vector notation, but are given here 'in longhand', without derivation, as a set of useful tools.

The volume of the reciprocal unit cell, V^* is given by:

$$V^* = 1/V$$

where V is the volume of the direct unit cell.

The direction $[uvw]$ lies in the plane (hkl) when:

$$hu + kv + lw = 0$$

Table 2.6 Interplanar angle, φ.

System	$\cos\varphi$
Cubic	$[h_1h_2 + k_1k_2 + l_1l_2] / \{[h_1^2 + k_1^2 + l_1^2][h_2^2 + k_2^2 + l_2^2]\}^{1/2}$
Tetragonal	$\dfrac{\dfrac{h_1h_2 + k_1k_2}{a^2} + \dfrac{l_1l_2}{c^2}}{\left[\left(\dfrac{h_1^2 + k_1^2}{a^2} + \dfrac{l_1^2}{c^2}\right)\left(\dfrac{h_2^2 + k_2^2}{a^2} + \dfrac{l_2^2}{c^2}\right)\right]^{1/2}}$
Orthorhombic	$\dfrac{\dfrac{h_1h_2}{a^2} + \dfrac{k_1k_2}{b^2} + \dfrac{l_1l_2}{c^2}}{\left[\left(\dfrac{h_1^2}{a^2} + \dfrac{k_1^2}{b^2} + \dfrac{l_1^2}{c^2}\right)\left(\dfrac{h_2^2}{a^2} + \dfrac{k_2^2}{b^2} + \dfrac{l_2^2}{c^2}\right)\right]^{1/2}}$
Monoclinic[a]	$d_1d_2\left(\dfrac{h_1h_2}{a^2\sin^2\beta} + \dfrac{k_1k_2}{b^2} + \dfrac{l_1l_2}{c^2\sin^2\beta} - \dfrac{(l_1h_2 + l_2h_1)\cos\beta}{ac\sin^2\beta}\right)$
Triclinic[a]	$\dfrac{d_1d_2}{V^2}[S_{11}h_1h_2 + S_{22}k_1k_2 + S_{33}l_1l_2 + S_{23}(k_1l_2 + k_2l_1) + S_{13}(l_1h_2 + l_2h_1) + S_{12}(h_1k_2 + h_2k_1)]$
Hexagonal	$\dfrac{h_1h_2 + k_1k_2 + \frac{1}{2}(h_1k_2 + h_2k_1) + \dfrac{3a^2l_1l_2}{4c^2}}{\left[\left(h_1^2 + k_1^2 + h_1k_1 + \dfrac{3a^2l_1^2}{4c^2}\right)\left(h_2^2 + k_2^2 + h_2k_2 + \dfrac{3a^2l_2^2}{4c^2}\right)\right]^{1/2}}$
Rhombohedral[a]	$\dfrac{d_1d_2\{(h_1h_2 + k_1k_2 + l_1l_2)\sin^2\alpha + [h_1(k_2 + l_2) + k_1(h_2 + l_2) + l_1(h_2 + k_2)\cos\alpha(\cos\alpha - 1)]\}}{a^2(1 - 3\cos^2\alpha + 2\cos^3\alpha)}$

[a] V = unit cell volume, d_1 is the interplanar spacing of $(h_1k_1l_1)$, and d_2 is the interplanar spacing of $(h_2k_2l_2)$.
$S_{11} = b^2c^2\sin^2\alpha$.
$S_{22} = a^2c^2\sin^2\beta$.
$S_{33} = a^2b^2\sin^2\gamma$.
$S_{12} = abc^2(\cos\alpha\cos\beta - \cos\gamma)$.
$S_{23} = a^2bc(\cos\beta\cos\gamma - \cos\alpha)$.
$S_{13} = ab^2c(\cos\gamma\cos\alpha - \cos\beta)$.

The intersection of two planes, $(h_1k_1l_1)$ and $(h_2k_2l_2)$, is the direction $[uvw]$, where:

$$u = k_1l_2 - k_2l_1$$
$$v = h_2l_1 - h_1l_2$$
$$w = h_1k_2 - h_2k_1$$

Three planes, $(h_1k_1l_1)$, $(h_2k_2l_2)$ and $(h_3k_3l_3)$, form a zone when:

$$h_1(k_2l_3 - l_2k_3) - k_1(h_2l_3 - l_2h_3) + l_1(h_2k_3 - k_2h_3) = 0$$

The plane $(h_3k_3l_3)$ belongs to the same zone as $(h_1k_1l_1)$ and $(h_2k_2l_2)$ when:

$$h_3 = mh_1 \pm nh_2;\ k_3 = mk_1 \pm nk_2;\ l_3 = ml_1 \pm nl_2$$

where m and n are integers.

The three directions $[u_1v_1w_1]$, $[u_2v_2w_2]$ and $[u_3v_3w_3]$ lie in one plane when

$$u_1(v_2w_3 - w_2v_3) - v_1(u_2w_3 - w_2u_3) + w_1(u_2v_3 - v_2u_3) = 0$$

Two directions $[u_1v_1w_1]$ and $[u_2v_2w_2]$ lie in a single plane (hkl) when:

$$h = v_1w_2 - v_2w_1$$

$k = u_2 w_1 - u_1 w_2$
$l = u_1 v_2 - u_2 v_1$

The reciprocal lattice vector

$$\mathbf{r} = u\mathbf{a}^* + v\mathbf{b}^* + w\mathbf{c}^*$$

lies perpendicular to the direct lattice planes (uvw), and the direct lattice vector

$$\mathbf{R} = h\mathbf{a} + k\mathbf{b} + l\mathbf{c}$$

lies perpendicular to the reciprocal lattice planes (hkl).

ANSWERS TO INTRODUCTORY QUESTIONS

How does a crystal lattice differ from a crystal structure?
Crystal structures and crystal lattices are different, although these terms are frequently (and incorrectly) used as synonyms. A crystal structure is built of atoms. A crystal lattice is an infinite pattern of points, each of which must have the same surroundings in the same orientation. A lattice is a mathematical concept. There are five different two-dimensional (planar) lattices and 14 different three-dimensional (Bravais) lattices.

All crystal structures can be built up from the Bravais lattices by placing an atom or a group of atoms at each lattice point.

What is a primitive unit cell?
A primitive unit cell is a lattice unit cell that contains only one lattice point. The four primitive plane lattice unit cells are labelled *p*: oblique (*mp*), rectangular (*op*), square (*tp*) and hexagonal (*hp*). There are five primitive Bravais lattices, labelled *P*: triclinic (*aP*), monoclinic primitive (*mP*), tetragonal primitive (*tP*), hexagonal primitive (*hP*) and cubic primitive (*cP*). In addition, the trigonal lattice, when referred to rhombohedral axes, has a primitive unit cell, although the lattice is labelled *hP*.

What are Miller-Bravais indices used for?
Miller indices, h, k and l, written in round brackets, (hkl), are used to specify the facets of a well-formed crystal or internal planes through a crystal structure. The terminology (hkl) represents not just one plane, but the set of all identical parallel planes.

The Miller indices of planes in crystals with a hexagonal unit cell can be ambiguous. In order to eliminate this confusion, four indices, ($hkil$), are often used to specify planes in a hexagonal crystal. These are called Miller-Bravais indices and are only used in the hexagonal system. The index i is given by:

$$h + k + i = 0, \text{or } i = -(h + k)$$

Because it is a redundant index, the value of i is sometimes replaced by a dot, to give indices ($hk.l$).

PROBLEMS AND EXERCISES

Quick Quiz

1. A lattice is:
   a. A crystal structure
   b. An ordered array of points
   c. A unit cell
2. The basis vectors in a lattice define:
   a. The crystal structure
   b. The atom positions
   c. The unit cell
3. The number of different two-dimensional plane lattices is:
   a. 5
   b. 6
   c. 7

4. A rectangular primitive plane lattice has lattice parameters:
   a. a, b, $\gamma = 90°$
   b. $a = b$, $\gamma = 90°$
   c. a, b, $\gamma \neq 90°$
5. The number of Bravais lattices is:
   a. 12
   b. 13
   c. 14
6. An orthorhombic body-centred Bravais lattice has lattice parameters:
   a. a, b, c, $\alpha = \beta = \gamma = 90°$
   b. a $(=b)$, c, $\alpha = \beta = \gamma = 90°$
   c. a $(=b)$, c, $\alpha = \beta = 90°$, $\gamma = 120°$
7. A face-centred (*F*) lattice unit cell contains:
   a. One lattice point
   b. Two lattice points
   c. Four lattice points
8. A unit cell with a lattice point at each corner and one at the centre of the cell is labelled:
   a. *B*
   b. *C*
   c. *I*
9. The notation [*uvw*] means:
   a. A single direction in a crystal
   b. A set of parallel directions in a crystal
   c. A direction perpendicular to a plane (*uvw*)
10. The notation {*hkl*} represents:
    a. A set of directions that are identical by virtue of the symmetry of the crystal
    b. A set of planes that are identical by virtue of the symmetry of the crystal
    c. Both a set of planes or directions that are identical by virtue of the symmetry of the crystal

Calculations and Questions

2.1. Several patterns of points are shown in the figure below. Assuming these to be infinite in extent, which of them are plane lattices? For those that are lattices, name the lattice type.

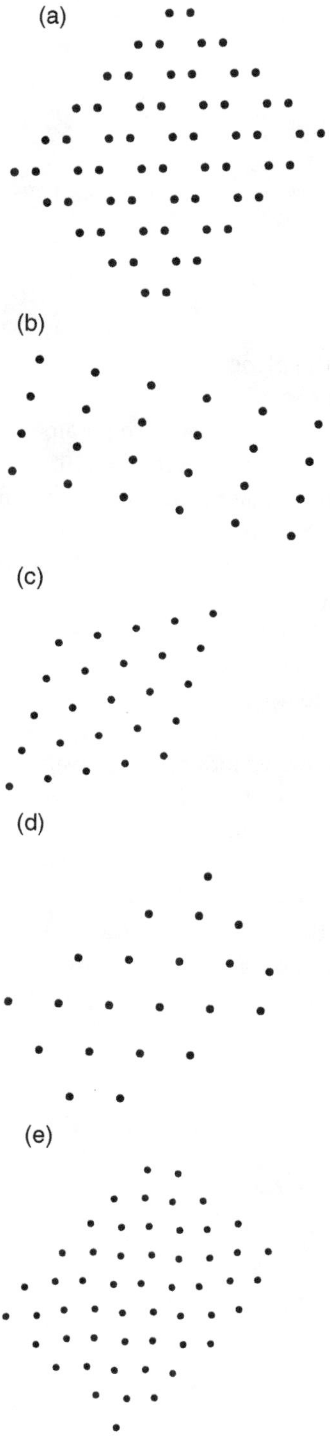

(a)

(b)

(c)

(d)

(e)

(f)

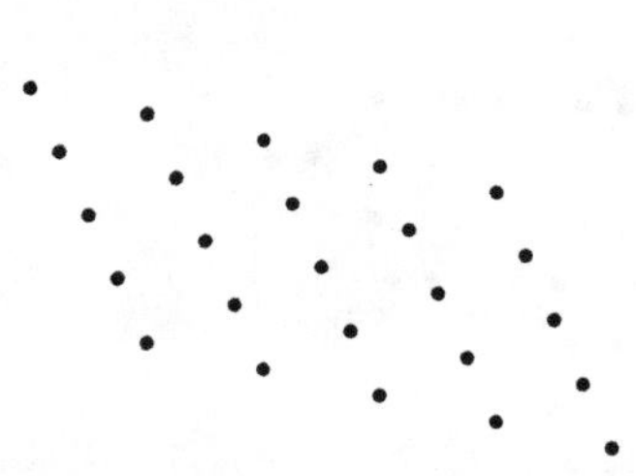

2.2. Draw the plane direct and reciprocal lattice for:
   a. an oblique lattice with parameters $a = 8$ nm, $b = 12$ nm, $\gamma = 110°$
   b. a rectangular centred lattice with parameters $a = 10$ nm, $b = 14$ nm
   c. the rectangular lattice in (b) drawn as a primitive lattice
   Confirm that the reciprocal lattices in (b) and (c) are identical and rectangular centred.

2.3. Sketch the direct and reciprocal lattice for:
   a. a primitive monoclinic Bravais lattice with $a = 15$ nm, $b = 6$ nm, $c = 9$ nm, $\beta = 105°$
   b. a primitive tetragonal Bravais lattice with $a = 7$ nm, $c = 4$ nm.

2.4. Index the lattice planes drawn in the figure below. The **c**-axis in all lattices is normal to the plane of the page and hence the index l is 0 in all cases.

(a)

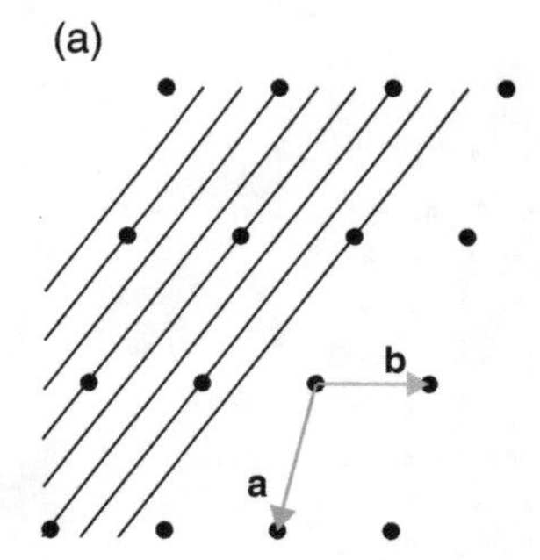

(b)

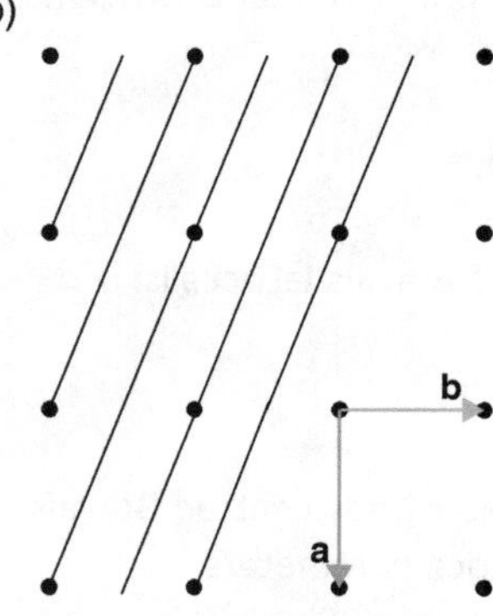

(c)

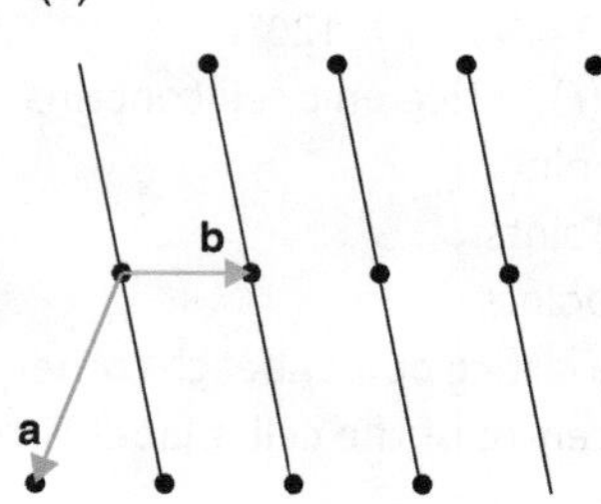

(d)

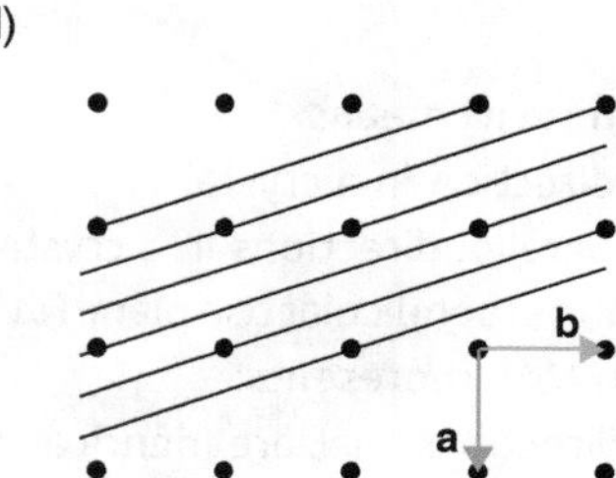

(e)

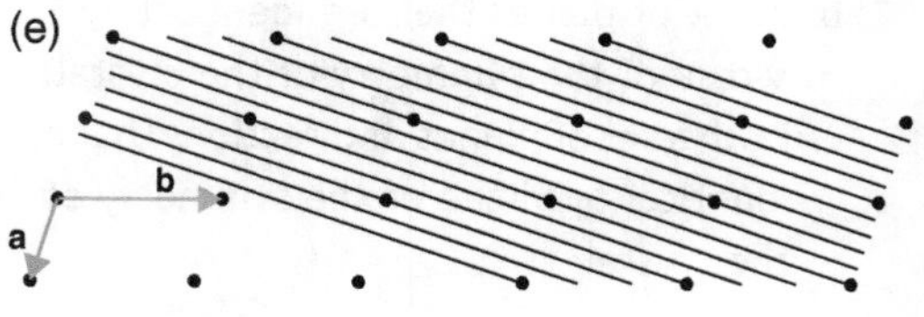

2.5. Index the lattice planes drawn in the figure below using Miller–Bravais (*hkil*) and Miller indices (*hkl*). The lattice is hexagonal with the **c**-axis normal to the plane of the page and hence the index l is 0 in all cases.

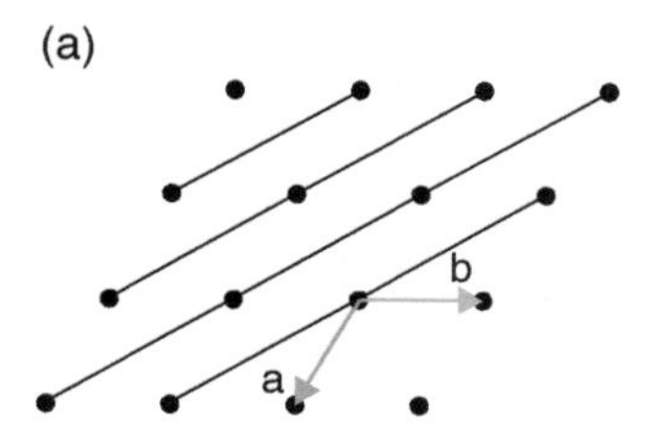

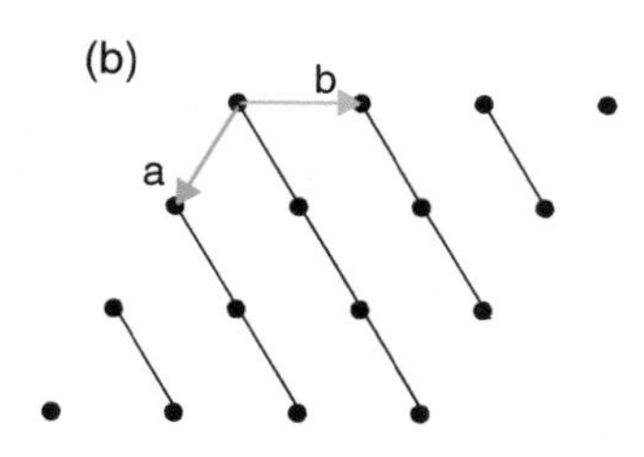

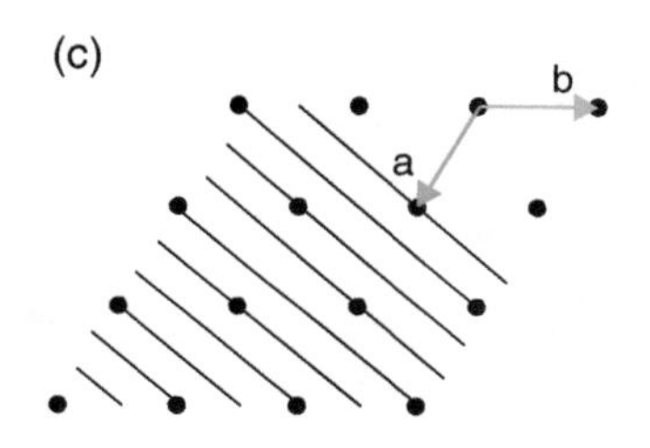

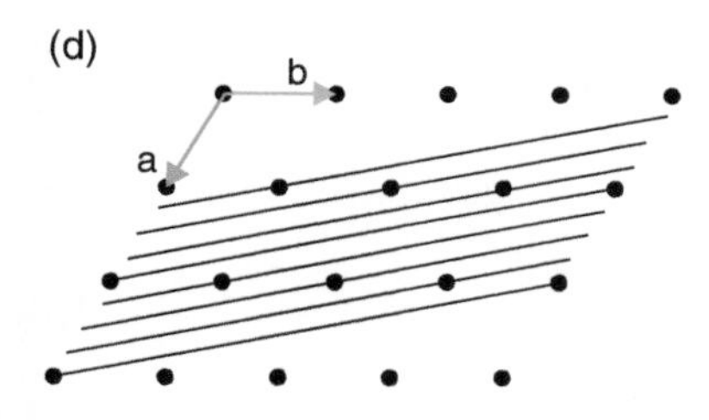

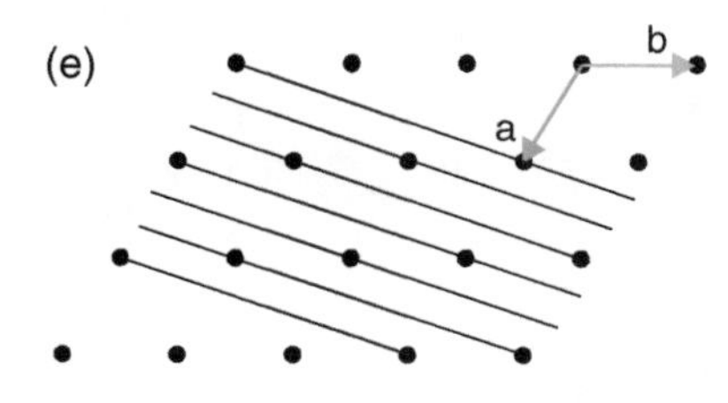

2.6. Give the indices of the directions marked on the figure below. In all cases the axis not shown is perpendicular to the plane of the paper and hence the index w is 0 in all cases.

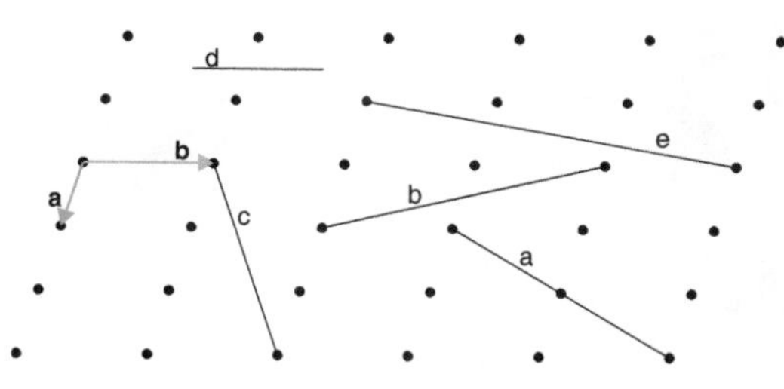

2.7. Along what direction [uvw] do the following pairs of planes $(h_1k_1l_1)$ and $(h_2k_2l_2)$ intersect?

(a) (110), $(1\bar{1}0)$; (b) $(\bar{2}10)$, (011);

(c) (111), (100); (d) (212), $(1\bar{2}1)$.

2.8. Calculate the interplanar spacing, d_{hkl}, for: (a) (111), cubic, $a = 0.365$ nm; (b) (210), tetragonal, $a = 0.475$ nm, $c = 0.235$ nm; (c) (321) orthorhombic, $a = 1.204$ nm, $b = 0.821$ nm, $c = 0.652$ nm; (d) (222), monoclinic, $a = 0.981$ nm, $b = 0.365$ nm, $c = 0.869$ nm, $\beta = 127.5°$; (e) (121), hexagonal, $a = 0.693$ nm, $c = 1.347$ nm.

2.9. Starting from the formula for a triclinic unit cell in Tables 2.4–2.6, by substitution of appropriate a, b, c, α, β, γ, confirm the other formulae in the tables.

2.10. Calculate the interplanar angles for the pairs of planes in cubic $SrCoO_3$, $a = 0.3855$ nm; (110)/(120); (111)/$(11\bar{1})$; (112)/(103).

2.11. Calculate the interplanar angles for the pairs of planes in tetragonal $BaTiO_3$, $a = 0.3991$ nm, $c = 0.40352$ nm; (110)/(211); (111)/$(11\bar{1})$; (111)/(121).

2.12. Calculate the interplanar angles for the pairs of planes in orthorhombic $LaCrO_3$, $a = 0.55133$ nm, $b = 0.54759$ nm, $c = 0.77585$ nm; (110)/(112); (111)/$(11\bar{1})$; (111)/(112); (111)/(211).

2.13. Calculate the interplanar angles for the pairs of planes in hexagonal $LaNiO_3$, $a = 0.54561$ nm, $c = 1.31432$ nm; (211)/(112); (111)/$(11\bar{1})$; (101)/(211); (110)/(121).

Chapter 3
Two-dimensional Patterns and Tiling

What is a point group?
What is a plane group?
What is an aperiodic tiling?

At the end of Chapter 1, an inherent difficulty became apparent. How is it possible to conveniently specify a crystal structure in which the unit cell may contain hundreds or even thousands of atoms? In fact, crystallographers make use of the symmetry of crystals to reduce the list of atom positions to reasonable proportions. However, the application of symmetry to crystals has far more utility than this accountancy task. The purpose of this chapter is to introduce the notions of symmetry, starting with two-dimensional patterns.

This aspect of crystallography, the mathematical description of the arrangement of arbitrary objects in space, was developed in the latter years of the nineteenth century. It went hand in hand with the observational crystallography sketched at the beginning of Chapter 1. They are separated here solely for ease of explanation.

3.1 THE SYMMETRY OF AN ISOLATED SHAPE: POINT SYMMETRY

Everyone has an intuitive idea of symmetry, and it seems reasonable to consider that the letters **A** and **C** are equally symmetrical, and both more so than the letter **G**. Similarly, a square seems more symmetrical than a rectangle. Symmetry is described in terms of transformations that leave an object apparently unchanged. These transformations are mediated by **symmetry elements** and the action of transformation by a symmetry element is called a **symmetry operation**. Typical symmetry elements include mirrors and axes, and the operations associated with them are reflection and rotation.

A consideration of the symmetry of an isolated object like the letter **A** suggests that it can be divided into two identical parts by a **mirror**, denoted *m* in text, and drawn as a continuous line in a diagram (Figure 3.1a). The same can be said of the letter **C**, although this

Crystals and Crystal Structures, Second Edition. Richard J. D. Tilley.

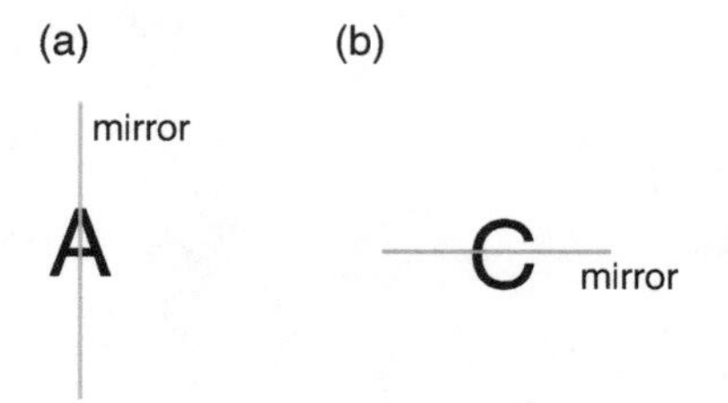

Figure 3.1 Mirror symmetry of letters. The symmetry element in each case is a mirror, drawn as a line.

time the mirror is horizontal (Figure 3.1b). An equilateral triangle contains three mirror lines (Figure 3.2a). However, it is also easy to see that the triangle can be thought to contain an **axis of rotation**, lying through the centre of the triangle and normal to the plane of the paper. The operation associated with this axis is rotation of the object in a **anticlockwise** manner[1] through an angle of (360/3)°, which regenerates the initial shape each time. The axis is called a **triad** axis or a **threefold** axis of rotation, and it is represented on a figure by the symbol ▲ (Figure 3.2b).

Note that a general (non-equilateral) triangle does not possess this combination of symmetry elements. An isosceles triangle, for example, only has one mirror line present (Figure 3.2c), and a scalene triangle has none. It is thus reasonable to say that the equilateral triangle is more symmetric than an isosceles triangle, which is itself more symmetrical than a scalene triangle. We are thus led to the idea that more symmetrical objects contain more symmetry elements than less symmetrical ones.

The most important symmetry operators for a planar shape consist of the mirror operator and an *infinite number* of rotation axes. Note that a mirror is a symmetry operator that can change the handedness or **chirality** of an object, that is, a left hand is transformed into a right hand by reflection. Two mirror image objects cannot be superimposed simply by rotation in the plane, just as a right-hand glove will not fit a left hand (Figure 3.3a). The only way in which the two figures can be superimposed is by lifting one from the page (i.e. using a third dimension), and turning it over. The chirality introduced by mirror symmetry has considerable implications for the physical properties of both isolated molecules and whole crystals. It is considered in more detail in later chapters.

The rotation operator 1, a **monad** rotation axis, implies that no symmetry is present, because the shape has to be rotated anticlockwise by (360/1)° to bring it back into register with the original (Figure 3.3b). There is no graphical symbol for a monad axis. A **diad** or twofold rotation axis is represented in text as 2 and in drawings by the symbol ⬮. Rotation in an anticlockwise manner by 360°/2 (180°) around a diad axis returns the shape to the original configuration (Figure 3.3c). A **triad** or threefold rotation axis is represented in text by 3, and on figures by ▲. Rotation anticlockwise by 360°/3 (120°) around a triad axis returns the shape to the original configuration (Figure 3.3d). A **tetrad** or fourfold rotation axis is represented in text by 4 and on figures by the symbol ◆. Rotation anticlockwise by 360°/4 (90°) about a tetrad axis returns the shape to the original configuration (Figure 3.3e). The **pentad** or fivefold rotation axis, represented in text by 5 and on figures by ⬟, is found in regular pentagons. Rotation anticlockwise by 360°/5 (72°) about a pentad axis returns the shape to the original configuration (Figure 3.3f). A **hexad** or sixfold rotation axis is represented in text by 6 and on figures by ⬢. Rotated anticlockwise

[1] Clearly whether the rotation takes place in a clockwise or anticlockwise direction is irrelevant in many cases. Conventionally an anticlockwise direction is always chosen.

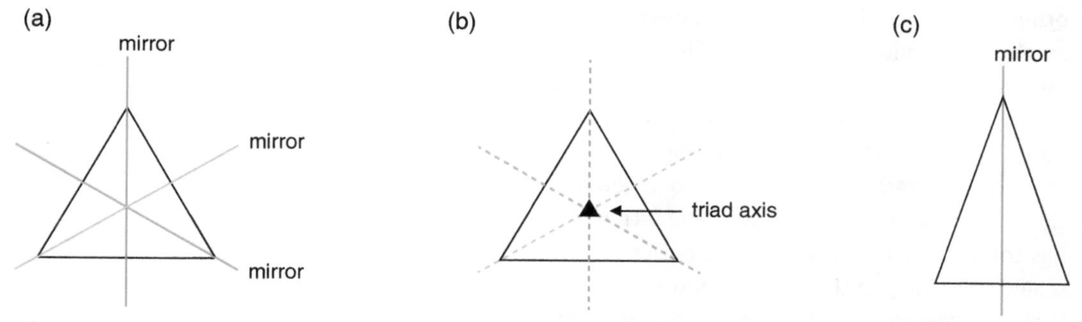

Figure 3.2 Symmetry of triangles: (a) mirrors in an equilateral triangle; (b) triad axis in an equilateral triangle; (c) mirror in an isosceles triangle.

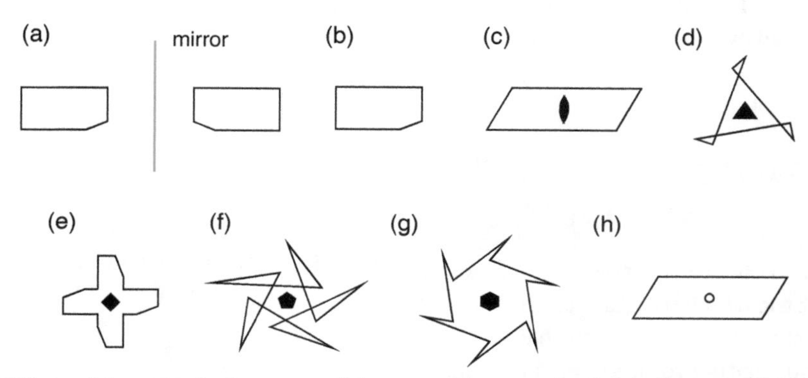

Figure 3.3 Mirror (a) and rotation axes: (b) monad (no symbol); (c) diad, twofold; (d) triad, threefold; (e) tetrad, fourfold; (f) pentad, fivefold; (g) hexad, sixfold; (h) centre of symmetry, equivalent to (c).

by 360°/6 (60°) around a hexad axis returns the shape to the original configuration (Figure 3.3g). In addition, there are an infinite number of other rotation symmetry operators, such as sevenfold, eightfold, and higher axes.

An important symmetry operator is the **centre of symmetry** or **inversion centre**, represented in text by $\bar{1}$ (pronounced 'one bar'), and in drawings by ∘. This operation is an inversion through a point in the shape, so that any object at a position (x, y) with respect to the centre of symmetry is paired with an identical object at ($-x$, $-y$), written ($\bar{x}$, $\bar{y}$) (pronounced 'x bar', 'y bar'). In two dimensions, the presence of a centre of symmetry is equivalent to a diad axis (Figure 3.3h).

When the various symmetry elements present in a shape are applied, it is found that one point is left unchanged by the transformations, and when they are drawn on a figure, they all pass through this single point. The collection of symmetry elements of an isolated shape is called the **point group** and the combination of elements is called the **general point symmetry** of the shape. (Here the word *group* is mathematically precise, and the results of symmetry operations can be related to each other using the formalism of group theory.) The symmetry elements that characterise the point group are collected together into a **point group symbol**. Crystallographers generally use **International** or **Hermann-Mauguin** symbols for this purpose, and this practice is followed here. The point symmetry is described by writing the rotation axes present, followed by the mirror planes. Thus, the equilateral triangle has the point symmetry, $3m$. Alternatively, it is said to belong to the **point group** $3m$. Note that the order in which the symmetry operators are written is specific and is described below. An alternative nomenclature for point group symmetry is that proposed by **Schoenflies**. This is generally used by chemists, and is described in Appendix C.

3.2 ROTATION SYMMETRY OF A PLANE LATTICE

As described in Chapter 1, a crystal is defined by the fact that the whole structure can be built by the regular stacking of a unit cell that is translated but neither rotated nor reflected. The same is true for two-dimensional 'crystals' or patterns. This imposes a limitation upon the combinations of symmetry elements that are compatible with the use of unit cells to build up a two-dimensional pattern or a three-dimensional crystal. To understand this, the rotational symmetry of the five unique plane lattices is described.

Suppose that a rotational axis of value n is normal to a plane lattice. It is convenient (but not mandatory) to locate it at a lattice point, O, part of the row of nodes A–O–B, each of which is separated by the lattice vector **t**, of length t (Figure 3.4a). The rotation operation will generate new rows of points, lying at angles with increments of $2\pi/n$ to the original row, containing points C and D (Figure 3.4b). As all points in a lattice are identical, C and D must lie on a lattice row parallel to that containing A and B.

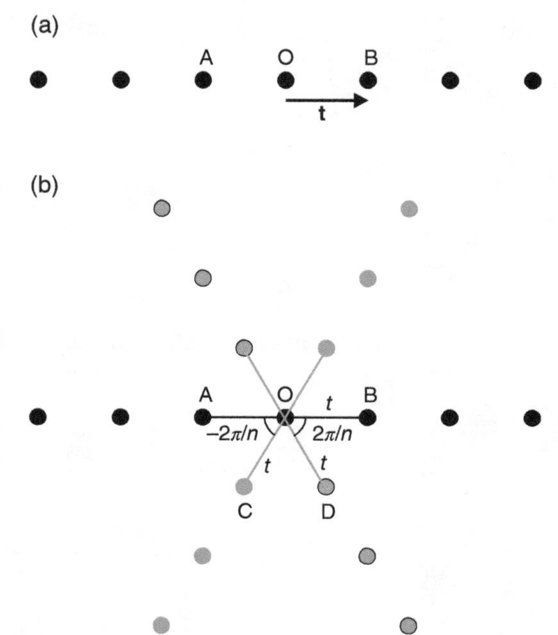

Figure 3.4 Lattice rotation: (a) a lattice row, A–O–B; (b) the same row rotated by $\pm 2\pi/n$.

The separation of C and D must be an integral number, m, of the distance t. By simple geometry, the distance CD is given by $2t\cos(2\pi/n)$. Hence, for a lattice to exist:

$$2t\cos(2\pi/n) = mt$$

$$2\cos(2\pi/n) = m \text{ (integer)}$$

The solutions to this equation, for the lowest values of n, are given in Table 3.1. This shows that the only rotation axes that can occur in a lattice are 1, 2, 3, 4 and 6. The fivefold rotation axis and all axes with n higher than 6 *cannot occur* in a plane lattice.

This derivation provides a formal demonstration of the fact mentioned in Chapters 1 and 2, that a unit cell in a lattice or a pattern cannot possess fivefold symmetry. The same will be found to be true for three-dimensional lattices and crystals, described in

Table 3.1 Rotation axes in a plane lattice

n	$2\pi/n$	$2\cos(2\pi/n) = m$
1	360°	2
2	180°	−2
3	120°	−1
4	90°	0
5	72°	0.618[a]
6	60°	1
7	51.43°	1.244[a]

[a]Not allowed.

Chapter 4. (Here it is pertinent to note that fivefold symmetry can occur in two- and three-dimensional patterns, if the severe constraints imposed by the mathematical definition of a lattice are relaxed slightly, as described in Section 3.7 and Chapter 9, Section 9.9.)

Note that the fact that a unit cell cannot show overall fivefold rotation symmetry does not mean that a unit cell cannot contain a pentagonal arrangement of atoms. Indeed, pentagonal coordination of metal atoms is commonly found in inorganic crystals, and pentagonal ring structures occur in organic molecules. However, the pentagonal groups must be arranged within the unit cell to generate an overall symmetry appropriate to the lattice type, described in the following section (see, e.g., Figure 2.4).

3.3 THE SYMMETRY OF THE PLANE LATTICES

The symmetry properties inherent in patterns are important in crystallography. The simplest of these patterns are the plane lattices described in Chapter 2. The symmetry elements that can be found in the plane lattices are the allowed rotation axes just described, together with mirror planes. The way in which these symmetry elements are arranged will provide an introduction to the way in which symmetry elements are disposed in more complex patterns.

Consider the situation with an oblique primitive (*mp*) lattice (Figure 3.5a). The symmetry of the unit cell is consistent with the presence of a diad axis, which can be placed conveniently through a lattice point, at the origin of the unit cell, without losing generality. This means that there must be a diad through every lattice point. However, the presence of this axis also forces diad axes to be placed at the centre of each of the unit cell sides, and also one at the centre of the cell (Figure 3.5b). The symmetry operations must leave every point in the lattice identical, and so the lattice symmetry is also described as the **lattice point symmetry**. The lattice point symmetry of the *mp* lattice is given the symbol 2.

The other lattice symmetries are described in a similar way. The rectangular primitive (*op*) lattice (Figure 3.5c) also has lattice symmetry consistent with a diad, located for convenience at the origin of the unit cell, which, of necessity, then repeats at each lattice point. This, as before, generates diads at the centre of each of the unit cell sides and one at the unit cell centre (Figure 3.5d). However, the lattice also shows mirror symmetry, not apparent in the oblique lattice. These lie along the unit cell edges and this necessitates the introduction of parallel mirror lines halfway along each unit cell side. The mirrors are represented as lines on Figure 3.5d and the point symmetry is given the symbol 2*mm*. (The order of the symbols for this and the other point symmetry groups listed is explained below.)

The rectangular centred (*oc*) lattice (Figure 3.5e) also has the same diads and mirrors as the *op* lattice, located in the same positions (Figure 3.5f). The presence of the lattice centring, however, forces the presence of additional diads between the original set.

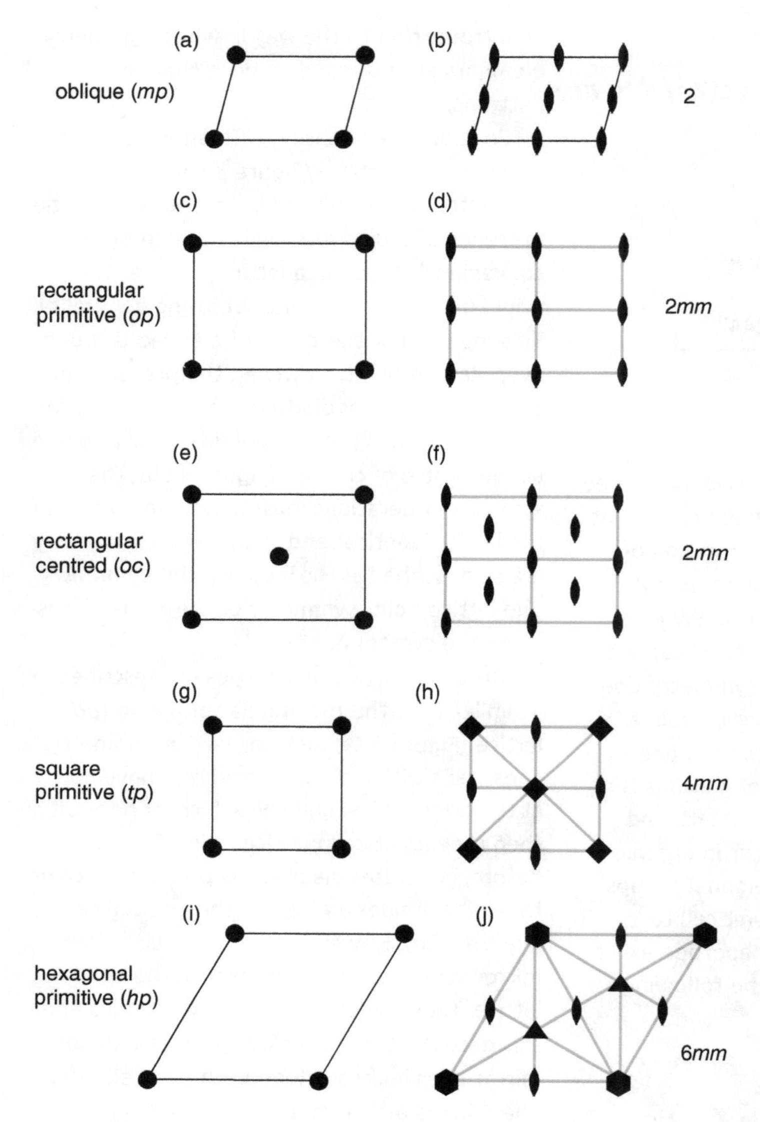

Figure 3.5 Symmetry of the plane lattices.

Nevertheless, the point symmetry does not change, compared with the *op* lattice, and remains 2*mm*.

The square (*tp*) lattice (Figure 3.5g) has, as principle symmetry element, a tetrad rotation axis through the lattice point at the unit cell origin, which necessitates a tetrad through each lattice point. This generates additional diads at the centre of each unit cell side, and another tetrad at the cell centre. These in turn necessitate the presence of mirror lines as shown (Figure 3.5h). The lattice point symmetry symbol is 4*mm*.

Finally, the hexagonal primitive (*hp*) lattice (Figure 3.5i) has a hexad rotation axis at each lattice point. This generates diads and triads as shown. In addition, there are six mirror lines through each lattice point. In other parts of the unit cell, two mirror lines intersect at diads and three mirror lines intersect at triads (Figure 3.5j). The lattice point symmetry is described by the symbol 6*mm*.

Note that this emphasis on the symmetry of the lattices has a profound effect on thinking about structures. Up to now, the lattice type has been characterised by the lattice parameters, the lengths of the lattice vectors and the inter-vector angles. Now, however, it is possible to define a lattice in terms of the symmetry rather than the dimensions. In fact, this is the norm in crystallography. As we will see, the crystal system of a phase is allocated in terms of symmetry and not lattice parameters. To pre-empt future chapters somewhat, consider the definition of a monoclinic unit cell in terms of the lattice parameters $a \neq b \neq c$, $\alpha = 90°$, $\beta \neq 90°$, $\gamma = 90°$ (see Chapter 1). The unit cell may still be regarded as monoclinic (*not* orthorhombic) even if the angle β is 90°, provided that the symmetry of the structure complies with that expected of a monoclinic unit cell.

3.4 THE TEN PLANE CRYSTALLOGRAPHIC POINT SYMMETRY GROUPS

The general point groups described in Section 3.1 are not subject to any limitations. The point groups obtained by excluding all rotation operations incompatible with a lattice are called the **crystallographic plane point groups**. These are formed, therefore, by combining the rotation axes 1, 2, 3, 4 and 6 with mirror symmetry. When the above symmetry elements are combined, it is found that there are just *ten* (two-dimensional) crystallographic plane point groups: 1, 2, *m*, 2*mm*, 4, 4*mm*, 3, 3*m*, 6 and 6*mm*.

The application of these symmetry elements is identical to that depicted in Figure 3.3. However, as we are ultimately aiming to explain patterns in two and three dimensions rather than solid shapes, it is convenient to illustrate the symmetry elements with respect to a unit of pattern called a **motif**, placed at a general position with respect to the rotation axis (Figure 3.6). In this example the motif is an asymmetric 'three atom planar molecule'. (In crystals, the motif is a group of atoms: see Chapter 5.)

The patterns obtained when the allowed rotation operations act on the motif are shown in Figure 3.6a–e. In Figure 3.6a the rotation axis 1 may be placed anywhere, either in the motif or outside of it. In Figures 3.6b–e the position of the rotation axis, 2, 3, 4 or 6, is obvious. The point groups of the patterns are 1, 2, 3, 4 and 6 respectively.

The action of the mirror alone is to form a chiral image of the motif (Figure 3.6f), with a point group *m*. The combination of a mirror line with a diad produces four copies of the motif and an additional mirror (Figure 3.6g), point group 2*mm*. The combinations of a mirror with a triad, tetrad or hexad, yielding point groups 3*m*, 4*mm* and 6*mm*, are illustrated in Figure 3.6h–j.

The order of the symbols in the point group labels are allocated in the following way. The first (**primary**) position gives the rotation axis if present. The second (**secondary**) and third (**tertiary**) positions record whether a mirror element, *m*, is present. An *m* in the secondary position means that the mirror is perpendicular to the [10] direction, in all lattices. If only one mirror is present it is always given with respect to this direction. An *m* in the secondary position

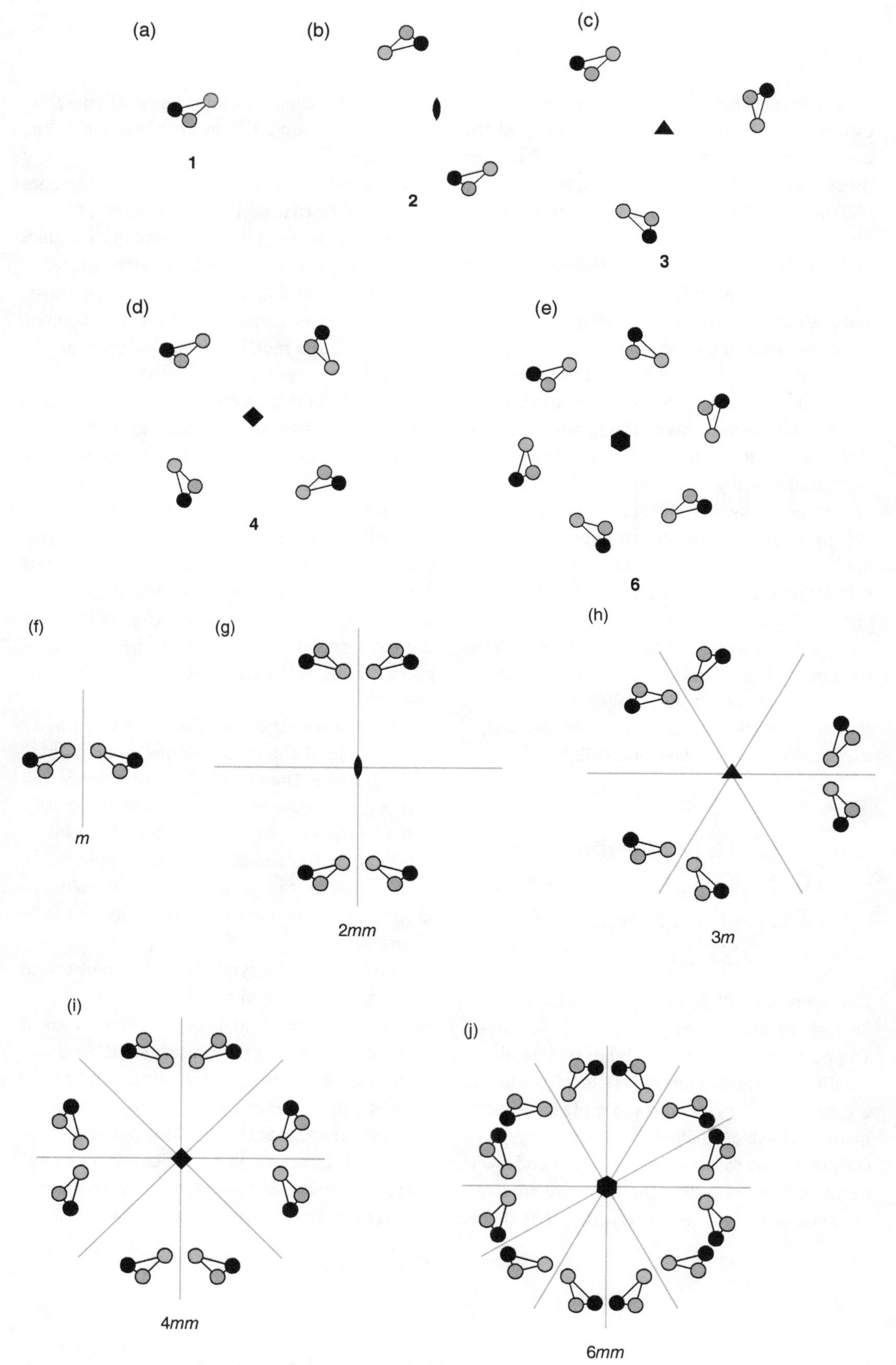

Figure 3.6 The ten crystallographic point groups.

Table 3.2 The order of two-dimensional point group and plane group symbols

Lattice	Position in Hermann-Mauguin symbol		
	Primary	**Secondary**	**Tertiary**
Oblique, *mp*	Rotation point	—	—
Rectangular, *op*, *cp*	Rotation point	[10]	[01]
Square, *tp*	Rotation point	[10] = [01]	$[1\bar{1}]$ = [11]
Hexagonal, *hp*	Rotation point	[10] = [01] = $[\bar{1}\bar{1}]$	$[1\bar{1}]$ = [12] = $[\bar{2}\bar{1}]$

is perpendicular to [01] in a rectangular unit cell, and to $[1\bar{1}]$ in both a square and a hexagonal unit cell (Table 3.2).

For example, objects belonging to the point group 4*mm*, such as a square, have a tetrad axis, a mirror perpendicular to the [10] direction, and an independent mirror perpendicular to the direction $[1\bar{1}]$. Objects belonging to the point group *m*, such as the letters **A** or **C**, have only a single mirror present. This would be perpendicular to, and thus define, the [10] direction in the object.

To determine the crystallographic point group of a planar shape, it is only necessary to write down a list of all of the symmetry elements present, order them following the rules set out, and then compare them to the list of ten given above.

3.5 THE SYMMETRY OF PATTERNS: THE 17 PLANE GROUPS

All two-dimensional patterns can be created by starting with one of the five plane lattices and adding a design, a single 'atom' say, or complex consisting of many 'atoms' (i.e. a motif), to each lattice point. Obviously, by varying the design, an infinite number of patterns can be created. However, the number of fundamentally different arrangements that can be created is much smaller and is limited by the symmetries that have been described above.

Perhaps the simplest way to start deriving these patterns is to combine the five lattices with the patterns generated by the ten point groups. For example, take as a motif the triangular 'molecule' represented by the point group 1 (Figure 3.6a), and combine this with the oblique *mp* lattice (Figure 3.5a). The 'molecule' can be placed anywhere in the unit cell and the rotation axis (1) can conveniently pass through a lattice point. The resultant pattern (Figure 3.7a and b) is labelled *p*1, and is called a **plane group**. The initial letter describes the lattice type (primitive), and the number part gives information on the symmetry operators present, in this case just the monad axis (1). In agreement with the designation of primitive, there is just one lattice point, and so one motif, per unit cell.

Remember that the motif can have any shape whatever. For example, a motif with a fivefold axis, say a regular pentagon, or a starfish, is allowed. However, the only pattern that can result, when this motif is combined with an *mp* lattice, is that given by the plane group *p*1 (see e.g. Figure 2.4). That is, the plane group is defined by the overall symmetry of the pattern, not by the symmetry present in the motif.

It is also possible to combine the same *mp* lattice with the arrangement representing point

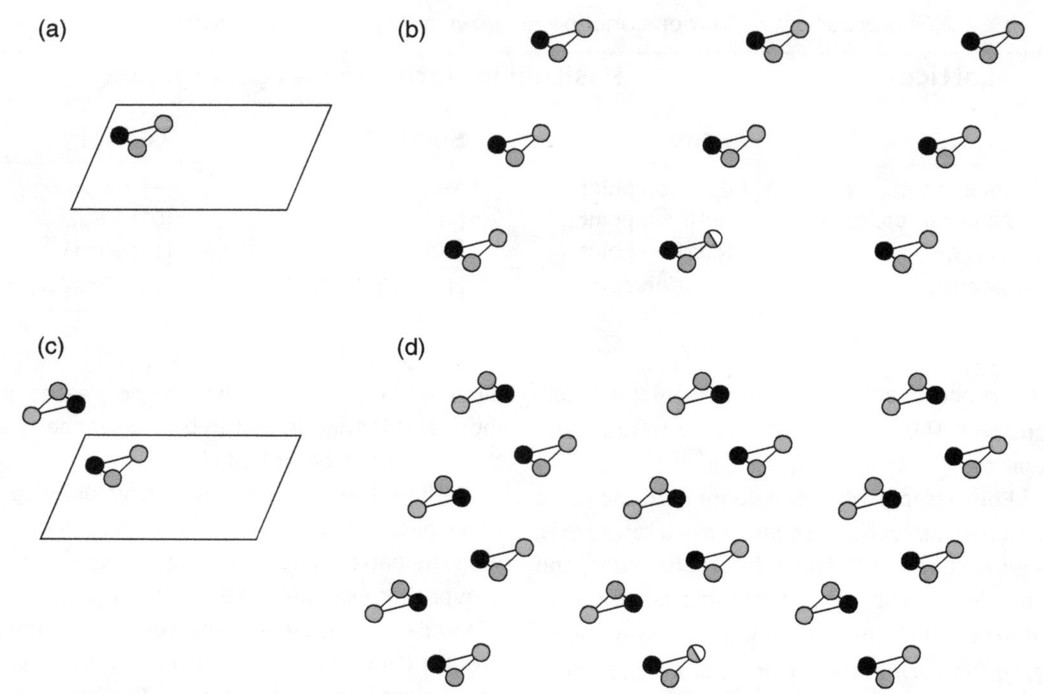

Figure 3.7 The plane groups $p1$ and $p2$: (a) the motif of point group 1 added to the plane lattice *mp*; (b) the pattern formed by repetition of (a), representative of plane group $p1$; (c) the motif in point group 2 added to the plane lattice *mp*; (d) the pattern formed by repetition of (c), representative of plane group $p2$.

group 2. In this case, it is convenient to place the diad axis through the lattice point at the origin (Figure 3.7c). Note that the presence of the diad makes the presence of two triangular 'molecules' mandatory, and it is convenient to consider the overall pattern to be generated by the addition of the whole group associated with point group 2, that is, two triangular 'molecules' and the rotation diad, to each lattice point. The pattern generated (Figure 3.7d) is labelled $p2$, where the initial letter relates to the primitive lattice and the label 2 indicates the presence of the diad axis. In agreement with the designation of primitive, there is only one lattice point in the unit cell, although this is associated with two triangular 'molecules' and a diad axis.

The arrangements equivalent to the next two point groups, m and $2mm$, cannot be combined with the *mp* lattice because the mirrors are incompatible with the symmetry of the lattice, and to form new patterns it is necessary to combine these with the *op* and *oc* lattices. When building up these patterns a new type of symmetry element is revealed. This is easy to see if the point group m is combined with firstly the *op* lattice and then with the *oc* lattice.

Consider the *op* lattice (Figure 3.8a). The symmetry operator m replicates the triangular motif associated with each lattice point to yield a design consisting of two mirror image 'molecules'. Note that the mirror line running through the origin automatically generates

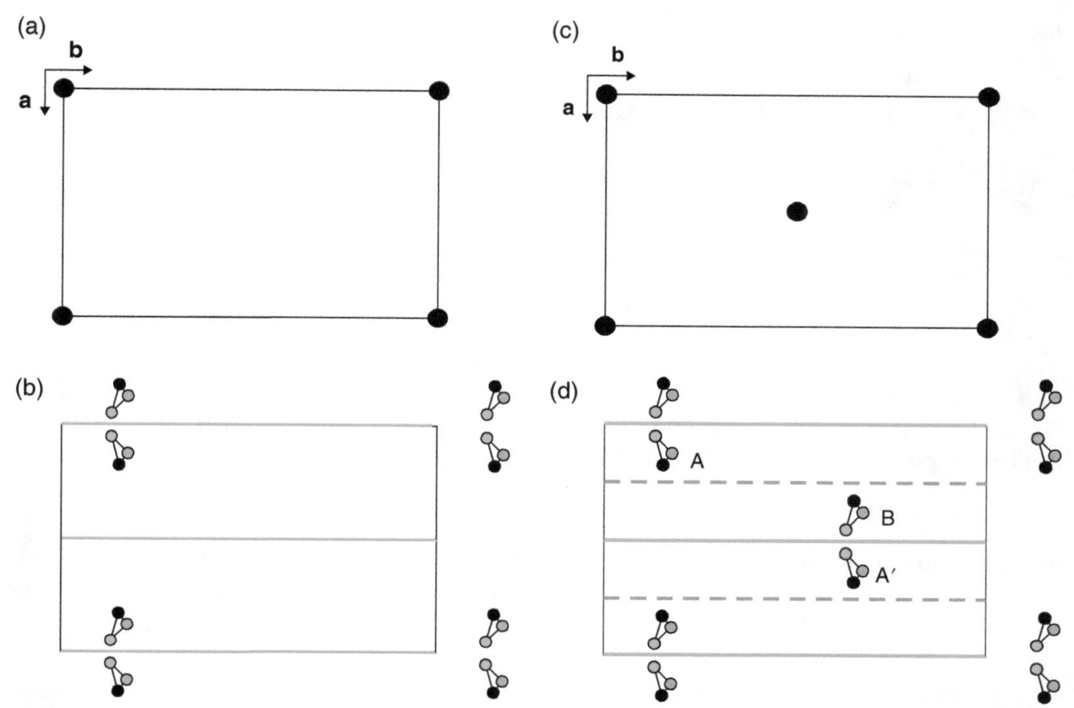

Figure 3.8 The plane groups *pm* and *cm*: (a) the plane lattice *op*; (b) the pattern formed by adding the motif of point group *m* to the lattice in (a), representative of plane group *pm*; (c) the plane lattice *oc*; (d) the pattern formed by adding the motif of plane group *m* to the lattice in (c), representative of plane group *cm*.

another parallel mirror through the centre of the unit cell, but the motif is not replicated further (Figure 3.8b). The same procedure applied to the *oc* lattice (Figure 3.8c) gives a different result, as the triangular 'molecule' labelled A associated with a lattice point at the origin must also be associated with the lattice point at the centre of the unit cell, to give A′ (which is identical to A) (Figure 3.8d). It is now apparent that the triangular 'molecules' A and B (Figure 3.8d) are related by the operation of a combination of a reflection and a translation, called **glide**. The glide operator appears as a result of combining the translation properties of the lattice with the mirror operator. The new symmetry element, drawn as a dashed line on figures, is the **glide line**.

The translation component of the glide must be subject to the limitations of the underlying lattice, just as the rotation axes are. Following initial reflection of the motif (Figure 3.9a) the glide line will lie parallel to a lattice row of the underlying plane lattice (Figure 3.9b). The translation of the motif, the glide vector **t**, will be parallel to the glide line and also to a lattice row of the underlying plane lattice. Suppose that the lattice repeat along the glide direction is the vector **T**. In order to fulfil the repeat constraints of the lattice, twice the vector **t** must bring the motif back to a new lattice position a distance of $p\mathbf{T}$ from the original lattice point (Figure 3.9b). It is possible to write:

$$2\mathbf{t} = p\mathbf{T}$$

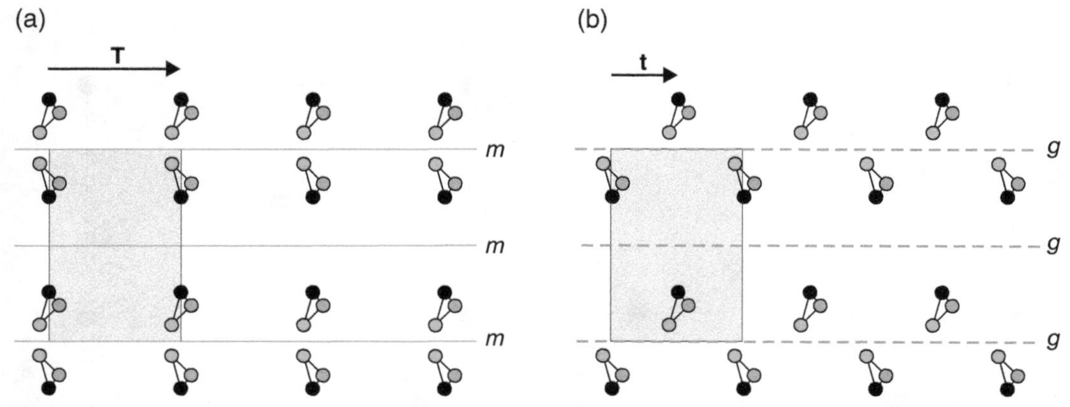

Figure 3.9 The glide operation: (a) reflection across a mirror line; (b) translation parallel to the mirror plane by a vector **t**. The lattice unit cell is shaded.

where p is an integer. Thus:

$$\mathbf{t} = (p/2)\mathbf{T} = 0\mathbf{T}, \mathbf{T}/2, 3\mathbf{T}/2, \ldots$$

Now, of these options, only a displacement of **T**/2 is distinctive. The glide displacement is therefore restricted to one half of the lattice repeat. Thus, displacements parallel to the unit cell **a**-axis is limited to **a**/2, and parallel to the **b**-axis is limited to **b**/2. Similar relationships hold for displacements in diagonal directions. Note that just as the operation of a mirror through the cell origin automatically generates a parallel mirror through the cell centre, a glide line through the cell origin automatically generates another parallel glide line through the unit cell centre.

In building up these patterns, it will become apparent that point groups with a tetrad axis can only be combined with a square (*tp*) lattice, and those with triad or hexad axes can only be combined with a hexagonal (hp) lattice.

When all the translations inherent in the five plane lattices are combined with the symmetry elements found in the ten point groups plus the glide line, just 17 different arrangements, called the 17 (two-dimensional) **plane groups**, are found (Table 3.3). All two-dimensional crystallographic patterns must belong to one or other of these groups and are described by one of 17 unit cells.

Each two-dimensional plane group is given a symbol that summarises the symmetry properties of the pattern. The symbols have a similar meaning to those of the point groups. The first letter gives the lattice type, primitive (*p*) or centred (*c*). A rotation axis, if present, is represented by a number, 1 (monad), 2 (diad), 3 (triad), 4 (tetrad) and 6 (hexad), and this is given second place in the symbol. Mirrors (*m*) or glide lines (*g*) along the directions specified in Table 3.3 are placed last.

The locations of the symmetry elements within the unit cells of the plane groups are illustrated in Figure 3.10. The first plane group, *p*1, is the least symmetrical, and has an oblique lattice containing only a monad axis of rotation through a lattice point at the origin of the unit cell (Figure 3.10a). The second plane group, *p*2, is also oblique, but with a diad axis through a lattice point at the origin of the unit cell (Figure 3.10b). The other plane groups with a single rotation axis are *p*4, *p*3 and *p*6. In all of these, the rotation axis passes through a lattice

Table 3.3 Plane (two-dimensional) symmetry groups

System	Lattice	Point group	Plane group	Number[a]	Figure
Oblique	*mp*	1	*p*1	1	3.10a
		2	*p*2	2	3.10b
Rectangular	*op*, *oc*	*m*, 2*mm*	*pm* (*p*1*m*1)	3	3.10c
			pg (*p*1*g*1)	4	3.10d
			cm (*c*1*m*1)	5	3.10e
			*p*2*mm*	6	3.10f
			*p*2*mg*	7	3.10g
			*p*2*gg*	8	3.10h
			*c*2*mm*	9	3.10i
Square	*tp*	4, 4*mm*	*p*4	10	3.10j
			*p*4*mm*	11	3.10k
			*p*4*gm*	12	3.10l
Hexagonal	*hp*	3, 3*m*, 6, 6*mm*	*p*3	13	3.10m
			*p*3*m*1	14	3.10n
			*p*31*m*	15	3.10o
			*p*6	16	3.10p
			*p*6*mm*	17	3.10q

[a]Each plane group has a number allocated to it, found in the *International Tables for Crystallography* (see Bibliography).

point at the origin of the unit cell (Figure 3.10j, m and p).

A number of plane groups do not include rotation axes. These are the groups with a mirror line (*m*) or glide line (*g*) through a lattice point at the origin of the unit cell running perpendicular to the **a**-axis of the unit cell, *pm*, *pg* and *cm* (Figure 3.10c–e). These plane groups are sometimes described by 'long' symbols, *p*1*m*1, *p*1*g*1 and *c*1*m*1, which stress the monad axis through the lattice point at the origin of the unit cell and the fact that the mirror or glide line runs perpendicular to the **a**-axis. The symbol *p*11*m* would mean that the mirror line was perpendicular to the **b**-axis, [01], as set out in Table 3.2.

The remaining plane groups are described by a string of four symbols, typified by *p2mm* and *p*2*mg*. The first two symbols give the lattice and the rotation axis present. The significance of the order of the last two symbols is given in Table 3.2. Thus, the meaning of the symbol *p*2*mm*, which is derived from the *op* lattice, is that the lattice is primitive, there is a diad axis through the lattice point at the origin of the unit cell, and a mirror line perpendicular to the **a**-axis, [10], and a mirror line perpendicular to the **b**-axis, [01] (Figure 3.10f). The meaning of the symbol *p*2*mg* is that the lattice is primitive, there is a diad axis through the lattice point at the origin of the unit cell, and a mirror line perpendicular to the **a**-axis, [10], and a glide line perpendicular to the **b**-axis, [01] (Figure 3.10g). The positions of the mirror and glide lines are often through the origin of the unit cell, but not always. They occur in positions that are compatible with the rotation axis at the origin. The addition of a monad axis, 1, is often used as a placeholder, to ensure that the mirror or glide line is correctly placed, as described above for

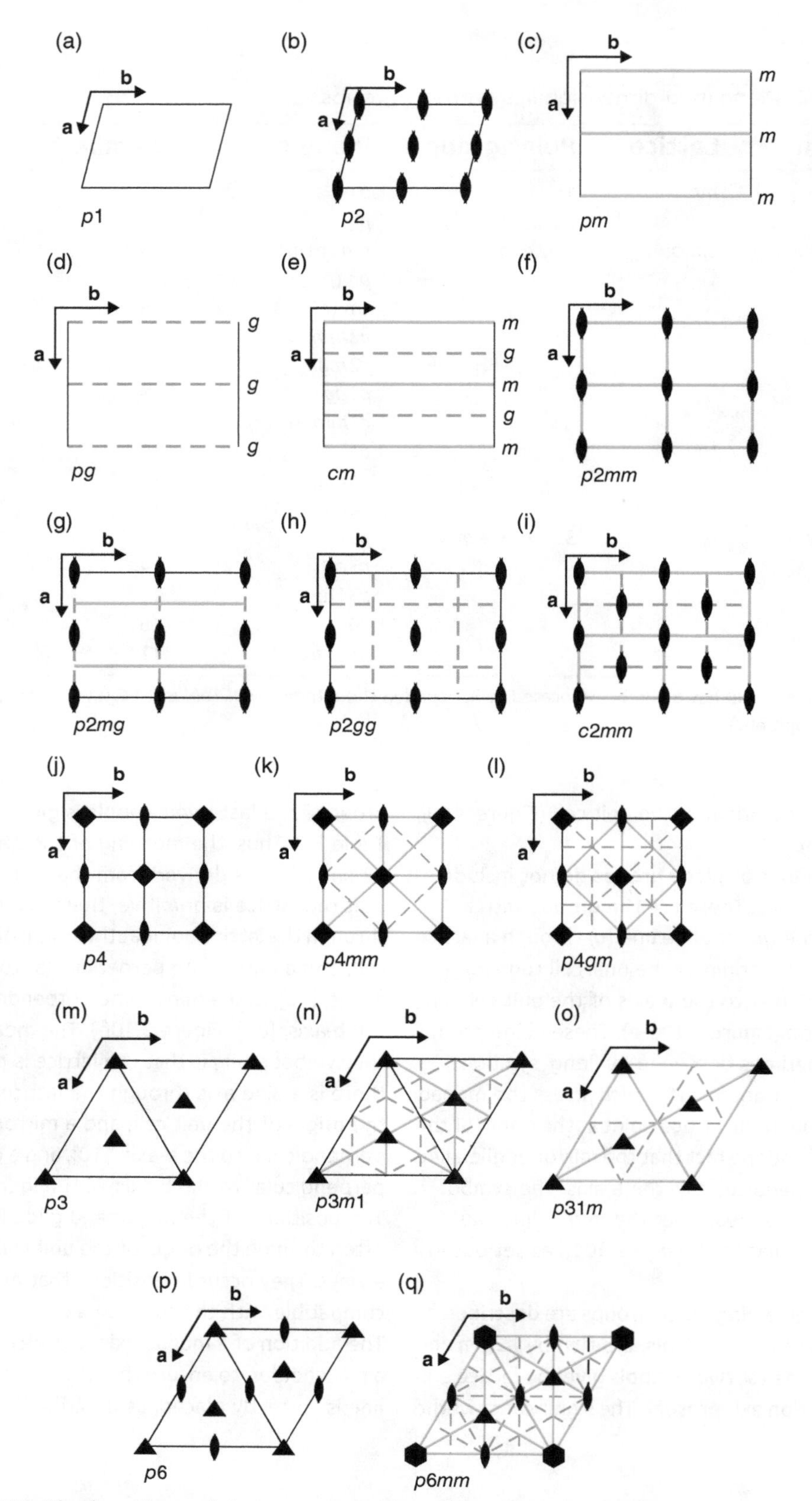

Figure 3.10 The locations of the symmetry elements within the unit cells of the 17 plane groups.

the groups $p1m1$ and $p11m$. Thus, the plane groups $p31m$ and $p3m1$ are both derived from $p3$ by the addition of mirrors. The group $p3m1$ has a mirror perpendicular to the **a**-axis, [10], while the group $p31m$ has a mirror along the **a**-axis, perpendicular to $[1\overline{1}]$.

To determine the plane group of a pattern, write down a list of all of the symmetry elements that it possesses (not always easy), order them, and then compare the list to the symmetry elements associated with the point groups given in Table 3.3.

3.6 TWO-DIMENSIONAL 'CRYSTAL STRUCTURES'

In describing a crystal structure, it is necessary to list the positions of all the atoms in the unit cell, and even in two-dimensional cases this can be a considerable chore. However, all structures must be equivalent to one of the 17 patterns described by the plane groups, and the task of listing atom positions can be reduced greatly by making use of the symmetry elements apparent in the appropriate plane group. To aid in this, each plane group is described in terms of a pair of standard diagrams. The first of these shows all of the symmetry elements in the unit cell, marked as in Figure 3.10. The accompanying diagram plots the various locations of a motif, initially placed in a **general position** in the unit cell, forced by the operation of the symmetry elements present (also see Section 3.7). Conventionally a motif is represented by a circle. Sometimes this diagram also shows the position of the motif in parts of the neighbouring unit cells, if that clarifies the situation.

Take the simplest plane group, $p1$ (number 1), as an example. Recall that there is only one lattice point in a primitive (p) unit cell. The motif can be placed anywhere in the unit cell. The unit cell has no symmetry elements present (Figure 3.11a), and hence the motif is not replicated (Figure 3.11b). Note that the cell is divided up into quarters, and motifs in the surrounding unit cells are also shown to demonstrate the pattern of repetition. Although the unit cell contains just one lattice point, the motif associated with the lattice point can contain any number of atoms, as in the triangular 'molecule' used previously (Figure 3.11c).

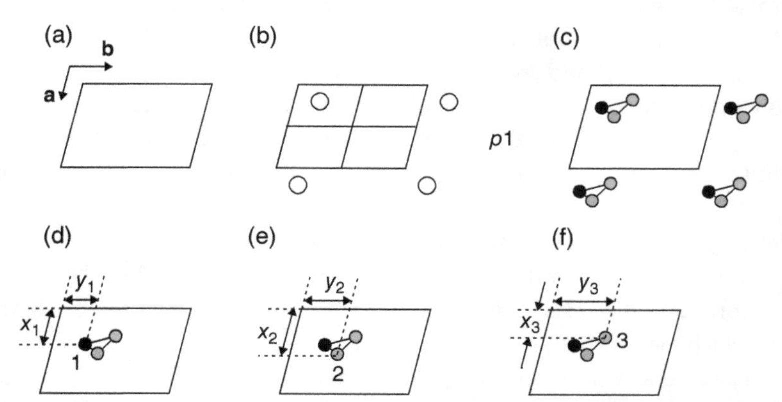

Figure 3.11 Space group diagrams for planar group $p1$: (a) symmetry elements; (b) equivalent general positions; (c) position of a triangular 'molecule' motif; (d–f) specification of the coordinates (x, y) of an atom.

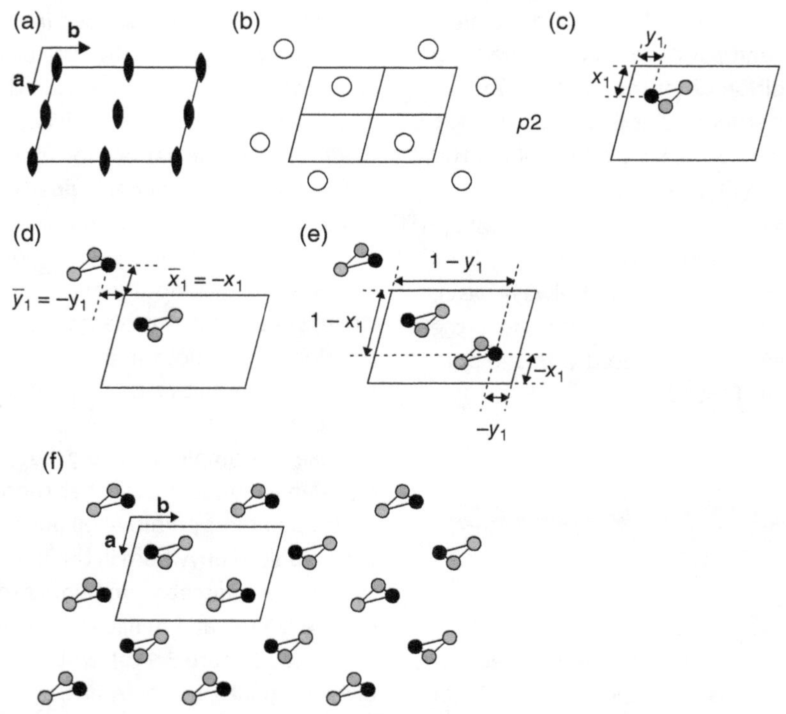

Figure 3.12 Space group diagrams for planar group $p2$: (a) symmetry elements; (b) equivalent general positions; (c–e) specification of the coordinates (x, y) of an atom in a general position; (f) the overall pattern.

The **atom positions** in *any* unit cell (not just in plane group $p1$), x and y, are always defined in directions parallel to the unit cell edges. The three atoms in the 'molecule' motif will have atom positions (x_1, y_1), (x_2, y_2) and (x_3, y_3), defined in this way (Figure 3.11d–f). Because of the absence of symmetry elements, there is no alternative but to list the (x, y) coordinates of each atom.

The standard crystallographic diagrams for plane group $p2$ are drawn in Figure 3.12a–e. The plane group $p2$ contains a diad axis at the origin of the unit cell which means that an atom at (x_1, y_1) will be repeated at a position, $-x_1, -y_1$, written $(\bar{x}, \bar{y})$, which is equivalent to the position $(1-x)$, $(1-y)$. If the motif is a group of atoms, the positions of each atom in the new group will be related to those of the starting group by the same rotation symmetry. Note that although there are more diad axes present, they do not produce more atoms in the unit cell. The pattern produced by extending the number of unit cells visible is drawn in Figure 3.12f.

The results of the operation of the symmetry elements present in the plane groups pm and pg are summarised in Figure 3.13a–f. In plane group pm, the mirror lines lie perpendicular to the **a**-axis (Figure 3.13a). The consequence of the mirror is to transform the motif into its mirror image. This form cannot be superimposed upon the original simply by moving it around on the plane of the page. In standard crystallographic diagrams, a chiral relationship with a normal motif is indicated by

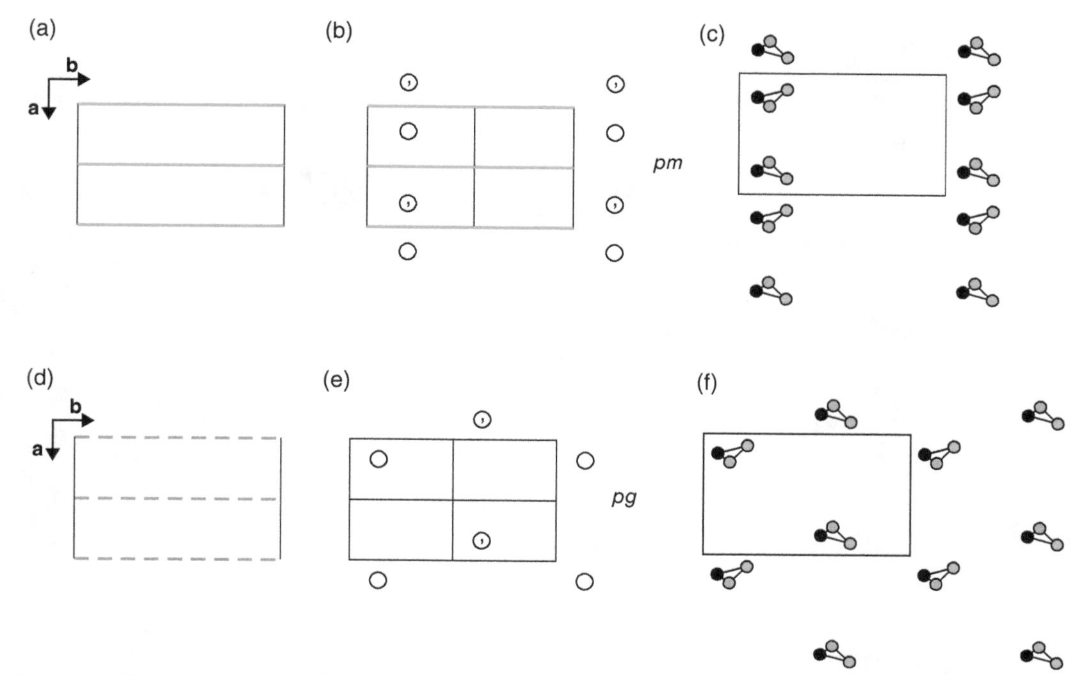

Figure 3.13 Space group diagrams for planar groups *pm* and *pg*: (a) symmetry elements present in group *pm*; (b) equivalent general positions for group *pm*; (c) an overall pattern consistent with group *pm*; (d) symmetry elements present in group *pg*; (e) equivalent general positions for group *pg*; (f) an overall pattern consistent with group *pg*.

placing a comma within the circular motif symbol (Figure 3.13b). The relationship between the two forms is clearer when the motif used is the triangular 'molecule' (Figure 3.13c). Plane group *pg* has glide lines lying perpendicular to the **a**-axis (Figure 3.13d). The glide operation also involves reflection, and so the two forms of the motif are chiral pairs (Figure 3.13e and f).

The defining symmetry element in the plane group *p*4 is a tetrad axis, which is conveniently located at the corner of a unit cell. The operation of the symmetry elements comprising the plane group *p*4 (Figure 3.14a) will cause a motif at a general position to be reproduced four times in the unit cell (Figure 3.14b). The positions of the atoms generated by the tetrad axis are derived as follows. An atom is placed at a general position (x, y) (Figure 3.14c). Successive anticlockwise rotations of 90° (Figures 3.13d–f), generate the remaining three. The equivalent atoms, transposed into just one unit cell, are shown in Figure 3.14g.

The positions are described in crystallographic texts as (x, y), $(\bar{y}, x)$, $(\bar{x}, \bar{y})$, $(y, \bar{x})$. This terminology can be misleading. It means this. If the first position chosen, (x, y) is, say, given by $x = 0.15$, $y = 0.35$, then the point $(\bar{y}, x)$, has an x coordinate of (−0.35) [which is equivalent to (1−0.35) or (+0.65)], and a y coordinate of 0.15. The four coordinates equivalent to the positions (x, y), $(\bar{y}, x)$, $(\bar{x}, \bar{y})$, $(y, \bar{x})$, are set out in Table 3.4 and can be visualised from Figure 3.14.

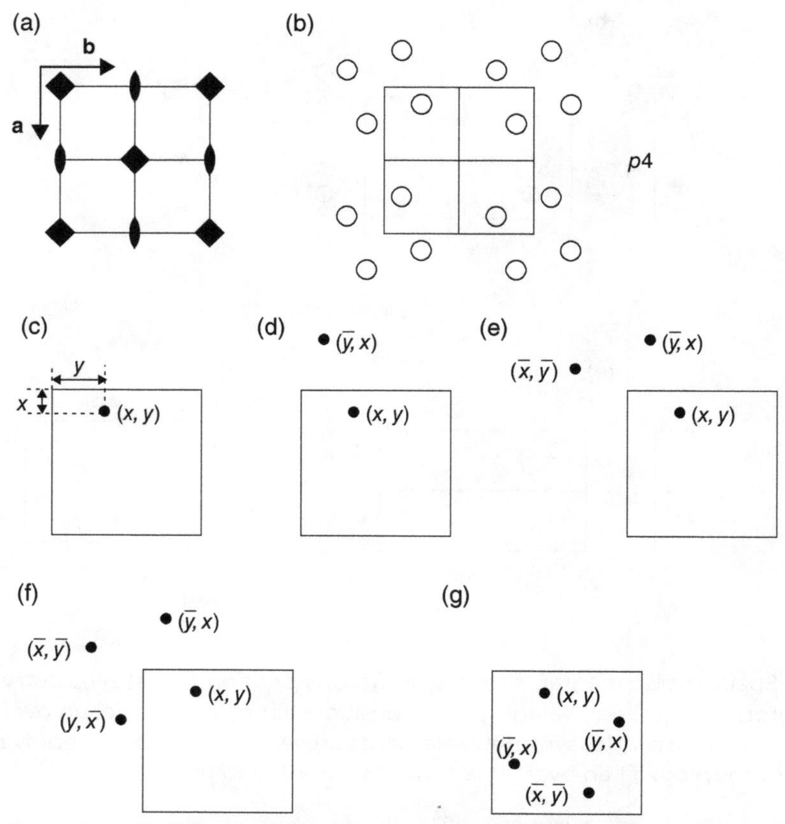

Figure 3.14 Space group diagrams for planar group *p*4: (a) symmetry elements; (b) equivalent general positions; (c–g) specification of the coordinates (*x*, *y*) of an atom in a general position.

Table 3.4 General equivalent atomic coordinates in plane group *p*4.

Position	*x* value	*y* value	Final *x*	Final *y*
(x, y)	0.15	0.35	0.15	0.35
$(\bar{y}, x)$	−0.35	0.15	(1−0.35) = 0.65	0.15
$(\bar{x}, \bar{y})$	−0.15	−0.35	(1−0.15) = 0.85	(1−0.35) = 0.65
$(y, \bar{x})$	0.35	−0.15	0.35	(1−0.15) = 0.85

3.7 GENERAL AND SPECIAL POSITIONS

In the previous section, an atom or a motif was allocated to a position, (*x*, *y*), which was anywhere in the unit cell. Such a position is called a **general position**, and the other positions generated by the symmetry operators present are called **general equivalent positions**. The number of general equivalent positions is given by the **multiplicity**. Thus, the general position (*x*, *y*) in the plane group *p*2 (number 2) has a multiplicity of 2 and in the plane group *p*4 (number 10) has a multiplicity of 4.

If, however, the (*x*, *y*) position of an atom falls upon a symmetry element, the multiplicity will decrease. For example, an atom placed at the unit cell origin in the group *p*2 will lie on the diad axis, and will not be repeated. This position will then have a multiplicity of 1, whereas a general point has a multiplicity of 2. Such a position is called **special position**. The special positions in a unit cell conforming to the plane group *p*2 are found at coordinates (0, 0), (½, 0), (0, ½) and (½, ½), coinciding with the diad axes. There are four such special positions, each with a multiplicity of 1.

In a unit cell conforming to the plane group *p*4, there are special positions associated with the tetrad axes, at the cell origin (0, 0), and at the cell centre (½, ½), and associated with diad axes at positions (½, 0) and (0, ½) (see Figure 3.14a). The multiplicity of an atom located on the tetrad axis at the cell origin will be one, as it will not be repeated anywhere else in the cell by any symmetry operation. The same will be true of an atom placed on the tetrad axis at the cell centre. If an atom is placed on a diad axis, say at (½, 0), another must occur at (0, ½). The multiplicity of an atom on a diad will thus be 2.

Note that the multiplicity of a diad in plane group *p*2 is different from that on a diad in plane group *p*4. The multiplicity of a special position will thus vary from one plane group to another and will depend upon the other symmetry elements present. Nevertheless, the multiplicity is always found to be a divisor of the multiplicity of a general position.

As well as the multiplicity of a site, its symmetry is often of importance. The **site symmetry** is specified in terms of the rotation axes and mirror planes passing through the site. The order of the symbols, when more than one is present, is the same as the order of the symmetry elements in the plane group symbol. A general position will always have a site symmetry of 1 (i.e. no symmetry elements are present). The site symmetry of an atom at a diad in plane group *p*2 or *p*4 will be 2, corresponding to the diad axis itself. The site symmetry of an atom on a tetrad axis in plane group *p*4 will be 4. When a position is not involved in site symmetry, the place is represented by a dot, as a placeholder. Thus in plane group *p*2*mm* (Figure 3.10f), the site symmetry of an atom on either of the two mirrors perpendicular to the **a**-axis is given as *.m.*, and an atom on a mirror perpendicular to the **b**-axis is given as *..m*. An atom on any of the diads is also on a mirror perpendicular to the **a**-axis and to a mirror perpendicular to the **b**-axis, hence the site symmetry is written as 2*mm*.

Both the general and special positions in a unit cell are usually listed in a table that gives the multiplicity of each position, the coordinates of the equivalent positions, and the site symmetry. In these tables, the general position is listed first, and the special positions are listed in order of decreasing multiplicity and increasing point symmetry. These are also accompanied by a **Wyckoff letter**, listed upwards from the site with the lowest multiplicity, in alphabetical order. The two following examples, Table 3.5 for the plane group *p*2, and Table 3.6 for plane group *p*4, give more detail.

From the foregoing, it is seen that in order to specify a two-dimensional crystal structure, all that needs to be specified are (i) the lattice parameters, (ii) the plane group symbol, and (iii) a list of atom types together with the Wyckoff letter and the atomic coordinates. For complex crystals, it simplifies the amount of information to be recorded enormously.

Table 3.5 Positions in the plane group *p*2

Multiplicity	Wyckoff letter	Site symmetry	Coordinates of equivalent positions
2	*e*	1	(1) x, y; (2) $\bar{x}, \bar{y}$
1	*d*	2	½, ½
1	*c*	2	½, 0
1	*b*	2	0, ½
1	*a*	2	0, 0

Table 3.6 Positions in the plane group *p*4

Multiplicity	Wyckoff letter	Point symmetry	Coordinates of equivalent positions
4	*d*	1	(1) x, y; (2) $\bar{x}, \bar{y}$; (3) $(\bar{y}, x)$; (4) $(y, \bar{x})$
2	*c*	2	½, 0; 0, ½
1	*b*	4	½, ½
1	*a*	4	0, 0

3.8 TESSELATIONS

A tesselation is a pattern of tiles that covers an area with no gaps or overlapping pieces. Obviously the unit cells associated with the five plane lattices will tesselate a plane periodically. In recent years, there has been considerable interest in the mathematics of such pattern-forming, much of which is of relevance to crystallography. There are an infinite number of ways in which a surface may be tiled, and although tiling theory is generally restricted to using identically shaped pieces (tiles), the individual pieces can be rotated or reflected, as well as translated, whereas in crystallography, only translation is allowed.

Regular tesselations or **tilings** are those in which every tile has the same shape, and every intersection of the tiles has the same arrangement of tiles around it. There are only three regular tesselations, derived from the three regular polygons: equilateral triangles, squares or hexagons (Figure 3.15). No other regular polygon will completely tile a plane without gaps appearing.

A **periodic tesselation** or **tiling** is one in which a region (i.e. a unit cell) can be outlined that tiles the plane by translation, that is by moving it laterally, but without rotating it or reflecting it. The rules for the production of a periodic tiling are less exacting than those to produce a regular tiling. This can be illustrated by a periodic tesselation using *equilateral pentagons* (Figure 3.16a), which has been known and used since ancient times. Note that the *equilateral* pentagon is distinct from the *regular* pentagon, because, although it has five sides of equal length, the internal angles differ from 72°. The tesselation can only be completed if some of the pentagonal tiles are rotated or reflected. The tesselation is periodic, however, because a unit of structure

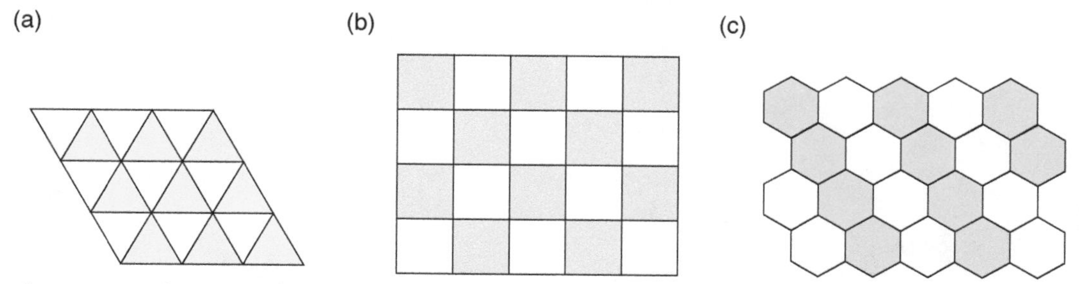

Figure 3.15 The regular tesselations: (a) equilateral triangles; (b) squares; (c) regular hexagons.

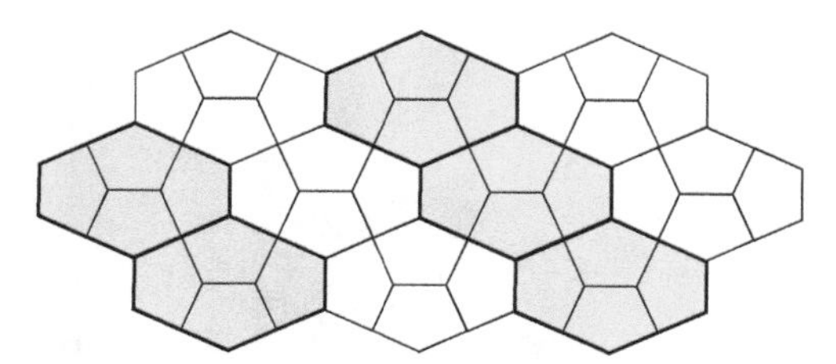

Figure 3.16 A periodic tiling using equilateral (not regular) pentagons.

(a unit cell) consisting of four pentagonal tiles (shaded) will tesselate the plane by translation alone.

A **non-periodic (aperiodic) tesselation or tiling** is one that cannot be created by the repeated translation of a small region (a unit cell) of the pattern. There are many of these, including, for example, spiral tilings. One family of non-periodic tesselation patterns of particular relevance to crystallography are called **Penrose tilings**. These can be constructed from two tile shapes that are called kites and darts, each with a particular geometry. Begin with a rhombus with each edge equal in length, with an acute apex angle of 72° (Figure 3.17a). The rhombus is divided into two along the long diagonal, in the golden ratio[2] of 1.618:1, and the point so found is joined to the obtuse corners of the rhombus. This creates the kite, with the longer edges of relative length equal to the golden ratio and the shorter edges equal to 1.0 (Figure 3.17b), and the dart, with similar relative edge lengths (Figure 3.17c). A kite is one-fifth of a decagon (Figure 3.17b). The rhombus itself tesselates periodically, because it is just an oblique unit cell. However, if ways of joining the two kites and darts are restricted, the pattern formed is non-periodic or aperiodic.

The way that the non-periodicity is forced upon the tesselation is to join the tiles in a particular way. An attractive way of labelling the tiles is to draw circular arcs of two colours or patterns, cutting the edges, again in the golden ratio (Figure 3.17e and f). The tesselation is then carried out by making sure that when tiles are joined, the circular arcs or patterns of the same type are continuous. A Penrose tiling constructed in this way, the 'infinite sun' pattern, is drawn in Figure 3.18. Any region of the pattern appears to show fivefold rotational symmetry, and this remains true for any area of the tiling, no matter how large the pattern becomes. However, extending the drawing reveals that the pattern itself never quite manages to repeat. Although, at first glance, it seems that a unit cell can be found, it turns out

[2] The 'golden ratio' is equal to $(1 + \sqrt{5})/2 = 1.61803398$, and is the ratio of any two successive terms in the Fibonacci number sequence.

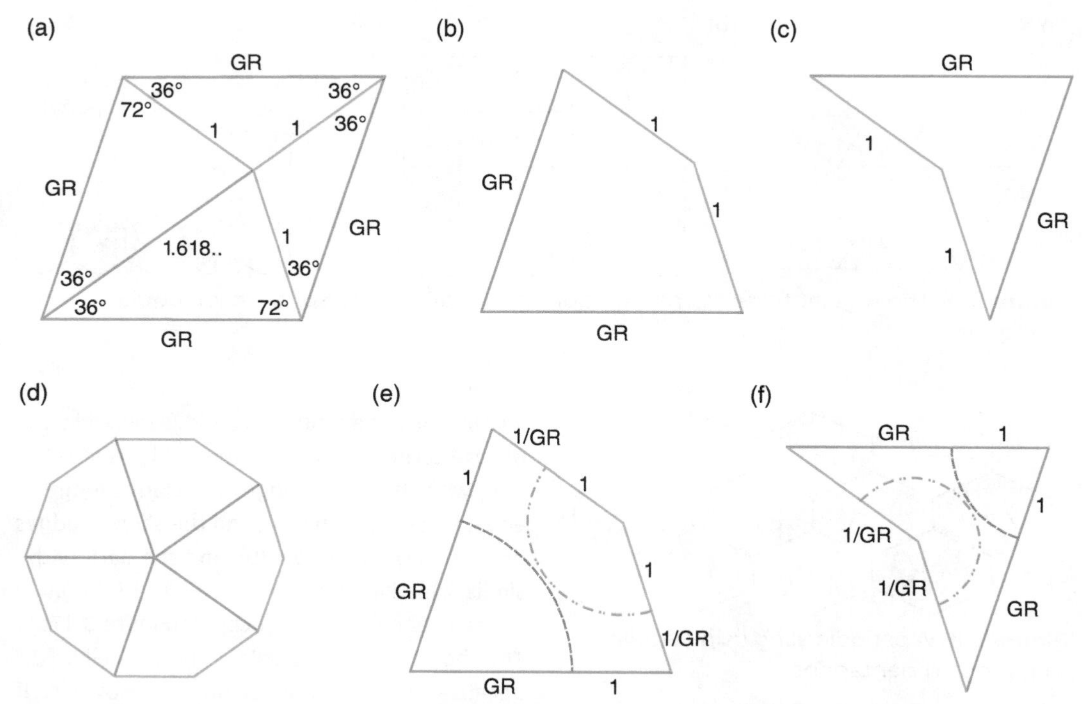

Figure 3.17 Kites and darts: (a) a rhombus with each edge equal to the golden ratio, GR; (b) a kite; (c) a dart; (d) five kites make up a decagon; (e and f) kite and dart with arcs that force an aperiodic Penrose tiling when matched.

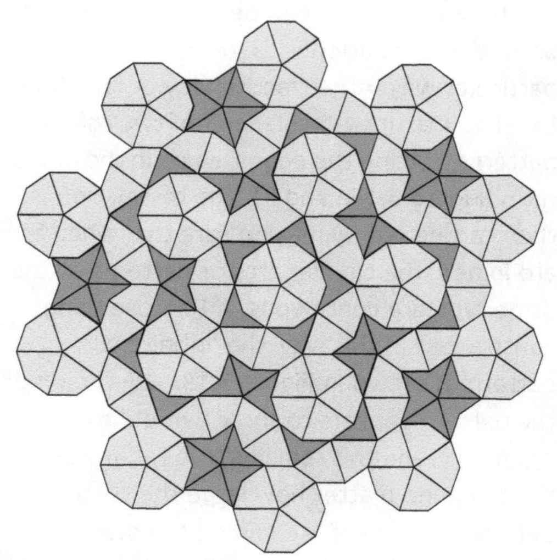

Figure 3.18 The 'infinite sun' pattern of Penrose tiles.

that this is never possible. References on how to construct Figure 3.18 and other Penrose tilings are given in the Bibliography.

This has startling consequences for crystallography. A Penrose tiling pattern appears to have a fivefold or perhaps 10-fold rotational symmetry, a feature not allowed in the 'classical' derivation of patterns given above. More surprisingly, solids that are analogous to three-dimensional tessellations and Penrose tilings have been discovered. These break the fundamental rules of crystallography by having unit cells with (apparently) fivefold and higher rotational symmetry. These enigmatic phases, now called **quasicrystals**, are described in Section 9.9.

ANSWERS TO INTRODUCTORY QUESTIONS

What is a point group?

A point group describes the symmetry of an isolated shape. When the various symmetry elements that are present in a two-dimensional (planar) shape are applied in turn, it is seen that one point is left unchanged by the transformations. When these elements are drawn on a figure, they all pass through this single point. (For this reason, the combination of operators is called the general point symmetry of the shape.) There are no limitations imposed upon the symmetry operators that are allowed, and in particular, fivefold rotation axes are certainly allowed, and are found in many natural objects, such as starfish or flowers. There are many general point groups.

The scaffolding that underlies any two-dimensional pattern is a plane lattice. The point groups obtained by excluding all rotation operators incompatible with a plane lattice are called the crystallographic plane point groups. These are formed by combining the rotation operators 1, 2, 3, 4 and 6, with mirror symmetry. There are just ten crystallographic point groups associated with two-dimensional shapes or 'crystals'.

What is a plane group?

A plane group describes the symmetry of a two-dimensional repeating pattern.

A way of making such a repeating pattern is to combine the ten (two-dimensional) crystallographic plane point groups with the five (two-dimensional) plane lattices. The result is 17 different arrangements, called the 17 (two-dimensional) plane groups. All two-dimensional crystallographic patterns must belong to one or other of these groups. These 17 are the only crystallographic patterns that can be formed in a plane. Obviously, by varying the motif, an infinite number of designs can be created, but they will all be found to possess one of the 17 unit cells described.

What is an aperiodic tiling?

A tesselation or tiling is a pattern of tiles that covers an area with no gaps or overlapping pieces. A periodic tiling is one in which a region (a unit cell), can be outlined that tiles the plane by translation, that is by moving it laterally, but without rotating it or reflecting it. An aperiodic (non-periodic) tiling is one that cannot be created by the translation of a small region of the pattern without rotation or reflection. In effect this means that the pattern does not possess a unit cell. Although any small area may suggest a unit cell, when the pattern is examined over large distances it is found that repetition does not occur. The 'unit cell' is infinitely long. The best-known family of aperiodic tesselation patterns is called Penrose tilings.

PROBLEMS AND EXERCISES

Quick Quiz

1. Which one of the following operations is ***not*** a symmetry element:
   a. Rotation
   b. Reflection
   c. Translation
2. A point group is another name for:
   a. The symmetry of a pattern
   b. A collection of symmetry elements
   c. A single symmetry element

3. A tetrad operator involves an anticlockwise rotation of:
   a. 180°
   b. 90°
   c. 60°
4. Which one of the following rotation axes is ***not*** compatible with a crystal:
   a. Fourfold
   b. Sixfold
   c. Eightfold
5. The number of two-dimensional crystallographic point symmetry groups is:
   a. 17
   b. 10
   c. 5
6. The number of plane groups is:
   a. 17
   b. 10
   c. 5
7. A glide operator consist of:
   a. Reflection plus rotation
   b. Rotation plus translation
   c. Reflection plus translation
8. If the position (x, y) = (0.25, 0.80), the position ($\bar{y}$, x) is:
   a. (−0.80, 0.25)
   b. (−0.25, 0.80)
   c. (0.80, −0.25)
9. If the (*x*, *y*) position of an atom falls upon a symmetry element, the multiplicity will:
   a. Increase
   b. Decrease
   c. Remain unchanged
10. A Penrose tiling can be described as:
   a. Regular
   b. Periodic
   c. Aperiodic

Calculations and Questions

3.1. Write out the symmetry elements present in the 'molecules' in the figure below, and the corresponding point group of each one. Assume that the molecules are planar, exactly as drawn, and not three-dimensional. The shapes are: (a) pentagonal C_5H_5, in ferrocene; (b) linear, CS_2; (c) triangular, SO_2; (d) square, XeF_4; (e) planar, C_2H_4, ethene.

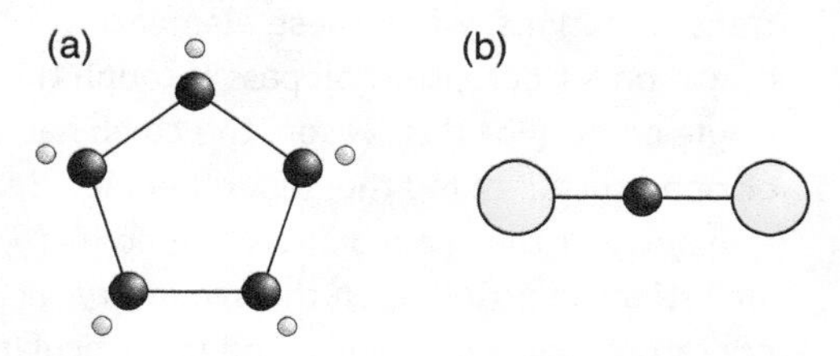

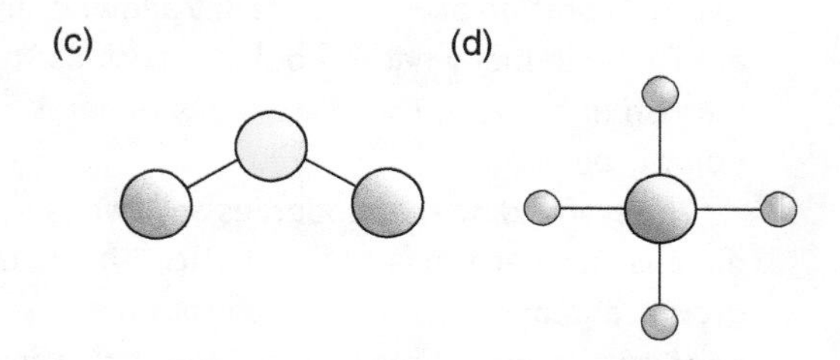

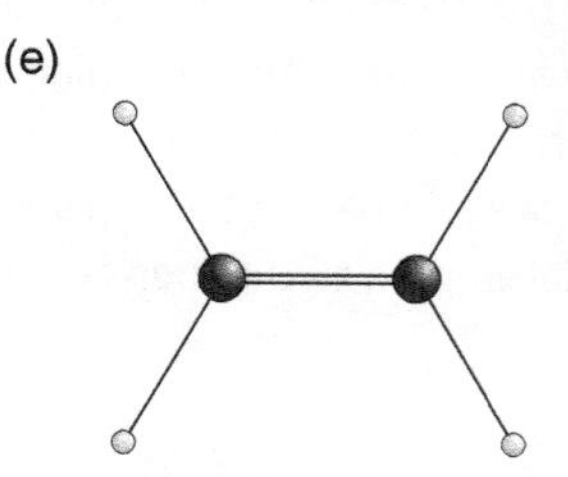

3.2. Write out the symmetry elements present in the two 'crossword' blanks drawn in the figure below, and the corresponding point group of each pattern.

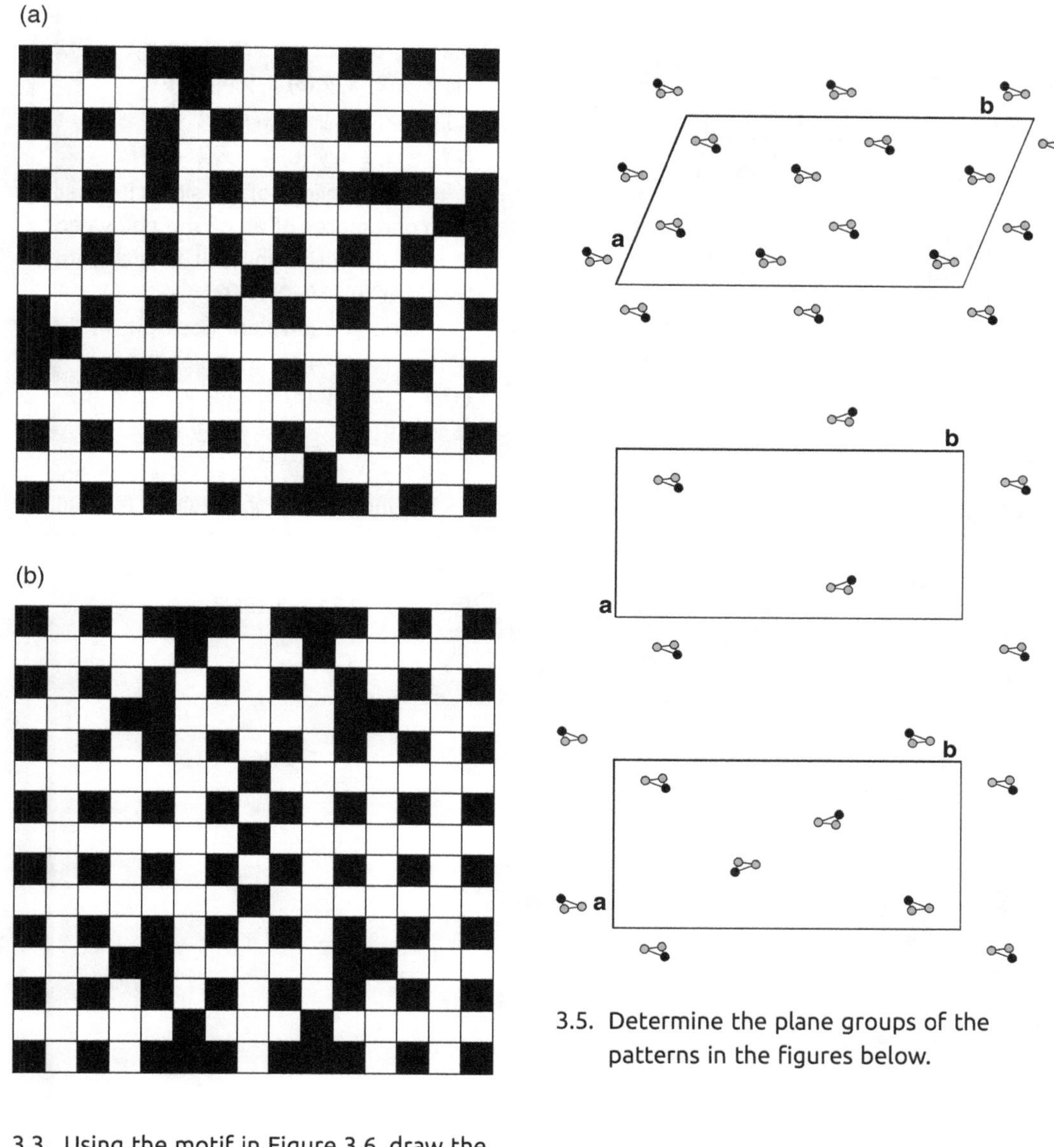

3.3. Using the motif in Figure 3.6, draw the diagrams for the point groups 5 and 5m.

3.4. Determine the plane groups of the patterns in the figures below.

3.5. Determine the plane groups of the patterns in the figures below.

(a)

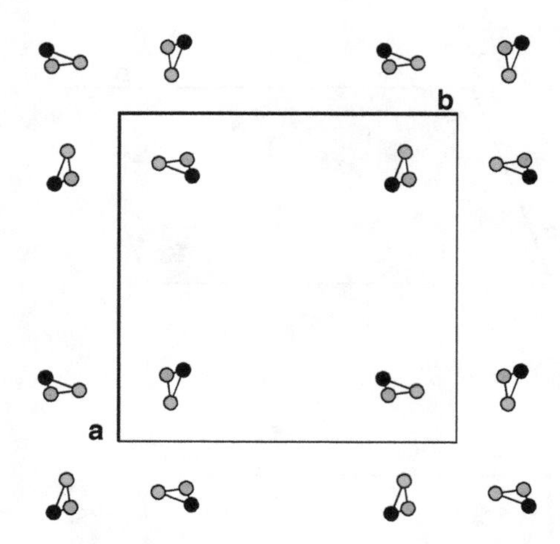

(b)

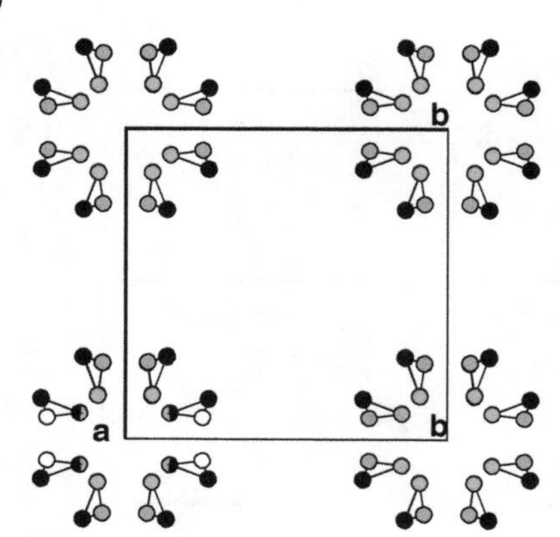

(c)

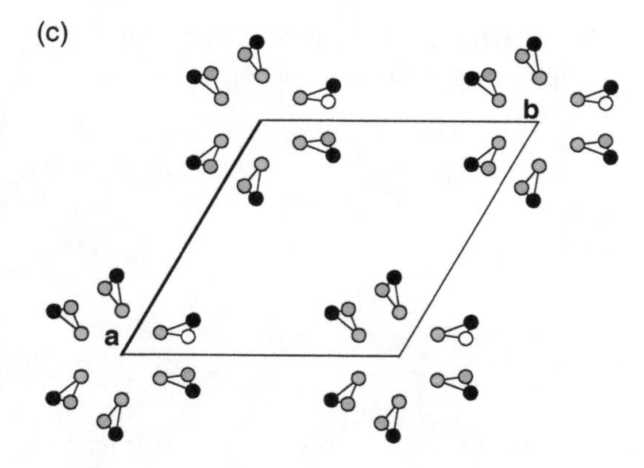

3.6. The general equivalent positions in the plane group *c*2*mm* are given by:
(1) x, y (2) $\overline{x}, \overline{y}$ (3) $\overline{x}, y$ (4) $x, \overline{y}$
(5) $\frac{1}{2} + x, \frac{1}{2} + y$ (6) $\frac{1}{2} + \overline{x}, \frac{1}{2} + \overline{y}$
(7) $\frac{1}{2} + \overline{x}, \frac{1}{2} + y$ (8) $\frac{1}{2} + x, \frac{1}{2} + \overline{y}$
The multiplicity of the site is 8. What are the coordinates of all of the equivalent atoms within the unit cell, if one atom is found at (0.125, 0.475)?

3.7. The general equivalent positions in the plane group *p*4*gm* are given by:
(1) x, y (2) $\overline{x}, \overline{y}$ (3) $\overline{y}, x$ (4) $y, \overline{x}$
(5) $\overline{x} + \frac{1}{2}, y + \frac{1}{2}$ (6) $x + \frac{1}{2}, \overline{y} + \frac{1}{2}$
(7) $y + \frac{1}{2}, x + \frac{1}{2}$ (8) $\overline{y} + \frac{1}{2}, \overline{x} + \frac{1}{2}$
The multiplicity of the site is 8. What are the coordinates of all of the equivalent atoms within the unit cell, if one atom is found at (0.210, 0.395)?

Chapter 4

Symmetry in Three Dimensions

What is a rotoinversion axis?
What relates a crystal class to a crystallographic point group?
What are enantiomorphic pairs?

This chapter follows, in outline, the format of Chapter 3. In this chapter, though, finite three-dimensional crystals form the subject matter, and the two-dimensional concepts already presented will be taken into this higher dimension. Any solid object can be classified in terms of the collection of symmetry elements that can be attributed to the shape. The combinations of allowed symmetry elements form the **general three-dimensional point groups** or **non-crystallographic three-dimensional point groups**. A smaller group, constrained by the translational properties of a lattice, make up the **crystallographic three-dimensional point groups**. The symmetry operators are described by here by the **International** or **Hermann-Mauguin** symbols. The alternative Schoenflies symbols, generally used by chemists, are given in Appendix C.

4.1 THE MIRROR PLANE AND AXES OF ROTATION

The most important symmetry operators for a solid object are identical to those described in Section 3.1, and so are only summarised here. The **mirror** operator, denoted *m* in text, is a symmetry operator that changes the **handedness** or **chirality** of the object. A **chiral species** is one that cannot be superimposed upon its mirror image. That is, a left hand is transformed into a right hand by reflection. Objects, particularly molecules or crystals that display handedness, are termed **enantiomorphs**, or an **enantiomorphic pair**. This phenomenon is particularly well explored in organic chemistry. It is found that a molecule containing a carbon atom tetrahedrally coordinated to four different groups will be a chiral species, the central C atom being labelled as a chiral C atom. The amino acid alanine, $CH_3CH(NH_2)COOH$, is an example of such a

Crystals and Crystal Structures, Second Edition. Richard J. D. Tilley.

chiral molecule, and can exist in both right-handed or left-handed forms (Figure 4.1). (The crystal structure of alanine is described in more detail in Chapter 6.) Enantiomorphic objects are structurally identical in every way except for the change from right-handed to left-handed. The physical properties of these compounds are also identical except that a chiral object displays the additional property of **optical activity**. The chemical properties of enantiomers, are, however, quite distinct one from the other, and lead to different biological roles in living systems (also see Section 5.7).

The rotation operator 1, a **monad** rotation axis, implies that no symmetry is present, because the shape has to be rotated anticlockwise by (360/1)° to bring it back into register with the original. A **diad** (twofold) rotation axis implies that rotation in an anticlockwise manner by 360°/2 (180°) around the axis returns the shape to the original configuration. A **triad** (threefold) rotation axis indicates anticlockwise rotation by 360°/3 (120°) around the axis returns the shape to the original configuration. A **tetrad** (fourfold) rotation axis indicates that rotation anticlockwise by 360°/4 (90°) about the axis returns the shape to the original configuration. The **pentad** (fivefold) rotation axis indicates that rotation anticlockwise by 360°/5 (72°) about the axis returns the shape to the original configuration. Finally, a **hexad** (sixfold) rotation axis indicates that rotation anticlockwise by 360°/6 (60°) around the axis returns the shape to the original configuration. In addition, there are an infinite number of other rotation symmetry operators, such as sevenfold, eightfold, and higher axes, many of which are found in nature (Figure 4.2).

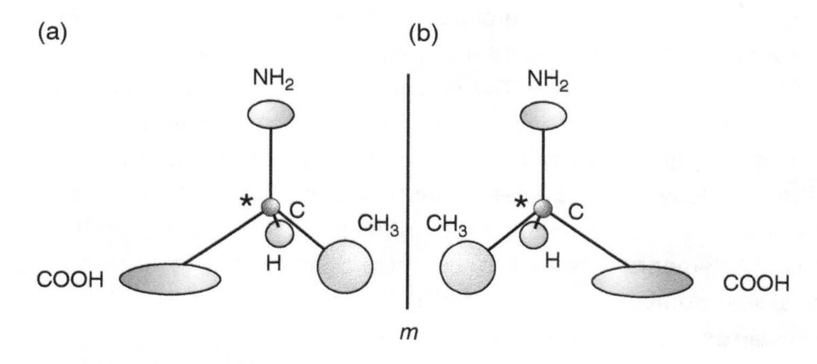

Figure 4.1 The enantiomorphic (mirror image) forms of the molecule alanine, $CH_3CH(NH_2)COOH$: (a) the naturally occurring form, (S)-(+)-alanine; (b) the synthetic form (R)-(-)-alanine. The chiral C atom is marked ∗.

Figure 4.2 Rotation axes in nature: (a) approximately fivefold, but really only a monad is present; (b) fivefold; (c) fivefold plus mirrors; (d) eightfold.

4.2 AXES OF INVERSION: ROTOINVERSION

Apart from the symmetry elements described above, an additional type of rotation axis occurs in a solid that is not found in planar shapes, the **inversion axis** $\bar{n}$. The operation of an inversion axis consists of a rotation combined with a centre of symmetry. These axes are also called **improper rotation axes**, to distinguish them from the ordinary **proper rotation axes** described above. The symmetry operation of an improper rotation axis is that of **rotoinversion**. Two solid objects related by the operation of an inversion axis are found to be enantiomorphous.

The simplest improper rotation axis, indicated by $\bar{1}$, and in drawings by , represents a centre of symmetry or inversion centre. An object possesses a centre of symmetry at position (0, 0, 0) if any point at position (x, y, z) is accompanied by an equivalent point located at ($-x$, $-y$, $-z$) written ($\bar{x}$, $\bar{y}$, $\bar{z}$) (Figure 4.3a). Note that a centre of symmetry inverts an object so that a lower face becomes an upper face, and vice versa (Figure 4.3b). The centre of symmetry of a regular octahedral MO_6 molecule lies at the central metal atom (Figure 4.3c). Any line that passes through the centre of symmetry of a solid connects equivalent faces or objects (Figure 4.3d).

The operation of a twofold improper rotation axis $\bar{2}$ transforms an initial atom position (Figure 4.4a) by a rotation of 180° anticlockwise (Figure 4.4b) then an inversion through the centre of symmetry (Figure 4.4c).

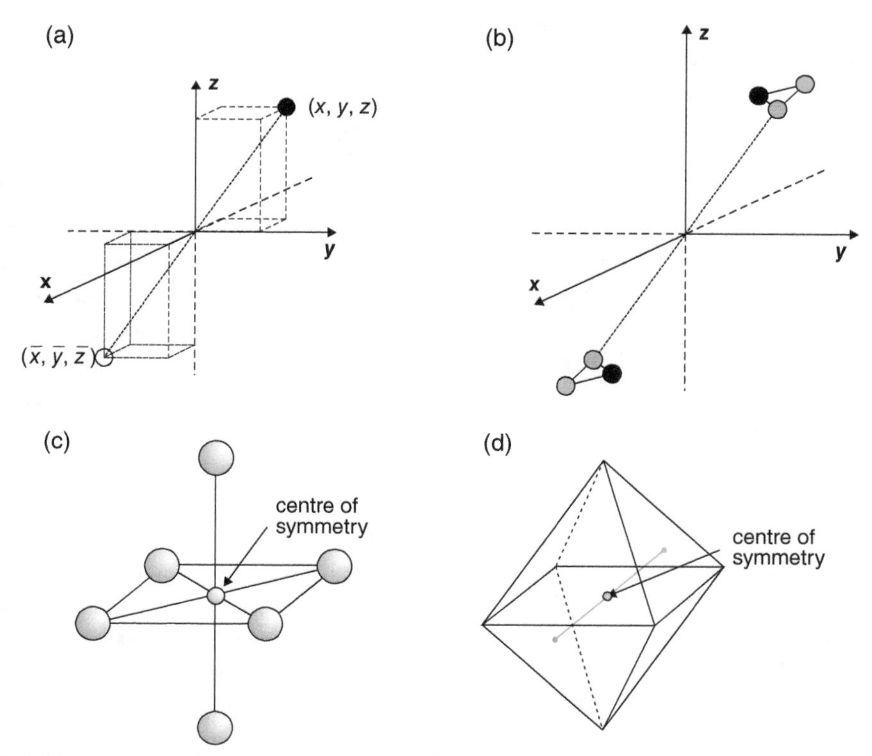

Figure 4.3 The centre of symmetry operator (see text).

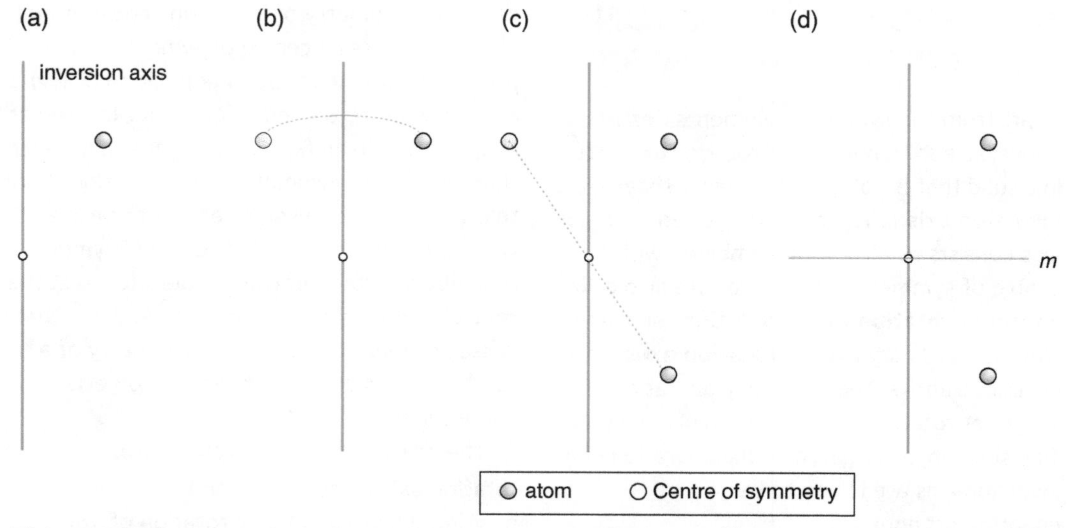

Figure 4.4 The operation of a twofold improper rotation axis $\bar{2}$, equivalent to that of a mirror plane, *m* (see text).

It is seen that the operation is identical to that of a mirror plane (Figure 4.4d), and this latter designation is used in preference.

The operation of some of the other improper rotation axes can be illustrated with respect to the five **Platonic solids**, the regular tetrahedron, regular octahedron, regular icosahedron, regular cube and regular dodecahedron. These polyhedra have regular faces and vertices, and each has only one type of face, vertex, and edge (Figure 4.5a–e). The regular tetrahedron, regular octahedron and regular icosahedron are made up from 4, 8 and 20 equilateral triangles respectively. The regular cube, or hexahedron, is composed of six squares, and the regular dodecahedron of 12 regular pentagons.

The tetrahedron illustrates the operation of a fourfold inversion axis $\bar{4}$. A tetrahedron inscribed in a cube allows the three Cartesian axes to be defined (Figure 4.6a). With respect to these axes, the fourfold inversion axes lie parallel to the *x*-, *y*- and *z*-axes. The operation of one such axis moves vertex A by a rotation of 90° in an anticlockwise direction, and then inversion through the centre of symmetry to generate vertex D (Figure 4.6b). In subsequent applications of the $\bar{4}$ operator, vertex D is transformed to vertex B, B to C, and C back to A.

Apart from the $\bar{4}$ axis, rotation triad (3) axes pass through any vertex and the centre of the opposite triangular face, along the <111> directions, (Figure 4.6c). In addition, the tetrahedron has mirror symmetry. The mirror planes are {110} planes that contain two vertices and lie normal to the other two vertices (Figure 4.6d).

The regular octahedron can be used to describe the threefold inversion axis $\bar{3}$. The highest order axis, is, however, a tetrad. Using Cartesian axes to locate the symmetry operators, these axes pass through a vertex and run along the *x*-, *y*- or *z*-axis. Each tetrad is accompanied by a mirror normal to it

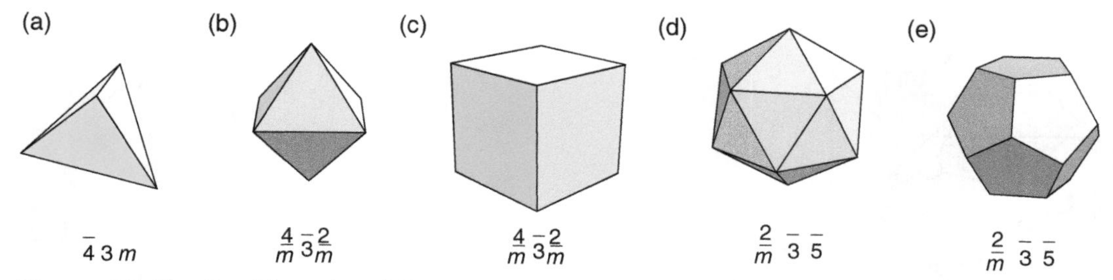

Figure 4.5 The five Platonic solids, together with the point group symbols: (a) regular tetrahedron; (b) regular octahedron; (c) regular cube; (d) regular icosahedron; (e) regular dodecahedron.

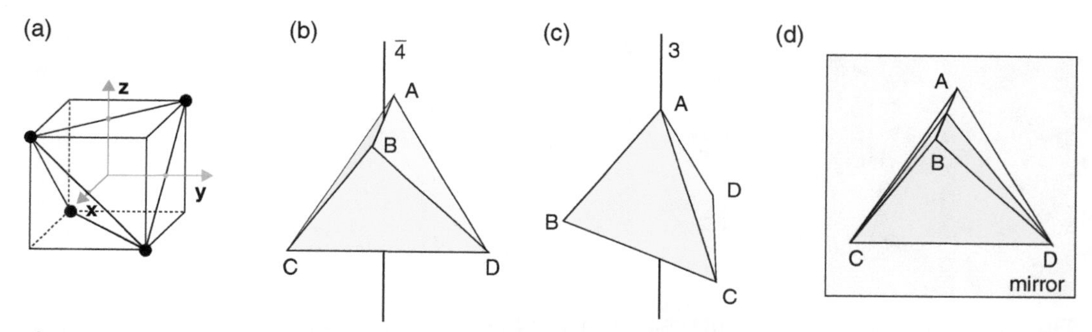

Figure 4.6 Symmetry elements present in a regular tetrahedron: (a) the three Cartesian axes; (b) fourfold inversion axes ($\bar{4}$); (c) rotation triad (3) axes; (d) mirror planes.

(Figure 4.7a and b). This combination is written $4/m$. The threefold inversion axes relate the positions of the vertices to each other. A $\bar{3}$ inversion axis runs through the middle of each opposite pair of triangular faces, along the cube <111> directions (Figure 4.7c). The operation of this symmetry element is to rotate a vertex such as A by 120° anticlockwise to B, and then invert it to generate the vertex F. The operation of the $\bar{3}$ axis is more readily seen from the view down the axis (Figure 4.7d). Successive 120° rotations in an anticlockwise direction, accompanied by an inversion, brings about the transformation of vertex A → (B) → F, F → (D) → B, B → (A) → E, E → (F) → C, C → (B) → D, D → (E) → A.

Finally, it is seen that diads (2) run through the middle of each edge of the octahedron, and run between any pair of the *x*-, *y*- or *z*-axes, along <110>. Each diad is accompanied by a perpendicular mirror (Figure 4.7e). The combination is written $2/m$.

The two most complex Platonic solids, the icosahedron and dodecahedron (Figure 4.5d and e), both have $\bar{5}$ (fivefold inversion) axes. In these a vertex is rotated by 72° (360°/5), and then translated by the height of the solid. In the case of a dodecahedron a $\bar{5}$ axis passes through each vertex, whilst in the case of a regular icosahedron, a $\bar{5}$ axis passes through the centre of each face.

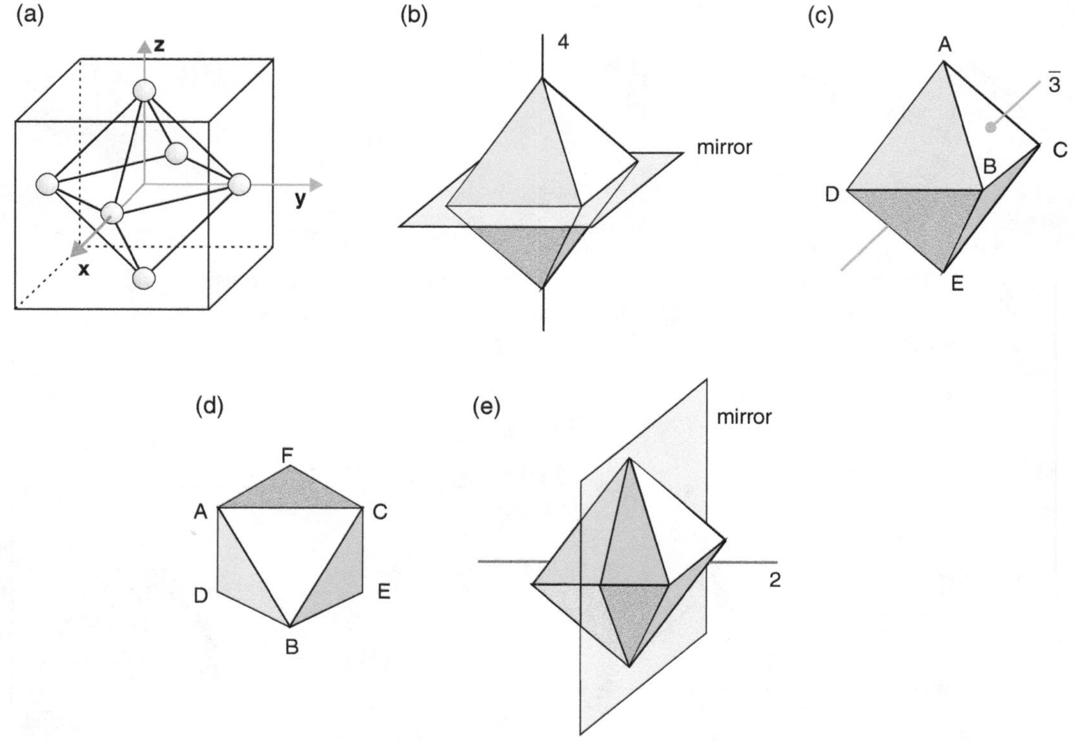

Figure 4.7 Symmetry elements present in a regular octahedron: (a) the three Cartesian axes; (b) a tetrad rotation axis; (c) a threefold inversion axis; (d) a triangular face viewed from above; (e) a diad axis.

4.3 AXES OF INVERSION: ROTOREFLECTION

The rotoinversion operator $\bar{n}$ is closely related to that of **rotoreflection**, and both are referred to as **improper rotation axes**. An axis of rotoreflection combines an anticlockwise rotation of $2\pi/n$ with reflection in a plane normal to the axis. These axes are given the symbol $\tilde{n}$. The point where the mirror plane intersects the rotation axis in a rotoreflection ($\tilde{n}$) operation is the centre of symmetry about which the inversion takes place during the operation of an $\bar{n}$ inversion axis. The operation of a rotoreflection axis has similar results to the operation of a rotoinversion axis, but the two sets of symbols *cannot* be directly interchanged. For example, the operation of a twofold improper rotoreflection axis $\tilde{2}$ on an initial atom position (Figure 4.8a) gives a rotation of 180° anticlockwise (Figure 4.8b), then reflected across a mirror normal to the axis (Figure 4.8c). The operation is identical to that of a centre of symmetry (Figure 4.8d), and this latter designation is used in preference to that of the improper axis.

The equivalence of the important rotoreflection axes with rotoinversion axes is given in Table 4.1. In crystallography the rotoinversion operation is always preferred, but

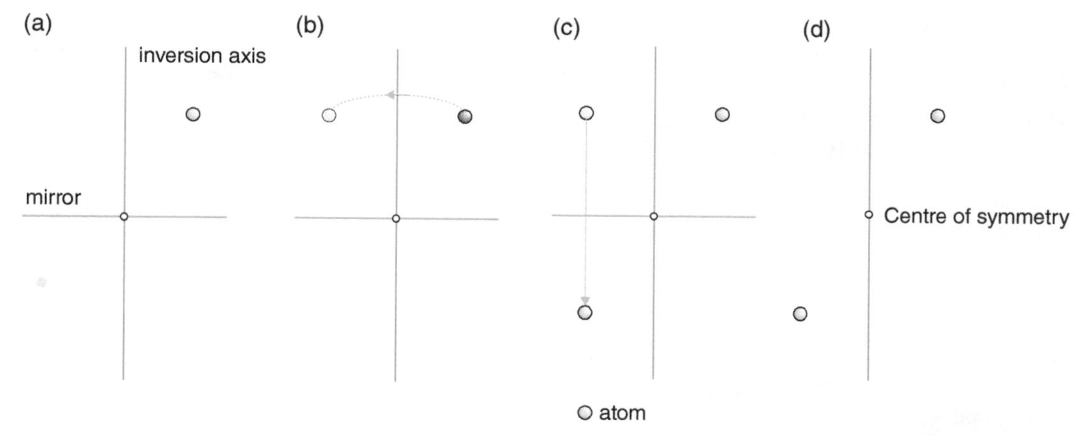

Figure 4.8 The operation of a twofold rotoreflection improper rotation axis $\tilde{2}$: (a) initial atom position; (b) rotation by 180° anticlockwise; (c) reflection across a mirror normal to the axis; (d) the operation of a centre of symmetry.

Table 4.1 Correspondence of rotoreflection and rotoinversion axes

Axis of rotoreflection	Axis of rotoinversion
$\tilde{1}$	$\bar{2} = m$
$\tilde{2}$	$\bar{1}$
$\tilde{3}$	$\bar{6}$
$\tilde{4}$	$\bar{4}$
$\tilde{6}$	$\bar{3}$

for molecular symmetry, which uses the Schoenflies system of symmetry nomenclature, the rotoreflection operations are chosen. Unfortunately, as mentioned above, *both* of these operations are often simply called *improper rotations*, without further specification, so care has to be exercised when comparing molecular symmetry descriptions (Schoenflies notation) with crystallographic descriptions (Hermann-Mauguin notation). A full correspondence between Schoenflies and Hermann-Mauguin nomenclature is given in Appendix C.

4.4 THE HERMANN-MAUGUIN SYMBOLS FOR POINT GROUPS

The symmetry operators that characterise a point group are collected together into the point group symbol. Crystallographers use the International or Hermann-Mauguin symbols. (The alternative Schoenflies symbols are given in Appendix C.) As in two dimensions, the order in which the symmetry operators are written is governed by specific rules, given in Table 4.2. The places in the symbol refer to **directions**. (When a number of directions are listed in Table 4.2 they are all equivalent, due to the symmetry of the object.) The first place or **primary** position is given to the most important or defining symmetry element of the group, which is often a symmetry axis. Symmetry axes are taken as parallel to the direction described. Mirror planes are taken to run **normal** to the direction indicated in Table 4.2. If a symmetry axis and the normal to a mirror plane are parallel, the two symmetry characters are separated by a slash, /. Thus the symbol $2/m$

Table 4.2 The order of the Hermann-Mauguin symbols in point groups

Crystal system	Required symmetry elements	Primary	Secondary	Tertiary
Triclinic	None	—	—	—
Monoclinic	1 diad or 1 mirror plane	[010], unique axis **b** [001], unique axis **c**	—	—
Orthorhombic	3 diads or 1 diad + 2 mirror planes	[100]	[010]	[001]
Tetragonal	1 tetrad	[001]	[100], [010]	$[1\bar{1}0]$, [110]
Trigonal, rhombohedral axes	1 triad	[111]	$[1\bar{1}0]$, $[01\bar{1}]$, $[\bar{1}01]$	
Trigonal, hexagonal axes	1 triad	[001]	[100], [010], $[\bar{1}\bar{1}0]$	
Hexagonal	1 hexad	[001]	[100], [010], $[\bar{1}\bar{1}0]$	$[1\bar{1}0]$, [120], $[\bar{2}\bar{1}0]$
Cubic	3 tetrads	[100], [010], [001]	[111], $[1\bar{1}\bar{1}]$, $[\bar{1}1\bar{1}]$, $[\bar{1}\bar{1}1]$	$[1\bar{1}0]$, [110], $[01\bar{1}]$, [011], $[\bar{1}01]$, [101]

indicates that a diad axis runs parallel to the normal to a mirror plane, that is, the diad is normal to a mirror plane.

In the example of the regular tetrahedron (Figures 4.5a and 4.6), the principle symmetry element is the $\bar{4}$ axis, which is put in the primary position in the point group symbol and defines the [100] direction of the cubic axial set. The rotation triad (3) lies along the <111> directions and so is given the secondary place in the symbol. In addition, the tetrahedron has mirror symmetry. The mirror planes are {110} planes that contain two vertices and lie normal to the other two vertices. They lie normal to some of the <110> directions, and are given the third place in the symbol, $\bar{4}3m$.

The same procedure can be used to collect the symmetry elements for a regular octahedron (Figures 4.5b and 4.7), described above, into the point group symbol $4/m\,\bar{3}\,2/m$. The regular cube (Figure 4.5c) can be described with exactly the same point group, $4/m\,\bar{3}\,2/m$. The two most complex Platonic solids, the icosahedron and dodecahedron, (Figure 4.5d and e), both have $\bar{5}$ (fivefold inversion) axes. The point group symbol for both of these solids is $2/m\,\overline{35}$, sometimes written as $\overline{53}\,2/m$.

To determine the point group of an object, it is necessary to write down a list of all of the symmetry elements that it possesses (not always easy), and then order them following the rules set out above.

4.5 THE SYMMETRY OF THE BRAVAIS LATTICES

The regular stacking of unit cells that can be translated but not rotated or reflected builds a crystal. This crystallographic limitation imposes a constraint upon the combinations of symmetry elements that are allowed in a lattice. In particular, the pentad axis and rotation axes higher than 6 are not allowed, for the reasons given in Chapter 3. It is also found that the pentad inversion axis $\bar{5}$, and all inversion axes higher than $\bar{6}$, are forbidden. The only operators allowed within the Bravais lattices are the centre of symmetry, $\bar{1}$, the mirror operator, *m*, the proper rotation axes 1, 2, 3, 4 and 6, and the improper rotation axes $\bar{2}$, $\bar{3}$, $\bar{4}$ and $\bar{6}$.

The operation of the allowed symmetry elements in the 14 Bravais lattices must leave each lattice point unchanged. The symmetry operators are thus representative of the point symmetry of the lattices. The most important lattice symmetry elements are given in Table 4.3. In all except the simplest case, two point group symbols are listed. The first is called the full Hermann-Mauguin symbol, and contains the most complete description. The second is called the short Hermann-Mauguin symbol, and is a condensed version of the full symbol. The order in which the operators within the symbol are written is given in Table 4.2.

The least symmetrical of the Bravais lattices is the triclinic primitive (*aP*) lattice. This is derived from the oblique primitive (*mp*) plane lattice by simply stacking other *mp*-like layers in the third direction, ensuring that the displacement of the second layer is not vertically above the first layer. Because the environment of each lattice point is identical with every other lattice point, a lattice point

Table 4.3 The symmetries of the Bravais lattices

Crystal system	Bravais lattice	Point group symbol	
		Full	Short[a]
Triclinic (anorthic)	*aP*	$\bar{1}$	
Monoclinic	*mP*	2/*m*	
	mC	2/*m*	
Orthorhombic	*oP*	2/*m* 2/*m* 2/*m*	*mmm*
	oC	2/*m* 2/*m* 2/*m*	*mmm*
	oI	2/*m* 2/*m* 2/*m*	*mmm*
	oF	2/*m* 2/*m* 2/*m*	*mmm*
Tetragonal	*tP*	4/*m* 2/*m* 2/*m*	4/*mmm*
	tI	4/*m* 2/*m* 2/*m*	4/*mmm*
Trigonal	*hR*	$\bar{3}$ 2/*m*	$\bar{3}$ *m*
Hexagonal	*hP*	6/*m* 2/*m* 2/*m*	6/*mmm*
Cubic	*cP*	4/*m* $\bar{3}$ 2/*m*	*m* $\bar{3}$ *m*
	cI	4/*m* $\bar{3}$ 2/*m*	*m* $\bar{3}$ *m*
	cF	4/*m* $\bar{3}$ 2/*m*	*m* $\bar{3}$ *m*

[a]Only given if different from the full symbol.

must lie at an inversion centre. This is the only symmetry element present, and the point group symbol for this lattice is $\bar{1}$.

The two monoclinic Bravais lattices, *mP* and *mC*, are generated in a similar way. The unique symmetry direction in monoclinic lattices is a diad (twofold) symmetry axis, conventionally labelled **b**; **a** and **c** lie in the lattice net perpendicular to **b** and include an oblique angle. Each layer of the lattice is similar to the plane *mp* lattice. The second and subsequent layers of the lattice are placed directly over the first layer, to generate the *mP* lattice, or displaced so that each lattice point in the second layer is vertically over the cell centre in the first layer to give the *mC* lattice. Because of this geometry, the diad axis of the plane lattice is preserved, and runs along the unique **b**-axis of the lattice. Moreover, the fact that the layers are stacked vertically means that a mirror plane runs perpendicular to the diad, through each plane of lattice points. No other symmetry elements are present. The point group symbol of both the *mP* and *mC* lattices is therefore 2/*m*.

The symmetry operators present in the other Bravais lattices can be derived in a similar way. The basis vector **c** for the tetragonal lattice is taken along the unique tetrad (fourfold) symmetry axis; **a** and **c** lie along diad (twofold) symmetry axes perpendicular to each other and **c**. The basis vectors **a**, **b** and **c** for an orthorhombic crystal lie along three mutually perpendicular diad (twofold) symmetry axes. The basis vector **c** for the hexagonal lattice lies parallel to the unique hexad (sixfold) symmetry axis; **a** and **b** are along diad (twofold) symmetry axes perpendicular to **c** and at 120° to each other.

As with the monoclinic lattices, face-centring or body-centring adds no new symmetry elements and the point group of the primitive lattice also applies to the others. Because of this there are only seven point groups, corresponding to the seven crystal systems, although there are 14 lattices (Table 4.3).

4.6 THE CRYSTALLOGRAPHIC POINT GROUPS

A solid can belong to one of an infinite number of general three-dimensional point groups. However, if the rotation axes are restricted to those that are compatible with the translation properties of a lattice, a smaller number, the **crystallographic point groups**, are found. The operators allowed within the crystallographic point groups are: the centre of symmetry $\bar{1}$, the mirror operator, *m*, the proper rotation axes 1, 2, 3, 4 and 6, and the improper rotation axes $\bar{1}$, $\bar{2}$, $\bar{3}$, $\bar{4}$ and $\bar{6}$. When these are combined together, 32 crystallographic point groups can be constructed, as listed in Table 4.4, of which 11 are centrosymmetric. The Schoenflies symbols for the crystallographic point groups are given in Appendix C.

There is no significant symmetry present in the triclinic system, and the symmetry operator 1 or $\bar{1}$ is placed parallel to any axis, which is then designated as parallel to the **a**-axis.

In the monoclinic system, the point group symbols refer to the unique axis, conventionally taken as the **b**-axis. The point group symbol 2 means a diad axis operates parallel to the **b**-axis. The improper rotation axis $\bar{2}$, taken to run parallel to the unique **b**-axis, is the same as a mirror, *m*, perpendicular to this axis. The symbol 2/*m* indicates that a diad axis runs parallel to the unique **b**-axis and a mirror lies perpendicular to the axis. Because there are mirrors present in two of these point groups, the possibility arises that crystals with this symmetry combination will be enantiomorphous, and show optical activity, (see Section 5.7). Note that in some

Table 4.4 Crystallographic point groups

Crystal system	Point group symbol		Centrosymmetric groups
	Full[a]	Short[b]	
Triclinic	1		
	$\bar{1}$		✓
Monoclinic	2		
	m ($\equiv \bar{2}$)		
	$\frac{2}{m}$ $2/m$		✓
Orthorhombic	222		
	$mm2$		
	$\frac{2}{m}\frac{2}{m}\frac{2}{m}$ $(2/m\ 2/m\ 2/m)$	mmm	✓
Tetragonal	4		
	$\bar{4}$		
	$\frac{4}{m}$ $(4/m)$		✓
	422		
	$4mm$		
	$\bar{4}2m$ or $\bar{4}m2$		
	$\frac{4}{m}\frac{2}{m}\frac{2}{m}$ $(4/m\ 2/m\ 2/m)$	$4/mmm$	✓
Trigonal	3		
	$\bar{3}$		✓
	32 or 321 or 312		
	$3m$ or $3m1$ or $31m$		
	$\bar{3}\frac{2}{m}$ or $\bar{3}\frac{2}{m}1$ or $\bar{3}1\frac{2}{m}$ $(\bar{3}\ 2/m$ or $\bar{3}\ 2/m\ 1$ or $\bar{3}1\ 2/m)$	$\bar{3}m$ or $\bar{3}m1$ or $\bar{3}1m$	✓
Hexagonal	6		
	$\bar{6}$ ($\equiv \frac{3}{m}$)		
	$\frac{6}{m}$ $(6/m)$		✓
	622		
	$6mm$		
	$\bar{6}2m$ or $\bar{6}m2$		
	$\frac{6}{m}\frac{2}{m}\frac{2}{m}$ $(6/m\ 2/m\ 2/m)$	$6/mmm$	✓
Cubic	23		
	$\frac{2}{m}\bar{3}$ $(2/m\ \bar{3})$	$m\bar{3}$	✓
	432		
	$\bar{4}3m$		
	$\frac{4}{m}\bar{3}\frac{2}{m}$ $(4/m\ \bar{3}\ 2/m)$	$m\bar{3}m$	✓

[a]The symbols in parenthesis are simply a more compact way of writing the full symbol.
[b]Only given if different from the full symbol.

cases the unique monoclinic axis is specified as the **c**-axis. In this case the primary position refers to symmetry operators running normal to the direction [001], as noted in Table 4.3.

In the orthorhombic system, the three places refer to the symmetry elements associated with the **a**-, **b**-, and **c**-axes. The most symmetrical group is $2/m\,2/m\,2/m$, which has diads along the three axes, and mirrors perpendicular to the diads. This is abbreviated to *mmm* because mirrors perpendicular to the three axes generate the diads automatically. In point group 2*mm*, a diad runs along the **a**-axis, which is the intersection of two mirror planes. Group 222 has three diads along the three axes.

In the tetragonal system, the first place in the point group symbol refers to the unique **c**-axis. This must always be a proper or improper rotation tetrad axis. The second place is reserved for symmetry operators lying along the **a**-axis, and due to the tetrad, also along the **b**-axis. The third place refers to symmetry elements lying along the cell diagonals, <110>. Thus, the point group symbol $\bar{4}2m$ means that an inversion tetrad lies along the **c**-axis, a rotation diad along the **a**- and **b**-axes, and mirrors bisect the **a**- and **b**-axes.

In the trigonal groups, the first place is reserved for the defining threefold symmetry element lying along the unique **c**-axis (hexagonal axes), and the second position for symmetry elements lying along the **a**-axis (hexagonal axes). Thus, the point group 32 has a rotation triad along the **c**-axis and a diad along the **a**-axis, together with two other diads generated by the triad.

In the hexagonal groups, the first place is reserved for the defining sixfold symmetry element lying along the unique **c**-axis, sometimes with a perpendicular mirror, and the second position for symmetry elements lying along the **a**-axis, the [100] (or equivalent) direction. The third position is allocated to symmetry elements referred to a [120] (or equivalent) direction. This direction is at an angle of 30° to the **a**-axis (Figure 4.9a). For example, the point group 6*mm* has a rotation hexad along the **c**-axis, a mirror with the mirror plane normal in a direction parallel to the **a**-axis, and a mirror with a normal parallel to the direction [120]. Operation of the hexad generates a set of mirrors, each at an angle of 30° to its neighbours (Figure 4.9b). The point group $6/m\,2/m\,2/m$ has a hexad along the **c**-axis, a mirror plane normal parallel to the **c**-axis (i.e. a mirror plane normal to the hexad), diads along [100] and [120], and mirror plane normals parallel to these directions (Figure 4.9c).

The cubic point groups show the greatest complexity in the arrangement of the symmetry elements. The first place in the symbol is reserved for symmetry elements associated with the **a**- (and hence **b**- and **c**-) axes. The second place refers to the type of triad lying along the cube body diagonal, <111>. The third place refers to symmetry elements associated with the face diagonals, <110>. Thus, the point group $4/m\ \bar{3}\ 2/m$ has four tetrad axes lying along the **a**- **b**- and **c**-axes, mirrors perpendicular to these axes, inversion triads along the cube body diagonals, and diads along the face diagonals <110> with mirrors perpendicular to them. The nine mirrors in this point group, three parallel to the cube faces and six parallel to the face diagonals, automatically generate the three tetrads and the six diads, so that the symbol is abbreviated to $m\bar{3}m$.

As remarked in Chapter 1, the external shape of a crystal can be classified into one of 32 crystal classes by making use of the symmetry elements that are present. These crystal classes correspond to the 32 crystallographic point groups. The two terms

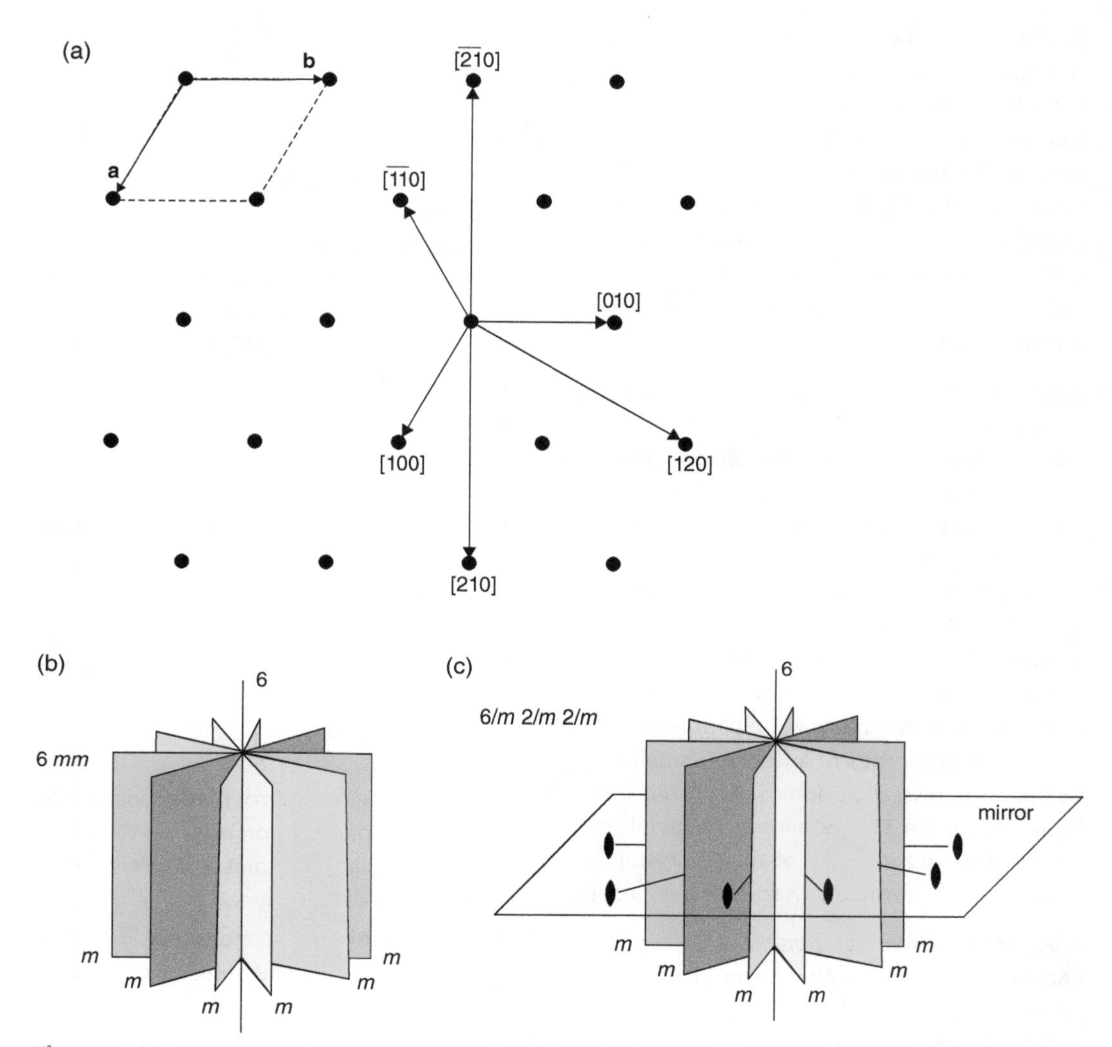

Figure 4.9 Symmetry elements present in a hexagonal crystal: (a) directions in a hexagonal lattice; (b) the point group 6*mm*; (c) the point group 6/*m* 2/*m* 2/*m*.

are often used completely interchangeably, and for practical purposes can be regarded as synonyms.

To determine the crystallographic point group of a crystal, write down a list of all of the symmetry elements present, order them following the rules set out above, and then compare them with the symmetry elements associated with the point groups listed in Table 4.4.

ANSWERS TO INTRODUCTORY QUESTIONS

What is a rotoinversion axis?

The inversion axis $\bar{n}$ depicts a symmetry operation that consists of a rotation combined with a centre of symmetry. (These axes are also called improper rotation axes, to distinguish them from the ordinary proper [or ordinary] rotation axes.) The symmetry operation of an improper rotation axis is that of rotoinversion. The initial atom position is rotated anticlockwise, by an amount specified by the order of the axis, and then inverted through the centre of symmetry. For example, the operation of a twofold improper rotation axis $\bar{2}$ is thus: the initial atom position is rotated 180° anticlockwise and then inverted through the centre of symmetry.

What relates a crystal class to a crystallographic point group?
The external shape of a crystal can be classified into one of 32 crystal classes dependent upon the symmetry elements present. The collection of symmetry elements present in a solid is referred to as the general point group of the solid. A solid can belong to one of an infinite number of general three-dimensional point groups. However, if the rotation axes are restricted to those that are compatible with the translation properties of a lattice, a smaller number, the crystallographic point groups, are found. There are 32 crystallographic point groups. They are identical to the 32 crystal classes, and the terms are used interchangeably.

What are enantiomorphic pairs?
Enantiomorphs are the two forms of optically active crystals; one form rotating the plane of polarised light in one direction, and the other rotating the plane of polarised light in the opposite direction. Such crystals are found existing as an enantiomorphic pair, consisting of right-handed and left-handed forms. These cannot be superimposed one upon the other, in the same way that a right-hand and left-hand glove cannot be superimposed one on the other.

The terms enantiomorphs and enaniomorphous pairs also apply to pairs of molecules, and as such may endow crystals with the ability to rotate the plane of polarised light in one direction or its opposite.

PROBLEMS AND EXERCISES

Quick Quiz

1. The operation of an improper rotation $\bar{n}$ axis involves:
   a. An anticlockwise rotation plus reflection
   b. An anticlockwise rotation plus inversion through a centre of symmetry
   c. An anticlockwise rotation plus translation
2. The number of regular Platonic solids is:
   a. 5
   b. 7
   c. 9
3. The places in the Hermann-Mauguin symbol for a point group refer to:
   a. Planes
   b. Axes
   c. Directions
4. The point group symbol $4/m$ means:
   a. A mirror plane runs parallel to a fourfold rotation axis
   b. The normal to a mirror plane runs parallel to a fourfold rotation axis
   c. A mirror plane contains a fourfold rotation axis
5. The number of three-dimensional crystallographic point symmetry groups is:
   a. 14
   b. 23
   c. 32
6. Which of the following tetrahedrally coordinated carbon (C) molecules will give rise to enantiomorphic pairs:
   a. C (H, OH, COOH, NH_2)
   b. C (H, OH, H, OH)
   c. C (COOH, NH_2, COOH, OH)
7. Two solids related by a rotoinversion axis, $\bar{n}$, are:
   a. Superimposable
   b. Identical
   c. Enantiomorphous

8. In a regular octahedron, a threefold inversion axis, $\overline{3}$, runs:
   a. Through the middle of a triangular face
   b. Through a triangular apex
   c. Along a triangular edge
9. The operation of a twofold improper rotoreflection axis, $\tilde{2}$, is identical to that of:
   a. Reflection
   b. A centre of symmetry
   c. A glide operation
10. The crystallographic point groups differ from the non-crystallogaphic point groups by taking into account:
   a. Lattice rotations
   b. Lattice translations
   c. Lattice reflections

Calculations and Questions

4.1. Write out the symmetry elements present and the corresponding point group for the molecules drawn in the figure below. The molecules are three-dimensional, with the following shapes: (a) $SiCl_4$, regular tetrahedron; (b) PCl_5, pentagonal bipyramid; (c) PCl_3, triangular pyramid; (d) benzene, hexagonal ring; (e) SF_6, regular octahedron. With reference to Appendix C, write the Schoenflies symbol for each point group.

(a)

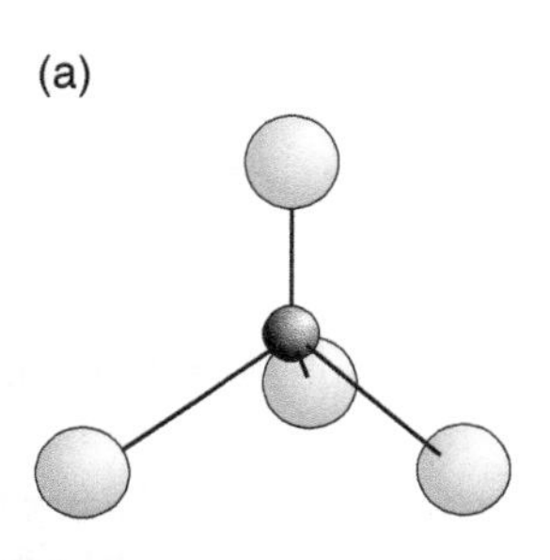

(b)

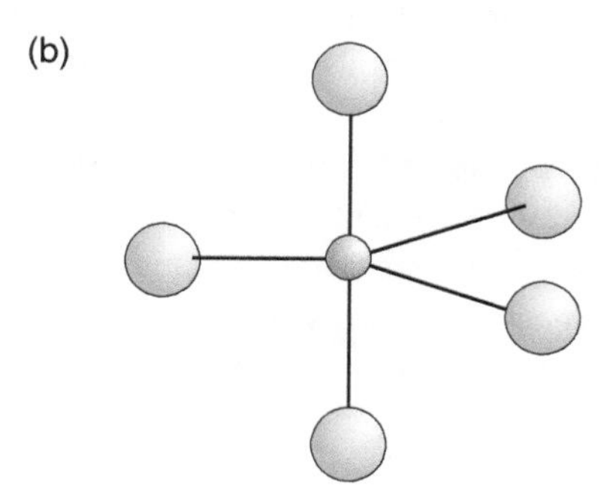

(c)

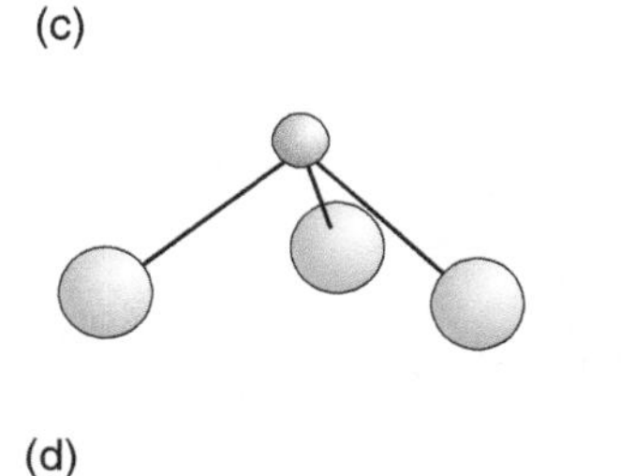

(d)

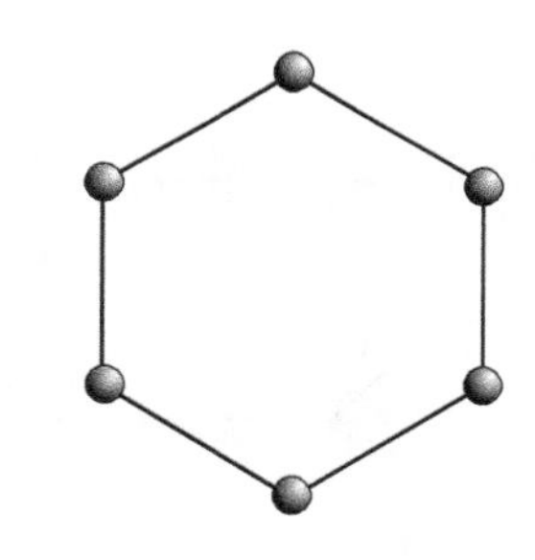

(e)

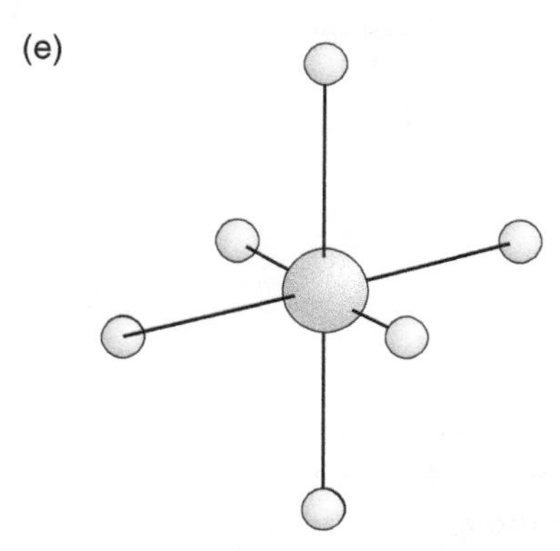

4.2. Determine the point group of each of the idealised 'crystals' in the figure below. (a) is a rhomb the shape of a monoclinic unit cell (see Chapter 1) whilst (b) and (c) are similar, but with corners replaced by triangular faces. Assume that the triangular faces are identical. The axes apply to all objects with **b** perpendicular to the plane containing **a** and **c**.

(a)

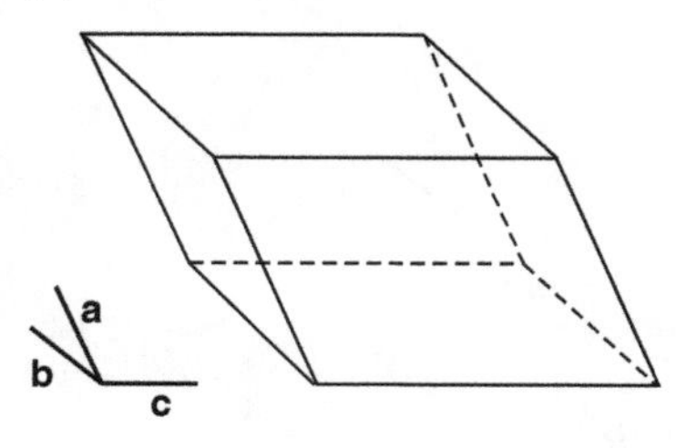

(b)

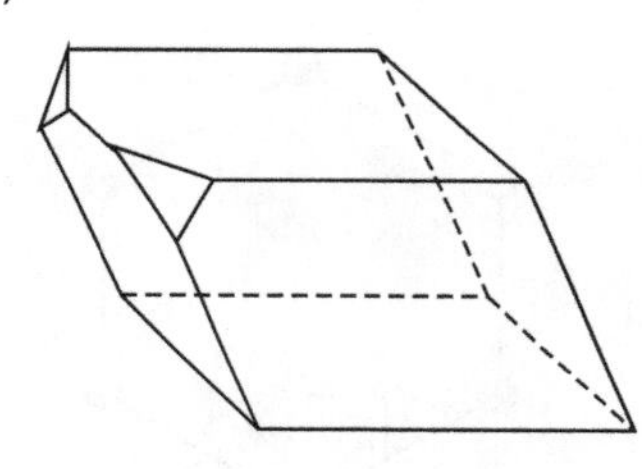

(c)

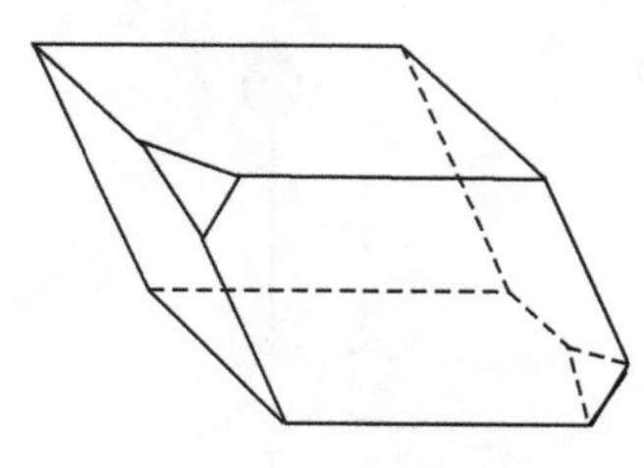

4.3. Determine the point group of each of the idealised 'crystals' in the figure below. (a) is a rhomb the shape of an orthorhombic unit cell (see Chapter 1) whilst (b) and (c) are similar, but with corners replaced by triangular faces. Presume that the triangular faces are identical. The axes apply to all objects.

(a)

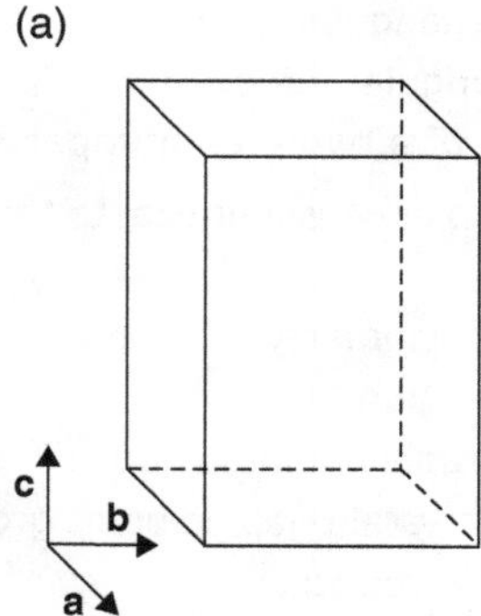

(b)

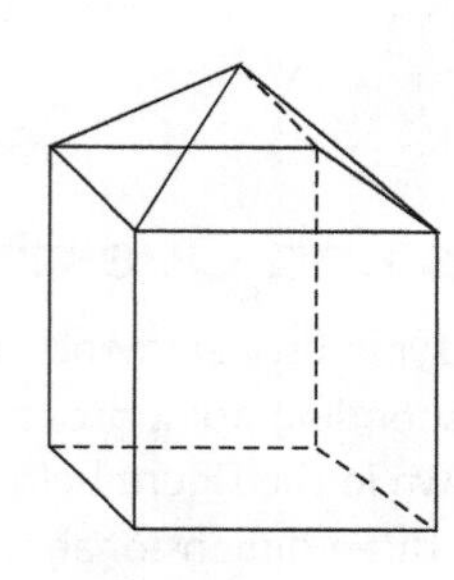

(c)

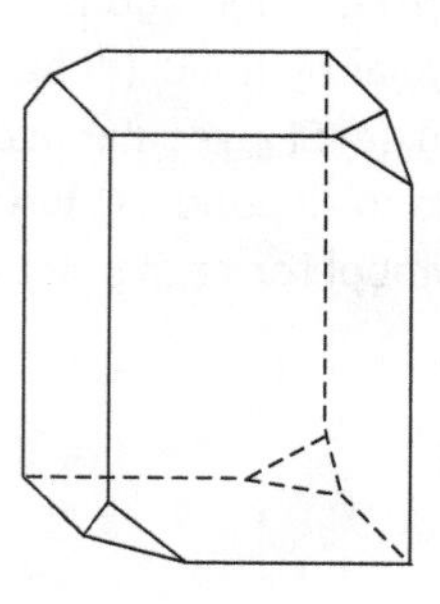

4.4. Write out the full Hermann-Mauguin symbols for the short symbols: (a) mmm; (b) $4/mmm$; (c) $\bar{3}m$; (d) $6/mmm$; (e) $m\bar{3}m$.

4.5. (a) Sketch a regular cube (short symbol $m\bar{3}m$) and mark the symmetry axes and planes; (b) mark the cube faces so that the point group is changed to the cubic group $2/m\bar{3}$ (short symbol $m\bar{3}$, see Table 4.4).

4.6. Determine the point groups of the unit cells of the crystal structures of the metals Cu, W and Mg given in Chapter 1.

4.7. Determine the point groups of the crystal structures of the compounds NaCl, CaF_2 and $SrTiO_3$ given in Chapter 1.

Chapter 5

Symmetry and Physical Properties

What are polar and axial vectors?
What are enantiomorphic pairs?
What are magnetic point groups?

In this chapter the relationship between the existence or absence of a physical property in a crystal and the symmetry of the crystal is outlined. These relationships are specified in terms of the point groups described in the preceding chapter and are generally treated in terms of tensor mathematics. Here, a few less complex examples are given, to illustrate the connection between properties and crystal structure, using optical, dielectric and magnetic properties as examples.

5.1 PROPERTIES AND SYMMETRY

The chemical and physical properties of a crystal are determined by a number of interlinked factors. Whilst the **magnitude** of the property usually reflects the atomic constituents present, the **presence** (or absence) of a physical property is frequently (but not always) dependent upon the underlying crystal **symmetry**. For a number of properties, the symmetry of the material is irrelevant. Density is simply dependent upon the number and type of atoms present in a unit cell, not their arrangement in space. Temperature, which is a measure of the vibration and rotation of the atoms that make up a material, is similar. Properties that do not reveal symmetry are called **non-directional**. Their magnitude is given by a scalar (number) together with the appropriate units.

On the other hand, **directional** properties, which need to be specified by a vector, can reveal aspects of the symmetry of the phase. For example, the symmetry of a crystal is partially revealed by the shape of the pits that form on surfaces when a crystal begins to dissolve in a solvent. Initial attack is at a point of enhanced chemical reactivity, often where a dislocation reaches the surface; the disruption

Crystals and Crystal Structures, Second Edition. Richard J. D. Tilley.

to the atomic packing and the resultant strain field enhances the local solubility. A pit forms as the crystal is corroded. The shapes of the pits, called **etch figures**, have a symmetry corresponding to one of the ten crystallographic plane point groups. This will be the point group that corresponds with the symmetry of the face. For instance, an etch pit on a (100) face of a cubic crystal will be square, whilst a pit on a (101) face of a tetragonal crystal will be rectangular.

5.2 POINT GROUPS AND PHYSICAL PROPERTIES

A directional property displays a symmetry that mirrors the underlying crystal symmetry. The broad relationship between the two is contained in **Neumann's principle**, which states that:

> *The symmetry elements of any physical property of a crystal must include the symmetry elements of the point group of the crystal.*

Note that this states that the symmetry elements of the physical property must **include** those present in the point group, and not that the symmetry elements are **identical** with those of the point group. This means that a physical property may show more symmetry elements than the point group, and not all properties are equally useful for revealing true point group symmetry.

One of the fundamental symmetry attributes of a point group is the possession or a lack of a centre of symmetry. Directions in crystals reflect this symmetry attribute. A **polar direction** in a crystal is a direction [*uvw*] that is not related by symmetry to the opposed direction $[\overline{u}\,\overline{v}\,\overline{w}]$. No polar directions can occur in a crystal belonging to any one of the 11 centrosymmetric point groups listed in Table 4.4. Polar directions are found in the 21 non-centrosymmetric crystal classes, leading to the names **polar point groups** or **polar crystal classes**. (Unfortunately, these terms are not exactly synonymous, and have slightly different meanings in the literature. In crystallographic usage, a polar class is simply one of the 21 non-centrosymmetric point groups listed in Table 4.4. In physics, a crystal class is considered to be a polar class if it gives rise to the pyroelectric effect, defined below.)

An important property of crystals belonging to 20 of the 21 non-centrosymmetric crystal classes (the exception being the cubic group 432) is **piezoelectricity**. In this phenomenon, the application of an applied stress will induce a polarisation in the crystal. The piezoelectric effect, described in more detail below, is widely exploited in many electrical and electronic applications. Some ten of these 20 non-centrosymmetric crystal classes that possess polar directions show a unique polar axis, which is a direction unrelated by symmetry to any other direction in the crystal. These crystals show an internal electric dipole, leading to an observable electric polarisation. These crystals show a change in the polarisation with temperature, the **pyroelectric effect**. In some of them, the polarisation direction can be changed (switched) by the application of an external electric field. This latter group form the **ferroelectrics**.

Another aspect of the lack of a centre of symmetry is the presence or absence of right-handed and left-handed configurations: **enantiomorphism** or **chirality**. In the enantiomorphic point groups, crystals may exist in one of two forms that are related to each other in the same way that right and left hands are related (see also Figure 4.1). Quartz, for example, exists in right-handed and left-handed crystals. Enantiomorphic crystals show **optical activity**, that is, they rotate the plane of polarised light as it traverses the crystal (Section 5.8). The occurrence of these properties is summarised in Table 5.1.

Table 5.1 Non-centrosymmetric crystal classes and physical properties

Crystal class	Polar directions	Unique polar axis	Piezo-electric effect	Pyro-electric effect	Enantiomorphic group	Optical activity possible
Triclinic						
1	All	None	✓	✓	✓	✓
Monoclinic						
2	All not $\perp$ 2	[010]	✓	✓	✓	✓
m	All not $\perp$ *m*	None	✓	✓		✓
Orthorhombic						
*mm*2	All not $\perp$ 2 or *m*	[001]	✓	✓		✓
222	All not $\perp$ 2	None	✓		✓	✓
Tetragonal						
4	All not $\perp$ 4	[001]	✓	✓	✓	✓
$\bar{4}$	All not $\parallel$ or $\perp$ $\bar{4}$	None	✓			✓
422	All not $\perp$ 4 or $\perp$ 2	None	✓		✓	✓
4*mm*	All not $\perp$ 4	[001]	✓	✓		
$\bar{4}2m$	All not $\perp$ $\bar{4}$ or 2	None	✓			✓
Trigonal						
3	All	[111]	✓	✓	✓	✓
32	All not $\perp$ 2	None	✓		✓	✓
3*m*	All not $\perp$ *m*	[111]	✓	✓		
Hexagonal						
6	All not $\perp$ 6	[001]	✓	✓	✓	✓
$\bar{6}$	All except $\bar{6}$	None	✓			
622	All not $\perp$ 6 or 2	None	✓		✓	✓
6*mm*	All not $\perp$ 6	[001]	✓	✓		
$\bar{6}m2$	All not $\perp$ 2	None	✓			
Cubic						
23	All not $\perp$ 2	None	✓		✓	✓
432	All not $\perp$ 4 or 2	None			✓	✓
$\bar{4}3m$	All not $\perp$ $\bar{4}$	None	✓			

$\perp$ = perpendicular to the axis following. $\parallel$ = parallel to the axis following.

Naturally, in crystals with a centre of symmetry, directional properties would be expected to be symmetrical with respect to the centre. However, some physical properties remain symmetrical even in crystals that lack a centre of symmetry. For example, thermal expansion occurs equally in any [*uvw*] and $[\bar{u}\,\bar{v}\,\bar{w}]$ direction, regardless of symmetry. Note that this does not mean that the thermal expansion is identical along all crystallographic directions, simply that opposite directions behave in an identical fashion under the applied stimuli of heating and cooling. Such physical properties, called **centrosymmetric physical properties**, will only differentiate between 11 of the point groups. This reduced assembly are called **Laue classes**. The 11 Laue classes, each of which contains the centrosymmetric and non-centrosymmetric crystallographic point groups that behave in the same way when investigated by centrosymmetric physical properties, are listed in Table 5.2.

Table 5.2 The Laue classes

Crystal system	Laue class (centrosymmetric group)	Non-centrosymmetric groups included in the class
Triclinic	$\bar{1}$	1
Monoclinic	$2/m$	2, m
Orthorhombic	mmm	222, $mm2$
Tetragonal	$4/m$	4, $\bar{4}$
	$4/mmm$	422, $4mm$, $\bar{4}2m$
Trigonal	$\bar{3}$	3
	$\bar{3}m$	32, $3m$
Hexagonal	$6/m$	6, $\bar{6}$
	$6/mmm$	622, $6mm$, $\bar{6}2m$
Cubic	$m\bar{3}$	23
	$m\bar{3}m$	432, $\bar{4}3m$

5.3 SPECIFICATION OF PHYSICAL PROPERTIES

Physical properties are quantified by imposing a **stimulus** on the material and then measuring the **response**. If it is important, the directional aspect of the stimulus needs to be represented by a **vector**. A well-known example is force, the effect of which depends upon both the magnitude and direction of application. Frequently both the stimulus and the response are vectors. For example, an electric field (i.e. a voltage) applied in a certain direction to a conductor will give rise to an electric current which may or may not run parallel to the applied field.

A stimulus vector **S** is specified by giving the components of the vector along three mutually perpendicular axes OX, OY, OZ (also written as Ox, Oy, Oz; Ox_1, Ox_2, Ox_3; 1, 2, 3), so that:

$$\mathbf{S} = [\mathbf{S}_1, \mathbf{S}_2, \mathbf{S}_3]$$

The magnitude (that is, the length) of a vector **S** is written S and the equation can be written in terms of the magnitudes along these axes as:

$$S = [S_1, S_2, S_3]$$

A response vector **R** of magnitude R can be specified in a similar way.

When the reference axes of the stimulus are chosen to coincide with the symmetry axes of the crystal, or to lie in a symmetry plane, they are referred to as **principal axes**. The reference axes for cubic, tetragonal and orthorhombic systems are chosen to lie parallel to the unit cell edges. In hexagonal systems one axis is taken to coincide with the crystallographic **c**-axis and the other two are normal to the **c**-axis. In monoclinic and triclinic crystals, whilst it is straightforward to define three Cartesian axes, the relationship between these and the crystallographic axes is not so simple and depends upon the precise unit cell chosen.

It is found that such vectors fall into two types from the point of view of symmetry operations, in particular, how they transform in terms of the mirror operation. These two classes are called **polar** and **axial** vectors.

The concept of a polar vector can be illustrated by the polarisation introduced into a molecule by an unsymmetrical distribution of charges. In general, an electric dipole moment, **p**, arising from a pair of charges $\pm q$ is given by:

$$\mathbf{p} = q\,\mathbf{r}$$

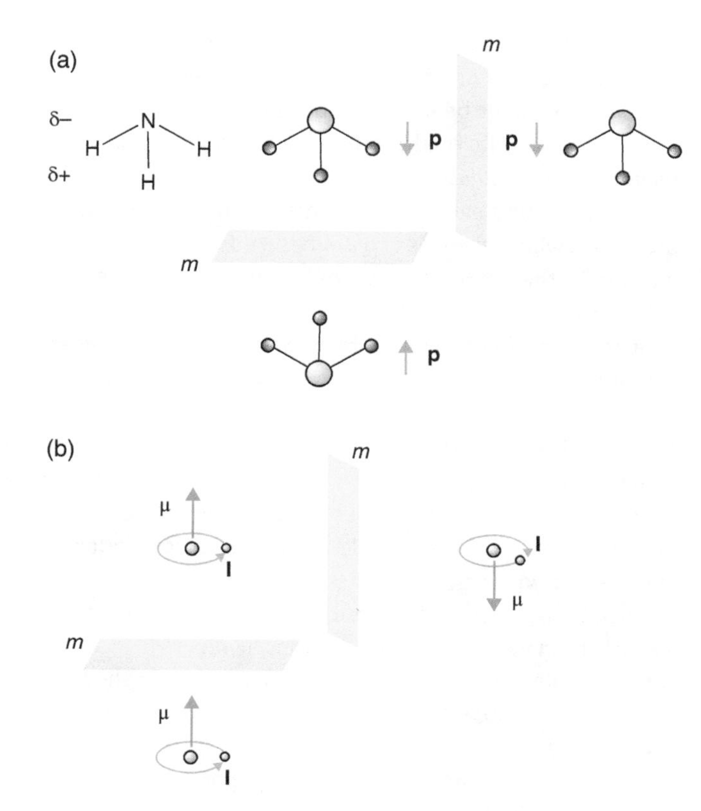

Figure 5.1 (a) Mirror transformations of a polar vector, **p**; (b) mirror transformations of an axial vector **μ**.

where **r** is the vector describing the location of the charges. If directions are not important the magnitude of the electric dipole moment is simply;

$$p = q\,r$$

For example, the molecule ammonia, NH_3, has a permanent dipole moment of 3×10^{-30} Cm, because the hydrogen atoms carry a small positive charge δ+, whilst the nitrogen atom, with a lone pair of electrons, has a small negative charge of δ– (Figure 5.1a). When a polar vector is reflected in a mirror parallel to **p**, there is no change in the direction of **p**. However, when the mirror is perpendicular to **p** the mirror image is reversed to **–p** (Figure 5.1a)

An axial vector can be illustrated by magnetic effects. A small closed loop of current, **I**, including an electron moving around an atomic nucleus, generates a magnetic dipole, **μ**, defined by:

$$\boldsymbol{\mu} = \mathbf{I}\,\pi\,r^2$$

where r is the radius of the orbit of the electron current. As before, if directions are not important, the magnetic dipole is:

$$\mu = I\pi\,r^2$$

An axial vector is reversed by a mirror operation parallel to the vector but is not inverted by a mirror perpendicular to the moment (Figure 5.1b).

5.4 REFRACTIVE INDEX

Light can be regarded as a wave of wavelength λ with electrical and magnetic field components, each of which is most properly described as a vector, although only the electric field vector is important for the optical properties described here. When a light beam enters a transparent crystal, the electric field component of a light wave, **E** (the *stimulus*) interacts with the electrons around each of the atoms in the structure, which gives rise to a measured slowing of the wave (the *response*). The refractive index of a crystal is the physical property that is the external manifestation of this stimulus–response pair. (Note that refractive index is sensitive to wavelength, but this aspect is ignored in the following descriptions.)

The refractive index experienced by the light is dependent upon the direction of the electric field vector with respect to the symmetry elements displayed by the structure. The electric field vector of a light beam is always perpendicular to the line of propagation of the light, but can adopt any angle otherwise (Figure 5.2a and b). The position of the electric vector defines the **polarisation** of the light wave. For ordinary light, such as that from the sun, the orientation of the electric vector changes in a random fashion every 10^{-8} seconds or so. Such light is said to be **unpolarised**. Light is **linearly** or **plane polarised** when the electric vector is forced to vibrate in a single plane.

A beam of linearly or unpolarised light can be resolved into two linearly polarised components with vibration directions perpendicular to each other. For convenience these can be called the horizontally and vertically polarised components (Figure 5.2c). When the beam enters a transparent crystal, each of these two linearly polarised components experiences its own refractive index. Because of these symmetry constraints, the refractive index of a crystal can have a single value or show multiple values (Table 5.3).

Crystals that possess cubic symmetry show a single refractive index for both the horizontally and vertically polarised light beams, and this is independent of the direction of the beam. Such crystals are described as optically **isotropic**. All other crystals are optically **anisotropic**.

Hexagonal, trigonal and tetragonal crystals show two refractive indices, called the **principal indices**, written n_o (also ω, n_ω, N_ω, O, n_O, N_O, N_o) and n_e (also ε, n_ε, N_ε, E, n_E, N_E, N_e). For tetragonal and hexagonal crystals, the refractive indices along the **a**- and **b**-axes are the same and different from the refractive index along the **c**-axis. In trigonal crystals the refractive index along the threefold rotation axis [111] is different from that normal to this direction. When a beam of light enters such a crystal along

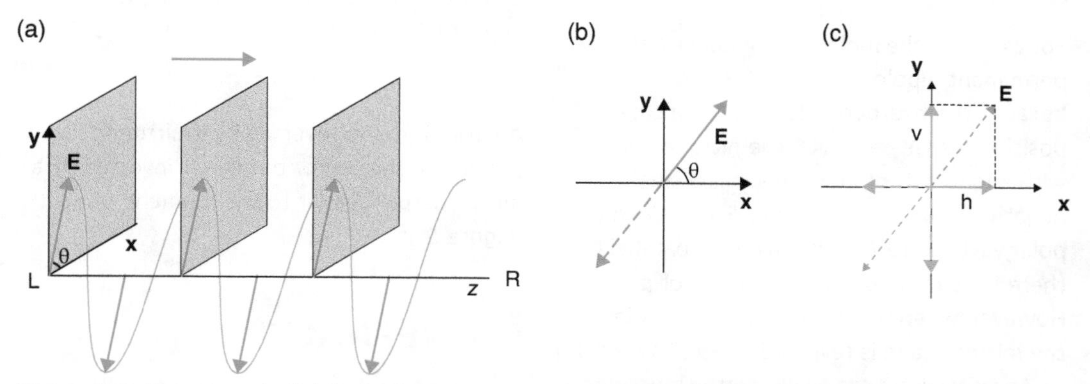

Figure 5.2 The electric field of a light wave: (a) perspective view; (b) looking into the wave; (c) the field resolved into horizontal (h) and vertical (v) components.

Table 5.3 Refractive indices of some mineral crystals[a]

Mineral	Formula	Symmetry	Unit cell parameters, nm	Refractive index[b]
Fluorite	CaF_2	Cubic, $m\bar{3}m$	$a = 0.5463$	1.441
Halite	NaCl	Cubic, $m\bar{3}m$	$a = 0.5640$	1.544
Periclase	MgO	Cubic, $m\bar{3}m$	$a = 0.4210$	1.740
Sylvite	KCl	Cubic, $m\bar{3}m$	$a = 0.6293$	1.4903
Spinel	$MgAl_2O_4$	Cubic, $m\bar{3}m$	$a = 0.80900$	1.719
Beryl	$Be_3Al_2Si_6O_{18}$	Hexagonal, $6/mmm$	$a = 0.9210$, $c = 0.9190$	1.570, 1.585
Calcite	$CaCO_3$	Trigonal, $\bar{3}m$	$a = 0.4990$, $c = 1.7061$	1.658, 1.485
Corundum	Al_2O_3	Trigonal, $\bar{3}m$	$a = 0.4750$, $c = 1.2982$	1.769, 1.761
Cassiterite	SnO_2	Tetragonal $4/mmm$	$a = 0.4738$, $c = 0.3187$	2.000, 2.097
Rutile	TiO_2	Tetragonal $4/mmm$	$a = 0.4594$, $c = 0.2959$	2.613, 2.909
Scheelite	$CaWO_4$	Tetragonal $4/m$	$a = 0.5249$, $c = 1.1374$	1.920, 1.937
Zircon	$ZrSiO_3$	Tetragonal, $4/mmm$	$a = 0.6607$, $c = 0.5982$	1.943, 1.997
Anhydrite	$CaSO_4$	Orthorhombic, mmm	$a = 0.6245$, $b = 0.6995$, $c = 0.6993$	1.571, 1.577, 1.614
Baryte	$BaSO_4$	Orthorhombic, mmm	$a = 0.8884$, $b = 0.5457$, $c = 0.7157$	1.636, 1.637, 1.647
Enstatite	$MgSiO_3$	Orthorhombic, mmm	$a = 1.8230$, $b = 0.8440$, $c = 0.5190$	1.659, 1.663, 1.669
Forsterite	Mg_2SiO_4	Orthorhombic, mmm	$a = 0.4754$, $b = 1.0197$, $c = 0.5981$	1.683, 1.695, 1.721
Gypsum	$CaSO_4.2H_2O$	Monoclinic, $2/m$	$a = 0.5679$, $b = 1.5202$, $c = 0.6523$, $\beta = 118.4$	1.520, 1.523, 1.530
Talc	$Mg_3Si_4O_{10}(OH)_2$	Monoclinic, $2/m$	$a = 0.5287$, $b = 0.9158$, $c = 1.895$, $\beta = 99.3$	1.554, 1.592, 1.595

[a]The measured refractive index of minerals varies: representative values are given.
[b]Cubic crystals: n, uniaxial crystals: n_o, n_e, biaxial crystals: $n_\alpha < n_\beta < n_\gamma$.

an arbitrary direction, the refractive index differs for the two polarisation components. One experiences a refractive index n_o and the other a refractive index n_e', which has a magnitude between n_o and n_e, dependent upon the direction of the incident beam.

In these crystals a beam of light directed along the direction of highest symmetry, the 4, 3 or 6 axis of rotation (the crystallographic **c**-axis), each polarisation component experiences only one refractive index, equal to n_o. This direction is called the **optic axis**. There is only one such unique direction, and these materials are called **uniaxial crystals**.

The birefringence of these crystals, Δn, is defined as:

$$\Delta n = (n_e - n_o)$$

Crystals in which n_e is greater than n_o are described as **uniaxial positive** and those with n_e less than n_o as **uniaxial negative**. In the tetragonal, trigonal and hexagonal systems, the unique optic axis lies along the direction of highest symmetry, the 4, 3 or 6 axis of rotation. If the beam is directed perpendicular to the optic axis the two refractive indices observed are n_o and n_e.

Orthorhombic, monoclinic and triclinic crystals exhibit three principal refractive indices, n_α which has the smallest value, n_γ, which has the greatest value, and n_β which is between the other two. The horizontally and vertically polarised components of a beam of light entering a crystal with one of these symmetries encounter different refractive indices, with magnitudes lying between the lowest, n_α and the highest, n_γ. Two optical axes exist in these crystals, and the refractive index encountered by both polarisation components of a light beam directed along either optic axis is n_β. These materials are known as **biaxial crystals**. Unlike uniaxial crystals, there is not an intuitive relationship between the optic axes and the symmetry, although one optical axis always lies along the direction of highest symmetry.

5.5 OPTICAL ACTIVITY

5.5.1 Specific Rotation

Optical activity is the ability of a material to rotate the plane of linearly polarised light to the right or the left as the beam passes through it (Figure 5.3). As with the refractive index, the stimulus is the electric vector **E** of the light wave, which interacts with the electron clouds around the atomic cores. The response includes a slowing of the wave, as in refractive index, but also a rotation of the vector **E**. Optical activity is found in many crystals, including some with cubic symmetry. It can arise in homogeneous phases and also those that contain enantiomorphic molecules, such as alanine (Figure 4.1). A material that rotates the plane of the light clockwise (when viewed into the beam of light) is called the **dextrorotatory**, *d*- or (+) form, whilst that which rotates the plane of the light anticlockwise is called the **laevorotatory**, *l*- or (−) form. The rotation observed, called the **specific rotation** of the crystal,[1] designated $[\alpha]_\lambda{}^t$, is measured under standard conditions which includes the wavelength, λ, of the light used, and the temperature, *t* °C. The specific rotation of a crystal decreases as the wavelength increases. The wavelength of light used to measure the specific rotation is usually the sodium D-line with a wavelength of 589.6 nm. The specific rotation is then written as, for example, $[\alpha]_D{}^{25}$. If the plane of polarised light is rotated clockwise when the observer looks towards the light source, the value of specific rotation is **positive**, and when the rotation is anticlockwise the specific rotation is **negative**. For crystals, the specific rotation is measured as the rotation per mm of crystal, with units ° mm^{-1}. For example, the specific rotation of α-quartz is $[\alpha]_D{}^{20}$ = +23.726° mm^{-1}, lying between the approximate values of 40° mm^{-1} at 400 nm and 14.5° mm^{-1} at 700 nm.

Optical activity in directions well away from the optic axes in uniaxial and biaxial birefringent crystals is usually masked by the much greater changes in the behaviour of the polarised light induced by the normal birefringent refractive indices. Indeed, optical activity is only a small perturbation of ordinary birefringence, and it only becomes noticeable when the ordinary birefringence is supressed, such as when the polarised light travels along an optic axis.

[1] Specific rotation is also termed the rotatory power ρ, but this is a poor descriptor as the units of this property are angle per unit length, not units of power.

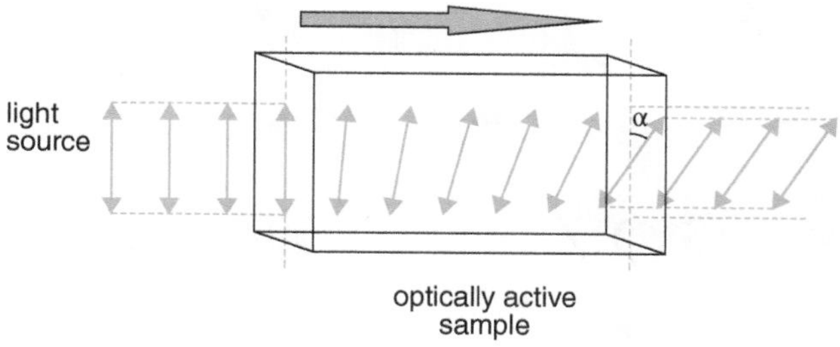

Figure 5.3 Schematic diagram of the rotation of the plane of linearly polarised light on passing through an optically active medium.

5.5.2 Crystal Symmetry and Optical Activity

The existence of left- and right-handed crystals implies an absence of a centre of symmetry in the crystal, and optical activity cannot exist in a crystal belonging to a centro-symmetric point group. The effect is, therefore, confined to crystals that belong to one of the 21 non-centrosymmetric point groups listed in Table 5.1. Crystals that belong to the 11 enantiomorphous point groups, 1, 2, 222, 4, 422, 3, 32, 6, 622, 23, 432, all of which can give crystals with a left- or right-handedness, display optical activity. From the other non-centrosymmetric point groups, optical activity can also occur in the four non-enantiomorphic classes *m*, *mm*2, $\bar{4}$, and $\bar{4}2m$, as directions of both left- and right-handed rotation occur in the crystal. There are thus 15 potentially optically active groups in all. In the remaining six non-centrosymmetric classes, optical activity is not possible because of the overall combination of symmetry elements present.

Optical activity is observable in any direction for crystals belonging to the two cubic enantiomorphic classes 23 and 432. In the non-enantiomorphic groups, no optical activity is found along an inversion axis or perpendicular to a mirror plane. Thus, no optical activity occurs along the optic axis in classes $\bar{4}$ or $\bar{4}2m$. In the remaining two groups, *m* and *mm*2, no optical activity is observed along the optic axes if these lie in a mirror plane.

5.5.3 Optical Activity in Homogeneous Crystals

The first instance of optical activity was noted in crystalline α-quartz, the room-temperature stable form of silicon dioxide, SiO_2, by Arago in 1811. This internal property is exhibited in the macroscopic crystalline habit. John Herschel, in 1822, reported that α-quartz occurs as left- or right-handed crystals that could be separated by eye. These pairs had the same relationship to each other as a left-hand glove to a right-hand glove or an object and its mirror image. Such right- and left-handed mirror image crystal forms are called enantiomorphs (also see Section 4.1).

Quartz is typical of homogeneous crystals, where the structure does not contain recognisable molecular building blocks. The room temperature form of quartz, low- or α-quartz, is trigonal, and both enantiomorphs share the point group 321. (Note that the space groups (Chapter 6) of these two forms differ.) At a temperature of 573°C low quartz transforms to high- or β-quartz, which is hexagonal. Again, there are two enantiomorphs, which share the point group 622. The right- and left-handed optical activity is maintained during the transition.

Both forms of quartz are built from corner-shared [SiO_4] tetrahedra. These form helices

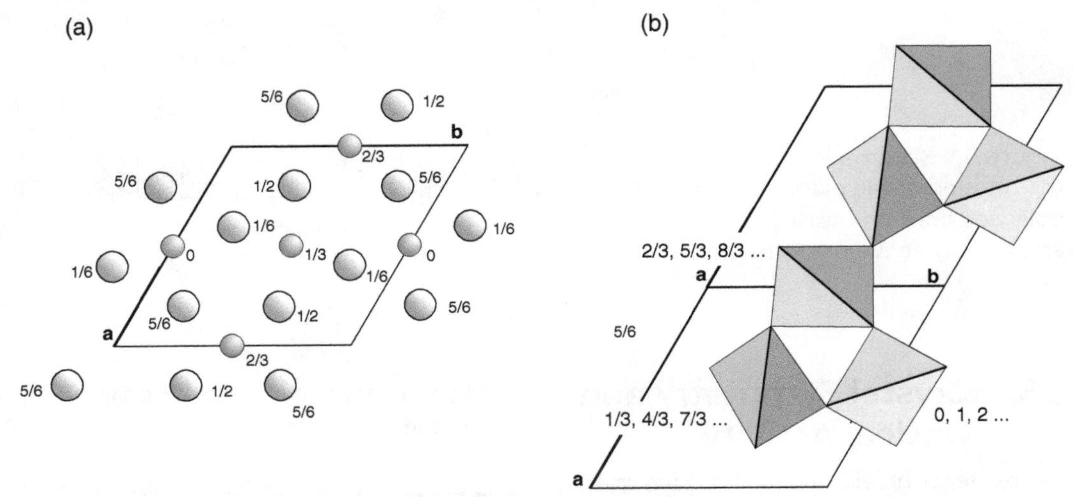

Figure 5.4 (a) The structure of right-handed high quartz; the **c**-axis height is given; (b) the helical configuration of the tetrahedra with the position of the central Si atom in each given.

(either right-handed or left-handed) along the optic axis, the crystallographic **c**-axis. This is illustrated for high quartz in Figure 5.4. The helices have Si atoms at the centres of tetrahedra at heights of 0/3, 1/3, 2/3, 3/3, 4/3, 5/3, 6/3…. The structure of the equivalent form of low quartz is similar, but the individual SiO_4 tetrahedra are distorted compared with those drawn. The plane of polarisation of linearly polarised light directed through a slice of crystal along the optic axis is rotated left or right, depending upon the handedness of the helices. Similar light directed normal to the optic axis shows no change.

5.5.4 Optical Activity in Crystals Containing Molecules

Optical activity in a crystal can also arise if the crystals contain enantiomorphic molecular components in the unit cell. The classical example of this phenomenon was described by Pasteur in 1848. At this epoch, crystals of the sodium salts of the two acids tartaric acid and racemic acid were well known and could be collected from old wine casks. These crystals, sodium tartrate and sodium racemate, seemed to be chemically and physically identical, except that when linearly polarised light was passed through a *solution* of sodium tartrate the plane of polarisation rotated to the right as viewed by an observer looking towards the light source. The corresponding sodium racemate solution acid was optically inactive and caused no rotation.

When crystals of sodium racemate were examined, Pasteur found that there were equal numbers of two forms, one right-handed and one left-handed, as with α-quartz. Solutions of the two crystal types rotated the plane of polarisation by equal amounts but in opposite directions. The compound sodium racemate should thus be described as a mixture of two forms of sodium tartrate, each of which rotated polarised light in equal and opposite directions. One of these was identical to natural sodium tartrate described above, whilst the other appeared not to occur in isolation.

The process of dissolution separates the solid crystals into molecules or ions. It was clear, therefore, that this example of optical activity was a feature that needed to be explained at a molecular level, but one which influenced the crystallographic morphology of the phases.

5.5.5 Optical Activity and Chiral Molecules

The optical activity of many crystals can be attributed to the presence of chiral molecules within the unit cell (Section 4.1). A chiral molecule is optically active. Although inorganic molecules with tetrahedral or octahedral bond geometry can form optically active pairs, optical activity has been explored in most detail with respect to organic molecules containing tetrahedrally coordinated carbon atoms. These mirror-image molecules are known as **optical isomers**. Optical isomers occur in organic compounds whenever four **different** atoms or groups of atoms are attached to a tetrahedrally coordinated central carbon atom. Such a carbon atom is called a **chiral carbon atom** or **chiral centre**. Often only one form of a molecule will be found in nature. For example, only one of the two enantiomers of the amino acid alanine (Figure 4.1), the dextrorotatory form, occurs naturally; the other can be made synthetically.

Tartaric acid crystals are optically active because they contain molecular species that are enantiomers of the tartrate anion. Tartrate anions show more complexity than simpler molecules such as alanine because they contain two chiral carbon atoms. These can 'cancel out' internally in the molecule so that three molecular forms actually exist: the two optically active mirror image structures that cannot be superimposed on each other, the laevorotatory (*l*-) and dextrorotatory (*d*-) forms, and the optically inactive (*meso*-) form, which can be superimposed on its mirror image[2] (Figure 5.5). Mixtures of enantiomers in equal proportions will not rotate the plane of linearly polarised light and are called **racemic** mixtures, after the 'racemic acid' of Pasteur. The nomenclature is thus that if the optical activity is cancelled internally by the action of more than one chiral centre, the form is labelled *meso*-, as in crystals of *meso*-tartaric acid. If optical activity is lost because equal numbers of *d*- (+) and *l*- (–) optically active enantiomers are present, the term used is *racemic* or *racemate*, as in *racemic*-tartaric acid.

5.5.6 Optical Activity, Chemical Reactivity and Symmetry

This chapter is predominantly concerned with the physical properties of solids. However, it is important to be aware that chemical properties can also be influenced in ways important to life. Enantiomers generally display identical chemical properties **except** when they react with other chiral molecules. This has profound effects for life because most biologically important molecules are chiral. Naturally occurring amino-acids are 'left-handed' whilst naturally occurring sugars are 'right-handed'.

---

[2] The nomenclature of optically active organic molecules has been revised since their absolute configuration has been determinable via X-ray diffraction (Chapter 7). The equivalent terms are (*d*)-tartaric acid = (+)-tartaric acid = (2R,3R)-tartaric acid; (*l*)-tartaric acid = (–)-tartaric acid = (2S,3S)-tartaric acid; mesotartaric acid = (2R,3S)-tartaric acid = (2S,3R)-tartaric acid.

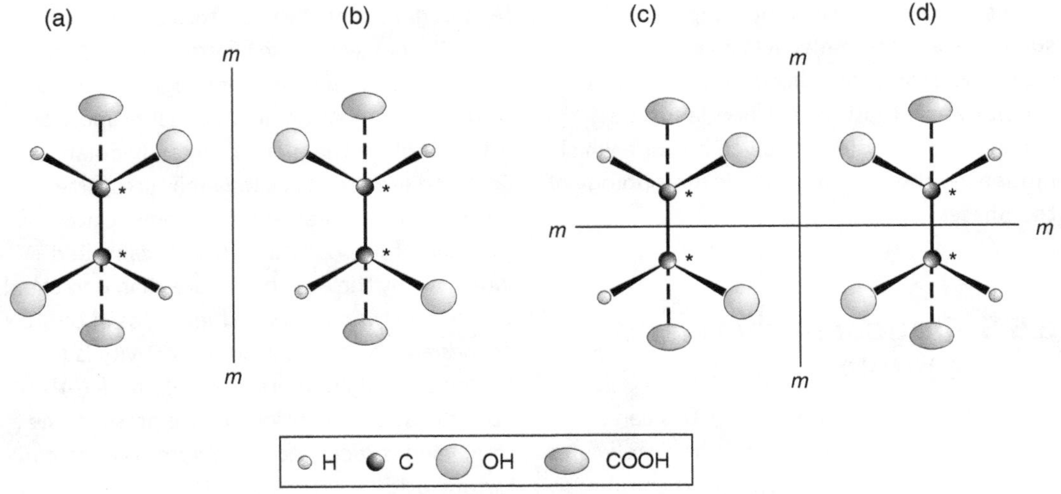

Figure 5.5 The structures of tartaric acid: (a) (*d*)-, (+)- or (2R,3R)-tartaric acid; (b) (*l*)-, (-)- or (2S,3S)-tartaric acid; (c) (2R,3S)-tartaric acid; (d) (2S,3R)-tartaric acid. The chiral carbon atoms are marked ∗. The four bonds around each of these are arranged tetrahedrally; bonds in the plane of the page are drawn as full lines, those receding into the page as dashed lines, and those pointing out of the page as triangles.

The molecules important to life on Earth are thus described as **homochiral**. (It seems that this bias is not restricted to life on Earth. Studies of the Murchison and Murray meteorites, reported from the late 1990s onwards, show that they also contain a preponderance of left-handed amino acids.)

The differences in biological and pharmacological activity between two enantiomers can be pronounced. Drugs and pharmaceuticals derived from natural products are often chiral and the two enantiomers differ considerably in activity, one perhaps being beneficial and one being non-active or even toxic. The sensation of the odour of caraway, for example, is triggered by the left-handed enantiomer of limonene, whilst that of mandarin oranges by the right-handed isomer. Vitamin C prevents the disease scurvy; the other enantiomer of this substance is biologically inactive. Such a list could be extended indefinitely.

5.6 THE PYROELECTRIC EFFECT

5.6.1 Pyroelectric and Ferroelectric Crystals

A crystal that is a pyroelectric is found to possess a spontaneous polarisation, $\mathbf{P}_s$, which means that a pyroelectric crystal shows a permanent polarisation that is present in the absence of an applied electric field. Polarisation is a polar vector (Section 5.3) which influences how it will transform under the symmetry operations present in the point group of the crystal. In a pyroelectric crystal, a change of temperature induces a polarisation change $\Delta\mathbf{P}$ in the already existing polarisation $\mathbf{P}_s$ – the **pyroelectric effect**. The change obtained on heating is reversed on cooling.

Ferroelectrics also possess a spontaneous polarisation, $\mathbf{P}_s$, in the absence of an electric field. They are distinguished from pyroelectrics

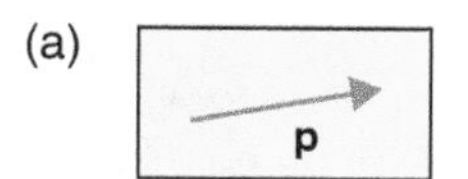

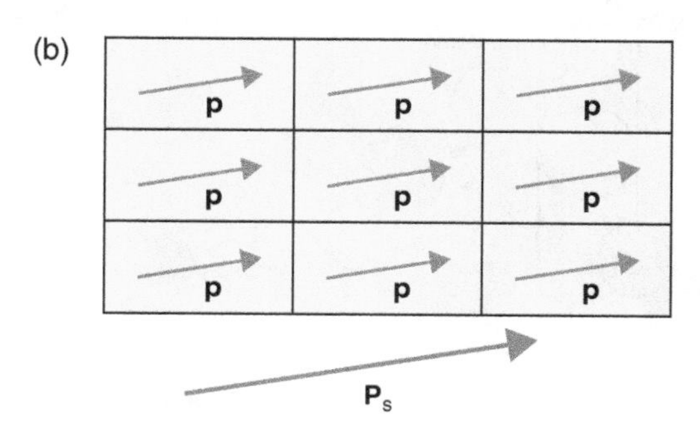

Figure 5.6 Schematic of a pyroelectric or ferroelectric crystal: (a) unit cell containing a dipole, **p**; (b) a pyroelectric crystal showing spontaneous polarisation $\mathbf{P}_s$.

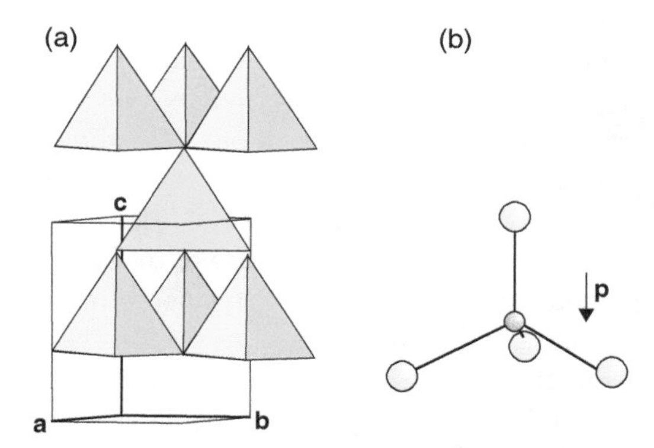

Figure 5.7 The wurtzite structure of ZnS: (a) the crystal structure of wurtzite; (b) an individual $[ZnS]_4$ tetrahedron.

by the fact that the direction of the spontaneous polarisation, $\mathbf{P}_s$, can be changed in direction (switched) by an externally applied electric field. They are, therefore, a subset of pyroelectrics and as such, all ferroelectrics are also pyroelectrics. Note that $\mathbf{P}_s$ is a vector, pointing from negative to positive. (The pyroelectric effect that is normally observed in a crystal is, in fact, composed of two separate effects called the **primary** (or **true**) pyroelectric effect and the **secondary** pyroelectric effect. If a crystal is fixed so that its size is constant as the temperature changes, the primary effect is measured. Normally, though, a crystal is unconstrained. An additional pyroelectric effect will now be measured, the secondary pyroelectric effect, due to strains in the crystal produced by the thermal change.)

The observed spontaneous polarisation $\mathbf{P}_s$ is the sum of existing elementary dipoles in the crystal. This implies that each unit cell in the crystal must also contain an overall permanent electric dipole, **p** (Figure 5.6). These elementary dipoles, which are also vectors, can arise in many ways. One obvious way is for a crystal to contain polar molecules, that is, molecules that carry a permanent electric dipole. There are many such inorganic and organic molecules, including ammonia (Figure 5.1) and water. For example, the H atoms in water are slightly positive (δ+) with respect to the O atom (δ-), and the molecule has a permanent electric dipole moment of 6.2×10^{-30} Cm. Thus, any crystalline hydrate is potentially pyroelectric.

An electric dipole can also arise simply as a consequence of the arrangement of the atoms in a non-molecular crystal. For example, ZnS with the wurtzite structure (see Appendix B) is built up of a 'parallel' arrangement of $[ZnS_4]$ tetrahedra, all of which are oriented similarly (Figure 5.7). In each tetrahedron, the centre of gravity of the negative charges located on the S atoms does not coincide with the positive charge located on the Zn atoms, so that each contains an internal electric dipole, pointing from the centre of gravity of the S atoms to the Zn atom.

Perovskite oxides, ABO_3, are analogous. In these compounds the structure is built of corner-sharing anion BO_6 octahedra, each of which contains a medium-sized B cation, with a large A cation at the corners of the unit cell

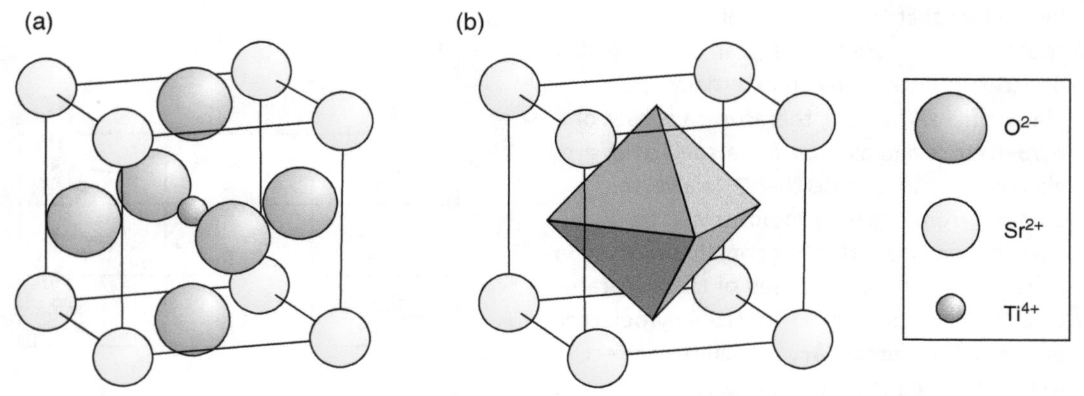

Figure 5.8 The ideal cubic perovskite structure: (a) atom locations; (b) the octahedral coordination of Ti^{4+} emphasised.

(Figure 5.8). This ideal structure is exhibited by cubic $SrTiO_3$ (Appendix B). In many members of the perovskite family the octahedra are distorted and the cations are displaced from the exact centre of each octahedron so that the centre of gravity of the negative charges from the octahedron of oxygen anions does not coincide with the location of the cation, giving rise to a dipole in each octahedron. The structural features that give rise to the spontaneous polarisation are temperature-sensitive and usually diminish as the temperature increases. Above a temperature called the **Curie temperature** the distortions vanish and the structure no longer has a spontaneous polarisation. The material is then no longer pyroelectric or ferroelectric.

$BaTiO_3$ provides a classic example (Figure 5.9). Above 398 K the structure is cubic, equivalent to the ideal $SrTiO_3$ structure. The phase is neither pyro- nor ferroelectric. Between the temperatures of 278 and 398 K the structure is tetragonal, with axial elongation along the **c**-axis and axial shortening along the **a**- and **b**-axes. The cause of the change is essentially because Ti^{4+} cations are displaced from the octahedral centres, resulting in the formation of an electric dipole along the tetragonal **c**-axis. This results in pyroelectric behaviour, and as the displacement of the Ti^{4+} cations and the direction of resulting electric dipoles can by changed (switched) by a high electric field, the phase also shows ferroelectric behaviour.

5.6.2 Crystallographic Aspects of Pyro- and Ferroelectric Behaviour

Although individual dipoles may exist in a structure, whether or not an **overall** dipole moment can be attributed to a unit cell, so resulting in an observable spontaneous polarisation $\mathbf{P}_s$, will depend upon the symmetry of the structure. As described above, each dipole is a polar vector. The operation of the symmetry elements present in the unit cell will generate a group of equivalent vectors, called a **form**. In a centrosymmetric crystal these vectors always add to zero, which immediately limits pyroelectricity to crystals belonging to the 21 non-centrosymmetric point groups.

However, another condition is also needed to produce a spontaneous polarisation: the

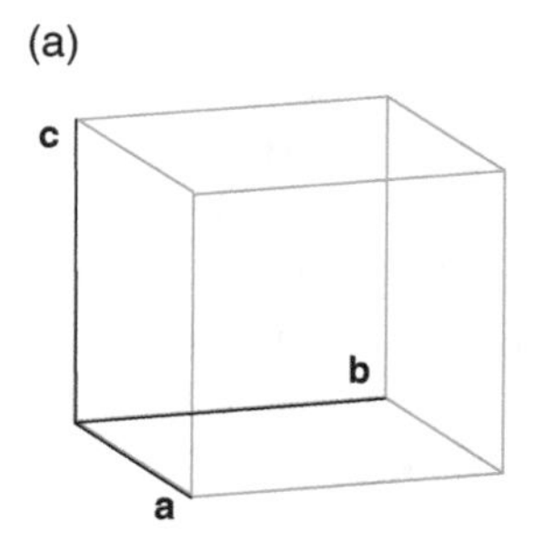

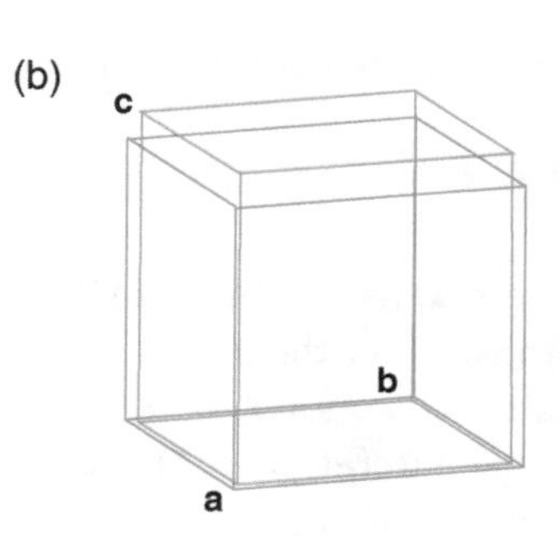

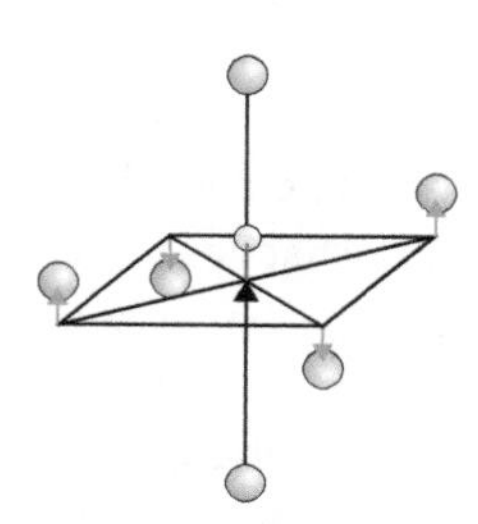

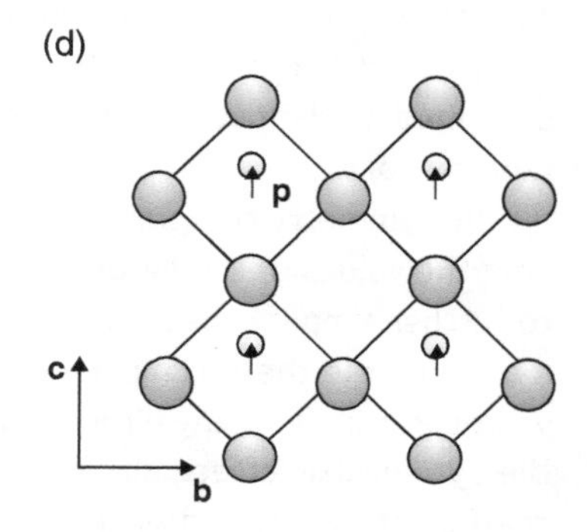

Figure 5.9 $BaTiO_3$:
(a) cubic form (T > 398 K);
(b) the tetragonal unit cell (398 > T > 278 K);
(c) ionic displacements (exaggerated) in each TiO_6 octahedron; (d) electric dipoles.

presence of a unique polar axis. This is a direction in a crystal unrelated by symmetry to any other direction, not even the antiparallel direction. You can understand this from the ZnS structure in Figure 5.7. The direction along the +**c**-axis (through the tetrahedra vertices) is not identical to that along the −**c**-axis (through the tetrahedra bases). Broadly speaking, the internal dipoles and hence $\mathbf{P}_s$ must lie parallel to the unique polar axis in order for spontaneous polarisation to be observed. This limits the number of possible centrosymmetric point groups still more. It is found that in 11 of the 21 non-centrosymmetric point groups, the vector sum of the form of vectors also adds to zero and a crystal belonging to these will not possess a unique polar axis and so will not show a pyroelectric effect. Thus, simply by a consideration of symmetry, it is possible to restrict the pyroelectric crystals to one of the remaining 10 classes, point groups 1, 2, 3, 4, 6, *m*, *mm*2, 3*m*, 4*mm* and 6*mm*, as these do permit a unique polar axis and exhibit a spontaneous polarisation, $\mathbf{P}_s$.

The pyroelectric effect is, however, not just the existence of a spontaneous polarisation, but the change in $\mathbf{P}_s$ with temperature, with units for the pyroelectric coefficient π (defined below) of C m^{-2} K^{-1}. A change in temperature, ΔT (the stimulus), will adjust the structure and hence alter the observed spontaneous polarisation (the response) by a measured amount ΔP_s. The equations describing the magnitudes of the effect, referred to orthogonal axes (Section 5.3) are:

$$\Delta P_1 = \pi_1 \ \Delta T$$
$$\Delta P_2 = \pi_2 \ \Delta T$$
$$\Delta P_3 = \pi_3 \ \Delta T$$

which can be condensed as

$$\Delta P_i = \pi_i \ \Delta T \qquad (5.1)$$

where ΔP_i is the change in the magnitude of the projection of the polarisation $\mathbf{P}_s$ along three Cartesian axes, OX, OY and OZ, with i taking the values 1 (referring to OX), 2 (referring to OY) or 3 (referring to OZ). Similarly, π_i with i = 1, 2 or 3 are the three **pyroelectric coefficients**, appropriate to axes OX, OY and OZ. The array of pyroelectric coefficients (π_1, π_2, π_3) is described mathematically as a first rank tensor.

The symmetry constraints described also limit the values of the pyroelectric coefficients, to further simplify the situation. Symmetry has least effect in the triclinic point group 1. The vector $\mathbf{P}_s$ can lie in any direction and to define the pyroelectric effect, all three values of the coefficients, π_1, π_2 and π_3 are required. In the monoclinic point groups, the polar axis lies parallel to OY, and in the point group m, can lie in the mirror plane perpendicular to this axis. The vector $\mathbf{P}_s$ can lie anywhere in this plane, and so two values of the pyroelectric coefficient are required to specify the vector, π_1 and π_3. In all the other groups, the direction of $\mathbf{P}_s$ must lie parallel to a symmetry axis (which is a principal axis). In these groups, the values of the pyroelectric coefficients not parallel to the allowed axis are zero. A consideration of the symmetry requirements thus simplifies the measurement of the pyroelectric effect by revealing that in the majority of crystals, only one pyroelectric coefficient is in operation, and that lies parallel to the main rotation axis of the group, as set out in Table 5.4.

Note that the value of polarisation falls with temperature, and so the coefficient π should be negative to represent this, but this feature is not always included in quoted values.

Table 5.4 Pyroelectric directions and coefficients

Crystal class	Allowed direction for $\mathbf{P}_s$	Components of $\mathbf{P}_s$	Coefficients of π
Triclinic			
1	Any	(P_1, P_2, P_3)	(π_1, π_2, π_3)
Monoclinic, OY parallel to diad			
2	Parallel to diad	(0, P_2, 0)	(0, π_2, 0)
m	Any direction in mirror plane	(P_1, 0, P_3)	(π_1, 0, π_3)
Orthorhombic, OX, OY, OZ parallel to crystallographic **a**, **b**, **c**			
*mm*2	Parallel to diads	(0, 0, P_3)	(0, 0, π_3)
Tetragonal, OZ parallel to crystallographic **c**			
4	Parallel to tetrad	(0, 0, P_3)	(0, 0, π_3)
4*mm*	Parallel to tetrad	(0, 0, P_3)	(0, 0, π_3)
Trigonal, OZ parallel to crystallographic **c**			
3	Parallel to triad	(0, 0, P_3)	(0, 0, π_3)
3*m*	Parallel to triad	(0, 0, P_3)	(0, 0, π_3)
Hexagonal, OZ parallel to crystallographic **c**			
6	Parallel to hexad	(0, 0, P_3)	(0, 0, π_3)
6*mm*	Parallel to hexad	(0, 0, P_3)	(0, 0, π_3)

5.7 DIELECTRIC PROPERTIES

5.7.1 Dielectrics

Insulating materials are generally described as **dielectrics** (a term used to characterise a polarizable insulator). When a dielectric material is exposed to an external electric field, **E** (the stimulus vector, pointing from positive to negative), the solid develops a surface charge resulting in an observable bulk polarisation, **P** (the response vector, pointing from negative to positive). The polarisation, **P**, need not be parallel to the external field, **E** (Figure 5.10).

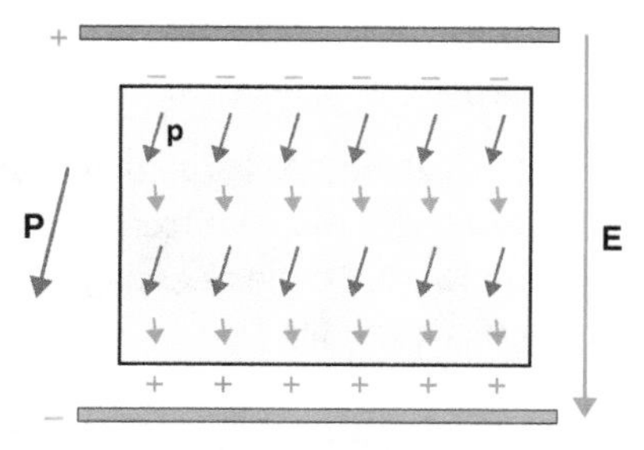

Figure 5.10 An applied external electric field, **E**, leads to an observable polarisation, **P**, in a dielectric.

Broadly speaking, the externally observed bulk polarisation is made up of the sum of internal dipoles ($\mathbf{p}_1$, $\mathbf{p}_2$, etc.) derived from displacement of various atomic and electronic constituents that make up the material by the imposed field. These internal dipoles are also vectors, pointing from the negative to the positive charge. In a normal dielectric, the observed polarisation of the material is zero in the absence of an electric field. as the dipoles giving rise to the effect are generated by the field itself.

The source of these induced dipoles is principally a distortion of the electron cloud around individual atoms, called **electronic polarizability**, displacement of ions from their equilibrium positions, called **ionic polarizability**, and a re-orientation of molecules that already carry a dipole, called **orientational polarizability** (Figure 5.11). In addition, there may be contributions from any

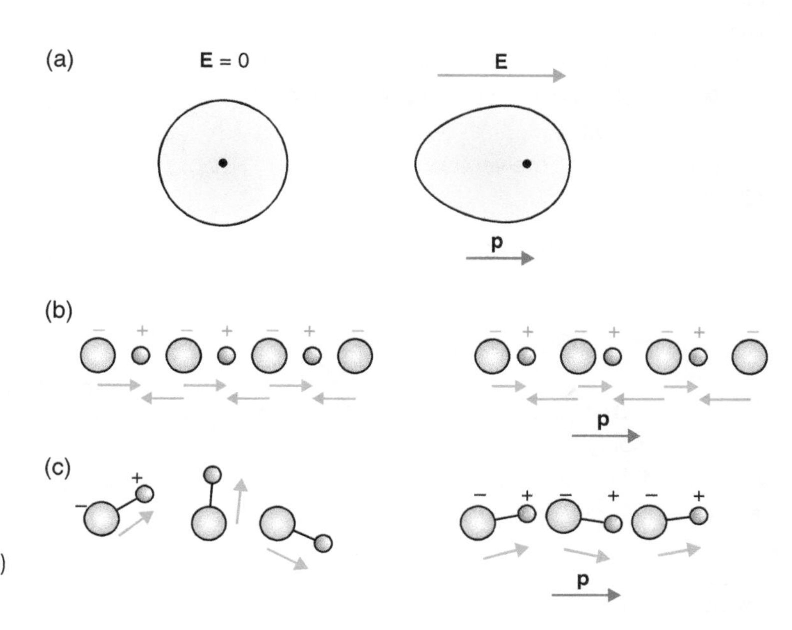

Figure 5.11 Polarisation: (a) electronic; (b) ionic; (c) orientational (see text).

mobile charge carriers or charged defects that may be present.

5.7.2 Isotropic Materials

When a linear, isotropic, homogeneous and nondispersive dielectric phase (examples can include gases, glasses or, in many instances, cubic crystals) is placed in an electric field, **E**, at ordinary field strengths, the induced polarisation, **P**, is **proportional and parallel** to the applied electric field vector, and is represented by:

$$\mathbf{P} = \varepsilon_0 \chi \mathbf{E} \qquad (5.2)$$

where **P** is the polarisation field density vector (the response) (units C m^{-2}), **E** is the electric field vector (the stimulus) (units Vm^{-1}), ε_0 is the permittivity of free space, (8.856×10^{-12} F m^{-1}), and χ is a unitless scalar constant called the **dielectric susceptibility** of the material. (At higher electric field strengths such as those found in laser beams, it is necessary to replace the right-hand side of the equation with a series, with $\varepsilon_0\chi\mathbf{E}$ as the first term). The directions of **E** and **P** are usually referred to a set of orthogonal axes, OX, OY, OZ (1, 2, 3), in the crystal, so that the components of the vectors **P** and **E** along OX, OY and OZ are:

$$\mathbf{P_1} = \varepsilon_0 \chi \mathbf{E_1}$$

$$\mathbf{P_2} = \varepsilon_0 \chi \mathbf{E_2}$$

$$\mathbf{P_3} = \varepsilon_0 \chi \mathbf{E_3}$$

The dielectric susceptibility is related to the more commonly listed relative permittivity ε_r (also called the dielectric constant, *K* or *k*) by the equation:

$$\chi = (\varepsilon_r - 1)$$

to give:

$$\mathbf{P} = \varepsilon_0 (\varepsilon_r - 1) \mathbf{E}$$

5.7.3 Non-isotropic Materials

In non-isotropic (but linear, nondispersive and homogeneous) media, which includes most crystals, Eq. (5.2) must be replaced by a more general formulation that takes into account the vector properties of both the applied electric field and the resultant polarisation. A field along any one axis may, in general, generate a polarisation along any of the axes. The resultant polarisation is written as:

$$\begin{aligned} \mathbf{P}_1 &= \varepsilon_0 \chi_{11} \mathbf{E}_1 + \varepsilon_0 \chi_{12} \mathbf{E}_2 + \varepsilon_0 \chi_{13} \mathbf{E}_3 \\ \mathbf{P}_2 &= \varepsilon_0 \chi_{21} \mathbf{E}_1 + \varepsilon_0 \chi_{22} \mathbf{E}_2 + \varepsilon_0 \chi_{23} \mathbf{E}_3 \\ \mathbf{P}_3 &= \varepsilon_0 \chi_{31} \mathbf{E}_1 + \varepsilon_0 \chi_{32} \mathbf{E}_2 + \varepsilon_0 \chi_{33} \mathbf{E}_3 \end{aligned} \qquad (5.3)$$

Thus, if $\mathbf{E}_1$, lying along OX, is the only electric field present, which can be written in the compact form [$\mathbf{E}_1$, 0, 0], the resulting polarisation **P** is given by:

$$\mathbf{P}_1 = \varepsilon_0 \chi_{11} \mathbf{E}_1$$

$$\mathbf{P}_2 = \varepsilon_0 \chi_{21} \mathbf{E}_1$$

$$\mathbf{P}_3 = \varepsilon_0 \chi_{31} \mathbf{E}_1$$

That is, a field only along OX ($\mathbf{E}_1$) in general gives rise to a polarisation along each of the axes OX, OY, OZ.

The array of nine coefficients relating response polarisation to electric field stimulus is a second rank tensor, written:

$$\begin{pmatrix} \chi_{11} & \chi_{12} & \chi_{13} \\ \chi_{11} & \chi_{12} & \chi_{13} \\ \chi_{11} & \chi_{12} & \chi_{13} \end{pmatrix}$$

The extended Eq. 5.3 for **P** is usually written in a compact notation as:

$$\mathbf{P}_i = \varepsilon_0 \chi_{ij} \mathbf{E}_j \qquad (i,\ j = 1, 2, 3) \qquad (5.4)$$

where the polarisation (the response) in a direction *i*, $\mathbf{P}_i$, is related to the applied field (the stimulus) in a direction *j*, $\mathbf{E}_j$, by a set of coefficients χ_{ij}, the order of the subscripts *ij*

Table 5.5 Coefficients for dielectric susceptibility

Symmetry	Dielectric susceptibility coefficients
Triclinic	$\begin{pmatrix} \chi_{11} & \chi_{12} & \chi_{13} \\ \chi_{21} & \chi_{22} & \chi_{23} \\ \chi_{31} & \chi_{32} & \chi_{33} \end{pmatrix}$
Monoclinic	$\begin{pmatrix} \chi_{11} & 0 & \chi_{13} \\ 0 & \chi_{22} & 0 \\ \chi_{31} & 0 & \chi_{33} \end{pmatrix}$
Orthorhombic	$\begin{pmatrix} \chi_{11} & 0 & 0 \\ 0 & \chi_{22} & 0 \\ 0 & 0 & \chi_{33} \end{pmatrix}$
Hexagonal, Trigonal, Tetragonal	$\begin{pmatrix} \chi_{11} & 0 & 0 \\ 0 & \chi_{11} & 0 \\ 0 & 0 & \chi_{33} \end{pmatrix}$
Cubic	$\begin{pmatrix} \chi_{11} & 0 & 0 \\ 0 & \chi_{11} & 0 \\ 0 & 0 & \chi_{11} \end{pmatrix}$

following the rule response–stimulus or r-s. Note that Eq. 5.4 is just a more compact way of writing the three Eqs (5.3) and contains neither more or less information.

As with pyroelectrics, the array of dielectric susceptibility coefficients can be simplified, with many being reduced to zero. In the case of polarisation, the main discriminator is the principal rotation axis. The resulting coefficients are listed in Table 5.5. It is seen that in many cases these reduce to zero, thus simplifying matters considerably.

5.8 MAGNETIC POINT GROUPS AND COLOUR SYMMETRY

When the atoms that make up a material have unpaired electrons in the outer electron orbitals, the spin of these manifests itself externally as magnetism. Each atom can be imagined to have an associated elementary magnetic dipole, $\boldsymbol{\mu}$, attached to it. The magnetic dipole is a vector quantity and behaves as an axial vector (thus differentiating its behaviour with respect to symmetry from that of the electric polarisation vector, which is a polar vector: Section 5.3).

The magnetic point groups specifically indicate the symmetry consequences of the presence of magnetic dipoles. These non-classical point groups are called **antisymmetrical crystallographic point groups, black and white groups, Shubnikov groups**, as well as a number of other names. The black and white point group terminology supposes that one colour, say black, is associated with one dipole orientation and the other, white, with the alternative dipole direction. The classical point groups are then termed the **neutral** groups.

The same considerations imply that the number of three-dimensional lattices should increase, and this is found to be so. There are 36 magnetic lattices, made up of 22 **antisymmetry lattices**, together with the 14 neutral Bravais lattices.

To illustrate this concept, we will briefly look at some simple magnetic materials. Paramagnetic compounds have the magnetic dipoles completely unaligned. These take random directions that are constantly changing in orientation. However, in some classes of magnetic materials, notably ferromagnetic, ferrimagnetic and antiferromagnetic solids, such as magnetic iron, Fe, or the ceramic magnet barium ferrite, $BaFe_{12}O_{19}$, the overall elementary magnetic dipoles on neighbouring atoms are aligned with each other. In ferromagnetic phases, all the magnetic dipoles are the same, and all point in a single crystallographic direction. In an antiferromagnetic phase, these are parallel across one crystallographic plane, as in a

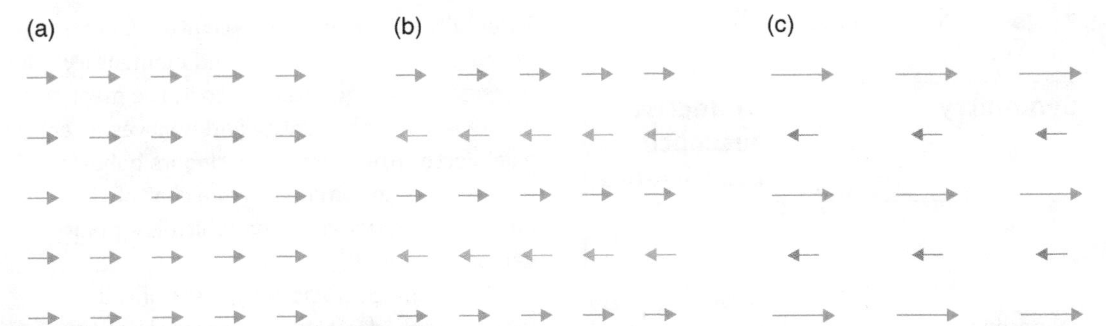

Figure 5.12 Magnetic ordering: (a) ferromagnetic; (b) antiferromagnetic; (c) ferrimagnetic. Magnetic dipoles, on individual metal atoms, are represented by arrows.

ferromagnetic, but on passing from one plane to a neighbouring plane the direction of the magnetic dipole is reversed. A ferrimagnetic crystal is similar, except that the magnitudes of the dipoles in adjacent planes are different (Figure 5.12).

From a macroscopic point of view, the magnetic dipole configuration is not apparent, because a crystal of a typical strongly magnetic material consists of a collection of domains (small volumes), in which all the elementary magnetic dipoles are aligned. Adjacent domains have the magnetic dipoles aligned along different directions, thus creating an essentially random array of magnetic dipoles. When searching for the point group of the crystal, this essentially random arrangement of dipoles can be safely ignored, as they can be in the case of paramagnetic solids. In both cases, the crystals can then be allocated to one of the 32 crystallographic point groups described above.

However, single magnetic domain crystals also exist. In such a case, the classical symmetry point group may no longer accurately reflect the overall symmetry of the crystal, because, in a derivation of the 32 (classical) point groups, the magnetic dipoles on the atoms were ignored. Clearly, in a single domain crystal of a magnetic compound, the magnetic dipoles ought to be taken into account when symmetry operators are applied.

For instance, the magnetic form of pure iron, α-iron, exists up to a temperature of 768°C. The non-magnetic form, once called β-iron, exists between the temperatures of 768°C and 912°C. Both of these forms of iron have the body-centred cubic structure, (Figure 5.13a). The point group of high temperature (β-iron), in which the magnetic dipoles are randomly oriented, or in multidomain samples of magnetic α-iron, is $m\bar{3}m$ $(\frac{4}{m}\bar{3}\frac{2}{m})$. If, however, a single domain crystal is prepared with all the magnetic dipoles aligned along [001], the **c**-axis is no longer identical to the **a**- or **b**-axes. The solid now belongs to the tetragonal system (Figure 5.13b). When the magnetic dipoles are aligned along [110] the appropriate crystal system is orthorhombic (Figure 5.13c), and when the dipoles are aligned along [111] the system is trigonal (Figure 5.13d). In each of these cases new magnetic point groups are required to describe the symmetry of the crystals.

The 32 classical point groups lead directly to the derivation of 58 further magnetic point groups, giving a total of 90 magnetic 'black and white' point groups. However, a further

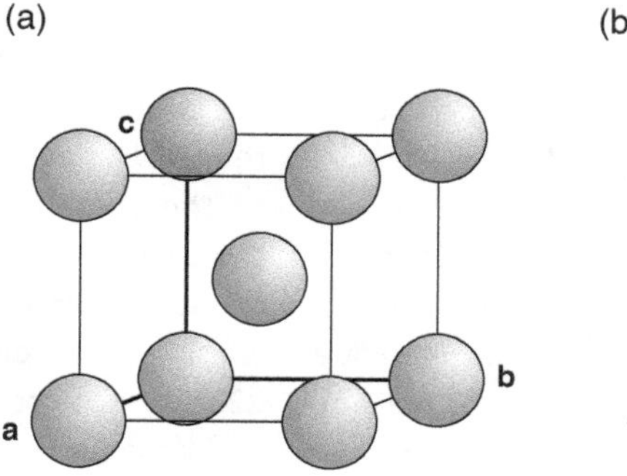

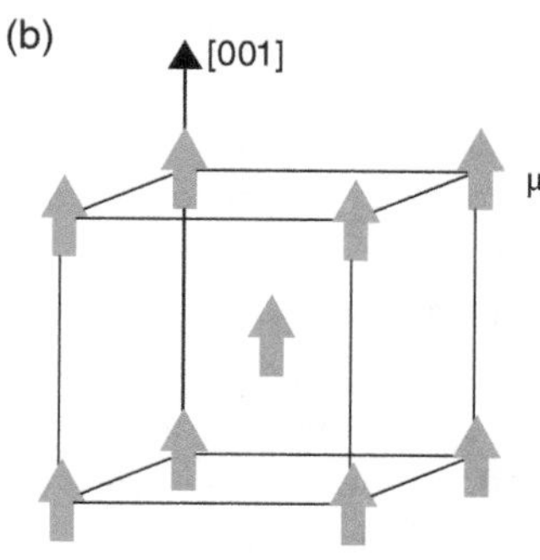

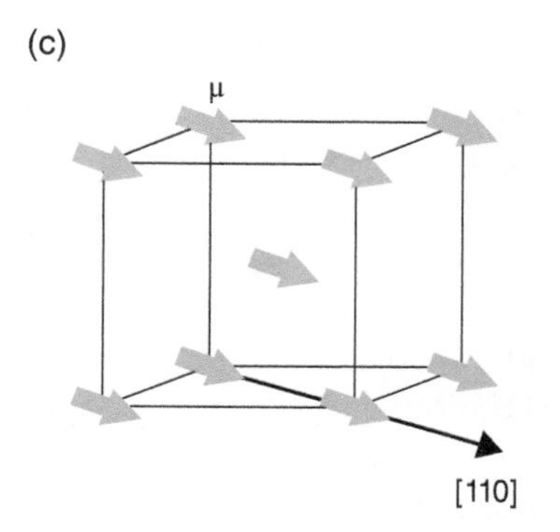

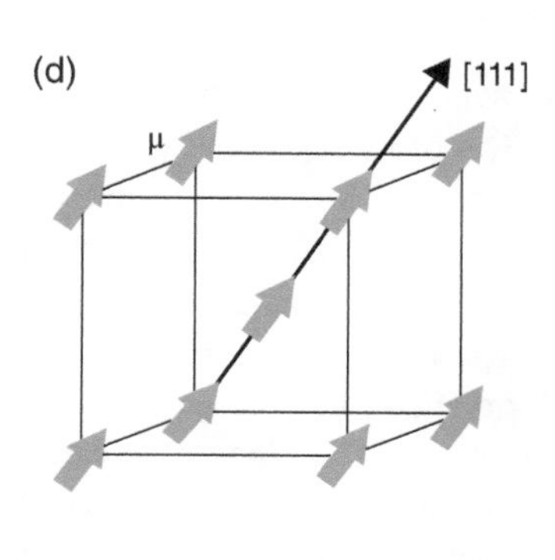

Figure 5.13 Magnetic structures of iron: (a) the structure of non-magnetic body-centred cubic iron; (b) magnetic dipoles (arrows) aligned along the cubic [001] direction; (c) aligned along the cubic [110] direction; (d) aligned along the cubic [111] direction.

32 magnetic groups, called grey groups, to differentiate them from the 90 'black and white' groups, are also needed to completely characterise these systems, giving a grand total of 122 magnetic point groups.

(The same increase in complexity is found when the point group of a pyroelectric or ferroelectric crystal with aligned electric dipoles is considered, as, from a symmetry point of view, the two cases are analogous. However, electric dipoles are polar vectors, and so differ from the symmetry behaviour of magnetic axial vectors with respect to mirror operations.)

More complex physical properties may require the specification of three or more 'colours'. In this case the general term 'colour symmetry' is used, and the lattices and point groups so derived are the **colour lattices** and **colour point groups**.

ANSWERS TO INTRODUCTORY QUESTIONS

What are polar and axial vectors?

Physical properties are quantified by imposing a stimulus on the material and then measuring the response. If either of these have a direction as well as a magnitude, then it needs to be represented by a vector. A well-known example is force, the effect of which depends upon both the magnitude and direction of application.

Vectors fall into two types from the point of view of symmetry operations, in particular, how they transform in terms of the mirror operation, polar and axial vectors. A polar vector is typified by the polarisation introduced into a molecule by an unsymmetrical distribution of charges, giving rise to an electric dipole moment,

p. When a polar vector is reflected in a mirror parallel to **p**, there is no change in the direction of **p**. However, when the mirror is perpendicular to **p** the mirror image is reversed to –**p**. An axial vector is typified by a magnetic moment **μ** introduced into a material by a closed loop of current **I**, including an electron moving around an atomic nucleus. Such a vector is reversed by a mirror operation parallel to the vector but not inverted by a mirror perpendicular to the moment.

What are enantiomorphic pairs?

The property of optical activity is the ability of a crystal to rotate the plane of linearly polarised light to the right or the left. Enantiomorphs are the two forms of optically active crystals, one form rotating the plane of polarised light in one direction, and the other rotating the plane of polarised light in the opposite direction. Such crystals are found to exist as an enantiomorphic pair, consisting of a right-handed and left-handed forms. These cannot be superimposed one upon the other, in the same way that right-hand and left-hand gloves cannot be superimposed one on the other.

The terms enantiomorphs and enantiomorphous pairs also apply to pairs of molecules that are optically active, such as alanine and the tartrate anion.

What are magnetic point groups?

In a magnetic crystal each atom can be imagined to have an associated elementary magnetic dipole, **μ**, attached to it. The magnetic point groups specifically indicate the symmetry consequences of the presence of magnetic dipoles. These non-classical point groups are called **antisymmetrical crystallographic point groups**, **black and white groups**, **Shubnikov groups**, as well as a number of other names. The black and white point group terminology supposes that one colour, say black, is associated with one dipole orientation and the other, white, with the alternative dipole direction. The classical point groups are then termed the **neutral** groups.

The 32 classical point groups lead directly to the derivation of 58 further magnetic point groups, giving a total of 90 magnetic 'black and white' point groups. However, a further 32 magnetic groups, called grey groups, to differentiate them from the 90 'black and white' groups, are also needed to completely characterise these systems, giving a total of 122 magnetic point groups.

PROBLEMS AND EXERCISES

Quick Quiz

1. A northwesterly wind should be represented by:
   a. An axial vector
   b. A polar vector
   c. A scalar
2. A monoclinic crystal will have:
   a. Two different principal refractive indices
   b. Three different principal refractive indices
   c. Four different principal refractive indices
3. An optically active crystal is one which is able to:
   a. Rotate the plane of polarised light as it traverses the crystal
   b. Split the light into two polarised components
   c. Combine two oppositely polarised beams of light
4. Optical activity is found in:
   a. All crystals with a centre of symmetry
   b. All crystals without a centre of symmetry
   c. Some crystals without a centre of symmetry

5. A molecule with two chiral centres:
   a. Is always optically active
   b. Is never optically active
   c. May show optical activity
6. A polar direction can occur only in:
   a. A centrosymmetric point group
   b. Any non-centrosymmetric point group
   c. A subset of the non-centrosymmetric point groups
7. A crystal with orthorhombic symmetry has:
   a. One pyroelectric coefficient
   b. Two pyroelectric coefficients
   c. Three pyroelectric coefficients
8. The polarisation in a cubic crystal produced by an external electric field is:
   a. Perpendicular to the field
   b. At an angle to the field
   c. Parallel to the field
9. An antiferromagnetic crystal contains:
   a. No magnetic dipoles
   b. One type of magnetic dipole
   c. Two types of magnetic dipoles
10. The number of magnetic point groups is:
    a. Less than the number of crystallographic point groups
    b. Equal to the number of crystallographic point groups
    c. Greater than the number of crystallographic point groups

Calculations and Questions

5.1. The oxides SnO_2 (cassiterite), refractive indices $n_0 = 2.000$, $n_e = 2.097$, and PbO_2 (platnerite), refractive indices $n_0 = 2.350$, $n_e = 2.250$, both adopt the rutile structure (Appendix B). A mixed crystal, $Sn_xPb_{1-x}O_2$ is found to have n_0 equal to n_e. Estimate the composition of the phase.

5.2. The variation of the specific rotation of quartz as a function of wavelength (the specific rotation dispersion) is given in the table below. (a) Plot the specific rotation dispersion curve. (b) What thickness of quartz crystal is needed to rotate the polarisation by 90° for red light, $\lambda = 670$ nm, and violet light, $\lambda = 420$ nm?

Wavelength, nm	Specific rotation, ° mm^{-1}
400	50.2
450	38.9
500	30.9
550	25.0
600	20.1
650	17.8
700	15.3

5.3. In the tetragonal form of $BaTiO_3$ (Figure 5.9), the Ti–O bond lengths along the **c**- (tetrad) axis are 0.22 and 0.18 nm. Estimate the electric dipole moment p and the spontaneous polarisation P_S.

5.4. The perovskite structure phase with a composition of $PbTi_{0.5}Zr_{0.5}O_3$, is tetragonal with approximate lattice parameters $a = b = 0.4000$ nm, $c = 0.4240$ nm. The pyroelectric coefficient is -5.0×10^{-4} C m^2 K^{-1}. (a) What will be the change in polarisation produced by a temperature rise of 75°C? (b) What is the direction of the polarisation?

5.5. Alumina, Al_2O_3, is hexagonal, with relative permittivity values of $\varepsilon_{11} = \varepsilon_{22} = 9.34$, $\varepsilon_{33} = 11.54$. If an electric field of 7500 V m^{-1} is applied in (a) the [001] and (b) the [100] or [010] directions, what will be the value of the induced polarisation P?

5.6. Spinel, $MgAl_2O_4$, is cubic with a relative permittivity of 8.6. If an electric field of 7500 V m^{-1} is directed along [111], what will be the polarisation along the **a**-, **b**- and **c**-axes and along [111]?

5.7. Chromium, Cr, has the A2 structure similar to that of iron, but instead of having the magnetic moments aligned in a ferromagnetic order, they are aligned in an antiferromagnetic array (Figure 5.12b). Assuming the possible arrangements are as depicted in the figure below, what is the new crystal system applicable to single domain crystals, (b), (c), and (d)?

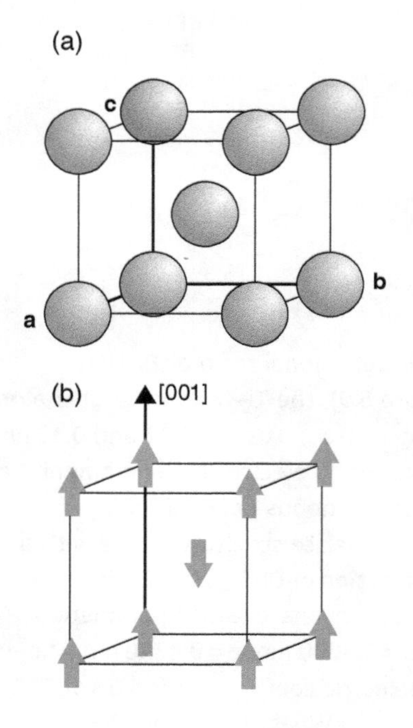

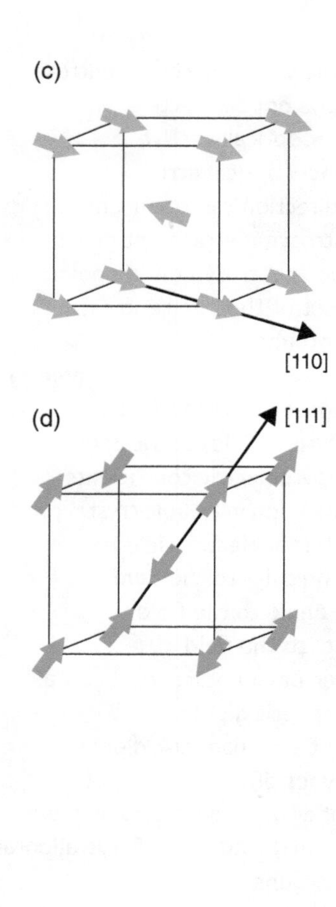

Chapter 6

Building Crystal Structures from Lattices and Space Groups

What symmetry operation is associated with a screw axis?
What is a crystallographic space group?
What are Wyckoff letters?

6.1 SYMMETRY OF THREE-DIMENSIONAL PATTERNS: SPACE GROUPS

The 17 plane groups, derived by combining the translations inherent in the five plane lattices with the symmetry elements present in the ten plane point groups, together with the glide operator, represent, in a compact way, all possible planar repeating patterns. In a similar way, a combination of the translations inherent in the 14 Bravais space lattices with the symmetry elements present in the 32 crystallographic point groups, together with a new symmetry element, the screw axis, described below, allows all possible three-dimensional repeating crystallographic patterns to be classified. The resulting 230 combinations are the **crystallographic space groups**.

There are parallels between the two- and three-dimensional cases. Naturally, mirror lines in two dimensions become **mirror planes** in three, and glide lines in two dimensions become **glide planes** in three. The glide translation vector, **t**, is constrained to be equal to half of

Crystals and Crystal Structures, Second Edition. Richard J. D. Tilley.

Table 6.1 Rotation, inversion, and screw axes allowed in crystals

Rotation axis, n	Inversion axis, $\bar{n}$	Screw axis, n_p				
1	$\bar{1}$ (centre of symmetry)					
2	$\bar{2}$ (m)	2_1				
3	$\bar{3}$	3_1	3_2			
4	$\bar{4}$	4_1	4_2	4_3		
6	$\bar{6}$	6_1	6_2	6_3	6_4	6_5

the relevant lattice vector, **T**, for the same reason that the two-dimensional glide vector is half of a lattice translation.

In addition, the combination of three-dimensional symmetry elements gives rise to a completely new symmetry operator, the **screw axis**. Screw axes are **rototranslational** symmetry elements, constituted by a combination of rotation and translation. A screw axis of order n operates on an object by (i) a rotation of $2\pi/n$ anticlockwise and then (ii) a translation by a vector **t** parallel to the axis, in a positive direction. The value of n is the **order** of the screw axis. For example, a screw axis running parallel to the **c**-axis in an orthorhombic crystal would entail an anticlockwise rotation in the **a–b** plane, (001), followed by a translation parallel to +**c**. This is a **right-handed screw rotation.** Now if the rotation component of the operator is applied n times, the total rotation is equal to 2π. At the same time, the total displacement is represented by the vector $n\mathbf{t}$, running parallel to the rotation axis. In order to maintain the lattice repeat, it is necessary to write:

$$n\mathbf{t} = p\mathbf{T}$$

where p is an integer, and **T** is the lattice repeat in a direction parallel to the rotation axis. Thus:

$$\mathbf{t} = (p/n)\,\mathbf{T}$$

For example, the repeat translations for a three-fold screw axis are:

$$(0/3)\mathbf{T}, (1/3)\mathbf{T}, (2/3)\mathbf{T}, (3/3)\mathbf{T}, (4/3)\mathbf{T}, \ldots$$

Of these, only $(1/3)\mathbf{T}$ and $(2/3)\mathbf{T}$ are unique. The corresponding values of p are used in writing the three-fold screw axis as 3_1 (pronounced 'three sub one') or 3_2 ('three sub two'). Similarly, the rotation diad 2 can only give rise to the single screw axis 2_1. The unique screw axes are listed in Table 6.1.

The operation of the screw axis 4_2 is described in Figure 6.1a–e. The first action is an anticlockwise rotation by $2\pi/4$, i.e. 90°. The rotated atom is then translated by vector $\mathbf{t} = \mathbf{T}/2$, which is half of the lattice repeat distance parallel to the screw axis, to generate atom B. Repetition of this pair of operations generates atom C from atom B. The total distance displaced by the operations is equal to $2\mathbf{t} = 2 \times \mathbf{T}/2 = \mathbf{T}$, so that the atom C is also replicated at the origin of the screw vector. The operation of the screw axis on atom C generates atom D. The total result of a 4_2 axis is thus to generate four atoms from the original one specified. This symmetry operation is often portrayed by a view along the axis (Figure 6.1f; see also Table 6.7). In this figure, the motif is represented by a circle, the + means that the motif is situated above the plane of the paper and ½ + indicates the position of a motif generated by screw operation. If the screw axis runs parallel to the **c**-axis, the heights can be written as $+z$ and $+z + ½c$, where c is the lattice parameter.

Screw axes are found in nature (Figure 6.2).

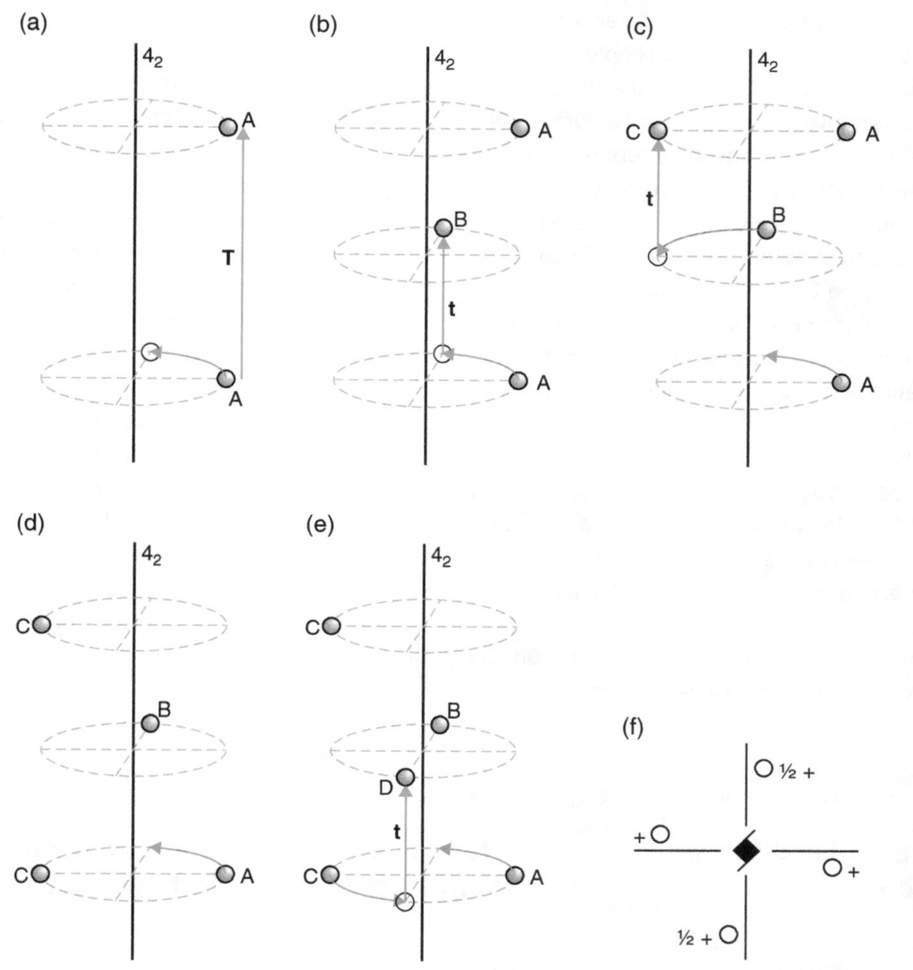

Figure 6.1 (a)–(e) The operation of a 4_2 screw axis parallel to the *z* direction; (f) standard crystallographic depiction of a 4_2 screw axis viewed along the axis.

Figure 6.2 Elm twigs with the leaves arranged following a 2_1 screw symmetry.

6.2 THE CRYSTALLOGRAPHIC SPACE GROUPS

The 230 crystallographic space groups summarise the total number of three-dimensional patterns that result from combining the 32 point groups with the 14 Bravais lattices and including the screw axes. Each space group is given a unique symbol and number (Appendix D). Naturally, just as the

17 plane groups can give rise to an infinite number of different designs, simply by altering the motif, so the 230 crystallographic space groups can give rise to an infinite number of crystal structures simply by altering the atoms and their relative positions in the three-dimensional motif. However, each resultant structure must possess a unit cell that conforms to one of the 230 space groups.

Each space group has a space group symbol that summarises the important symmetry elements present in the group. (As in previous chapters, the symbols used in crystallography, [International or Hermann-Mauguin symbols] are used throughout. The space group symbols are written in two ways, a more explicit 'full' form, or abbreviated to the 'short' form, both of which are listed in Appendix D. The alternative Schoenflies symbols are also given in Appendix D.) The space group symbol consists of two parts: (i) a capital letter, indicating the lattice that underlies the structure, and (ii) a set of characters that represent the symmetry elements of the space group.

Table 6.2 lists the letter symbols used in the space groups to describe the underlying lattice, and the coordinates of the associated lattice points. It is easy to become confused here. For example, a space group derived from a primitive hexagonal Bravais lattice is prefixed by the letter symbol *P*, for primitive, not *H*, which means a centred hexagonal lattice. Note that there are two ways of describing a trigonal lattice. The simplest is in terms of a primitive cell with rhombohedral axes. As with all primitive lattices, there is only one lattice point associated with the unit cell. However, a more convenient description of a trigonal structure can often be made by using a hexagonal unit cell. When this option is chosen, the coordinates of the lattice points in the unit cell can be specified in two equivalent ways, the **obverse**

Table 6.2 The crystallographic space group letter symbols

Letter symbol	Lattice type	Number of lattice points per unit cell	Coordinates of lattice points
P	Primitive	1	0, 0, 0
A	A-face centred	2	0, 0, 0; 0, ½, ½
B	B-face centred	2	0, 0, 0; ½, 0, ½
C	C-face centred	2	0, 0, 0; ½, ½, 0
I	Body-centred	2	0, 0, 0; ½, ½, ½
F	All-face centred	4	0, 0, 0; ½, ½, 0; 0, ½, ½; ½, 0, ½
R	Primitive (rhombohedral axes)	1	0, 0, 0
	Centred rhombohedral (hexagonal axes)	3	0, 0, 0; 2/3, 1/3, 1/3; 1/3, 2/3, 2/3 (obverse setting) 0, 0, 0; 1/3, 2/3, 1/3; 2/3, 1/3, 2/3 (reverse setting)
H	Centred hexagonal	3	0, 0, 0; 2/3, 1/3, 0; 1/3, 2/3, 0

setting or the **reverse setting**. The coordinates of the lattice points in each of these two arrangements are given in Table 6.2.

6.3 SPACE GROUP SYMMETRY SYMBOLS

The second part of a space group symbol consists of one, two, or three entries after the initial letter symbol described above. At each position, the entry consists of one or two characters, describing a symmetry element, either an axis or a plane. Many space groups contain more symmetry elements than that given in the space group full symbol. The standard notation gives only the essential symmetry operators, which, when applied, allows all of the other symmetry operations to be recovered. The symbols used to represent the symmetry operations are set out in Table 6.3.

As described previously, the position of the symmetry element in the overall symbol has a

Table 6.3 Symmetry elements in space group symbols

Symbol	Symmetry operation	Comments
m	Mirror plane	Reflection
a	Axial glide plane ⊥ [010], [001]	Glide vector **a**/2
b	Axial glide plane ⊥ [001], [100]	Glide vector **b**/2
c	Axial glide plane ⊥ [100], [010]	Glide vector **c**/2
	⊥ $[1\bar{1}0]$, [110]	Glide vector **c**/2
	⊥ [100], [010], $[\bar{1}\bar{1}0]$	Glide vector **c**/2, hexagonal axes
	⊥ $[1\bar{1}0]$, [120], $[\bar{2}\bar{1}0]$ $[\bar{1}\bar{1}0]$	Glide vector **c**/2, hexagonal axes
n	Diagonal glide plane ⊥ [001]; [100]; [010]	Glide vector ½(**a** + **b**); ½(**b** + **c**); ½(**a** + **c**)
	Diagonal glide plane ⊥ $[1\bar{1}0]$; $[01\bar{1}]$; $[\bar{1}01]$	Glide vector ½(**a** + **b** + **c**)
	Diagonal glide plane ⊥ [110]; [011]; [101]	Glide vector ½(−**a** + **b** + **c**); ½(**a**-**b** + **c**); ½(**a** + **b**-**c**)
d	Diamond glide plane ⊥ [001; [100]; [010]	Glide vector ¼ (**a** ± **b**); ¼(**b** ± **c**); ¼(±**a** + **c**)
	Diamond glide plane ⊥ $[1\bar{1}0]$; $[01\bar{1}]$; $[\bar{1}01]$	Glide vector ¼(**a** + **b** ± **c**); ¼(±**a** + **b** + **c**); ¼(**a** ± **b** + **c**)
	Diamond glide plane ⊥ [110]; [011]; [101]	Glide vector ¼(−**a** + **b** ± **c**); ¼(±**a**-**b** + **c**); ¼(**a** ± **b**-**c**)
1	None	—
2, 3, 4, 6	*n*-fold rotation axis	Rotation anticlockwise of 360°/*n*
$\bar{1}$	Centre of symmetry	—
$\bar{2}(=m), \bar{3}, \bar{4}, \bar{6}$	$\bar{n}$-fold inversion (rotoinversion) axis	360°/*n* rotation anticlockwise followed by inversion
2_1, 3_1, 3_2, 4_1, 4_2, 4_3, 6_1, 6_2, 6_3, 6_4, 6_5	*n*-fold screw (rototranslation) axis, n_p	360°/*n* right-handed screw rotation anticlockwise followed by translation by (*p*/*n*)**T**

Table 6.4 Order of the Hermann-Mauguin symbols in space group symbols

Crystal system	Primary	Secondary	Tertiary
Triclinic	—	—	—
Monoclinic	[010], unique axis **b** [001], unique axis **c**	—	—
Orthorhombic	[100]	[010]	[001],
Tetragonal	[001]	[100], [010]	$[1\bar{1}0]$, [110]
Trigonal, rhombohedral axes	[111]	$[1\bar{1}0]$, $[01\bar{1}]$, $[\bar{1}01]$	
Trigonal, hexagonal axes	[001]	[100], [010], $[\bar{1}\bar{1}0]$	
Hexagonal	[001]	[100], [010], $[\bar{1}\bar{1}0]$	$[1\bar{1}0]$, [120], $[\bar{2}\bar{1}0]$
Cubic	[100], [010], [001]	[111], $[1\bar{1}\bar{1}]$, $[\bar{1}1\bar{1}]$, $[\bar{1}\bar{1}1]$	$[1\bar{1}0]$, [110], $[01\bar{1}]$, [011], $[\bar{1}01]$, [101]

Table 6.5 The meaning of some space group symbols

Space group number	Short symbol	Full symbol	Meaning
3	*P*2	*P*121	Diad along unique **b**-axis (conventional)
		*P*112	Diad along unique **c**-axis
6	*Pm*	*P*1*m*1	Mirror plane normal to the unique **b**-axis (conventional)
		*P*11*m*	Mirror plane normal to the unique **c**-axis
10	*P*2/*m*	*P*1 2/*m* 1	Diad along and mirror plane normal to unique **b**-axis (conventional)
		*P*1 1 2/*m*	Diad along and mirror plane normal to unique **c**-axis

structural significance that varies from one crystal system to another, as set out in Table 6.4.

Note that symmetry planes are always designated by their normals. If a symmetry axis and the normal to a symmetry plane are parallel, the two symbols are divided by a slash, as in the space group *P* 2/*m* (pronounced 'P two over m'). Full symbols carry placemarkers to make the position of the symmetry element in the structure unambiguous. For example, the unique axis in the monoclinic system is conventionally taken as the **b**-axis. The short symbol for the monoclinic space group number 3 is *P*2. Using the conventional axes, this means that the only symmetry element of significance is a diad along the **b**-axis. To make this clear, the full symbol is *P*121. If the unique axis is taken as the **c**-axis, the full symbol would be *P*112, but the short symbol remains *P*2. More examples from the monoclinic system are given in Table 6.5.

In some instances, the space group symbol used does not appear to totally conform with the previous description. As an example, the orthorhombic space group No. 62 is illustrated. The standard space group symbol is (No. 62) *Pnma*, $P2_1/n\ 2_1/m\ 2_1/a$. However, the setting *Pbnm* and sometimes *Pcmn* are also found. In order to understand the implications of these alternatives, it is necessary to see how one is converted to another.

We start with the standard setting. From the previous discussion:

P indicates that the lattice is primitive.

2_1 (repeated three times) means that there is a screw axis of order 2 parallel to [100], [010] and [001].

n means that there is a diagonal glide plane with its normal running parallel to [100], so that the plane itself lies perpendicular to the *x*-axis.

m means that there is a mirror plane with its normal running parallel to [010], so that the plane itself lies perpendicular to the *y*-axis.

a means there is an axial glide plane with its normal running parallel to [001], so that the plane itself lies perpendicular to the *z*-axis. The glide vector runs parallel to **a**.

The unit cell indicating these symmetry features can now be sketched (Figure 6.3a). (Remember that the standard way of drawing these unit cells is that the **a**-axis points out of the page towards the reader, **b**-axis points in the plane of the page from left to right, and the **c**-axis points up the page from bottom to top. In a plane projection down the **c**-axis the a-axis points down the page from top to bottom and the **b**-axis points from left to right.) There are six possible permutations of *x*, *y* and *z* (or **a**, **b**, **c**), each one leading to a different arrangement of axes and hence a different symbol (Figure 6.3b–f).

Take the non-standard setting symbol *Pbnm* (Figure 6.3b). This space group is frequently used instead of the standard setting, especially when describing perovskite oxides. The basis vectors in the new setting are derived from those of the old setting by using the transformation described by the setting symbol (**cab**):

Suppose that the basis vectors of the unit cell axes in the standard setting is **a b c** *Pnma* (standard), (**abc**). The basis vectors for the new setting *Pbnm*, **a'**, **b' c'** are given by the transformation written (**cab**):

$$\mathbf{a}' \to \mathbf{c}, \mathbf{b}' \to \mathbf{a}, \mathbf{c}' = \mathbf{b}$$

That is to say, the new **a'** basis vector replaces the old **c** basis vector; the new **b'** vector replaces the old **a** vector and the new **c'** vector replaces the old **b** vector (Figure 6.4a). We can now describe the other four possibilities rather more quickly:

Pmcn, (**bca**)

$$\mathbf{a}' \to \mathbf{b}, \mathbf{b}' \to \mathbf{a}, \mathbf{c}' = \mathbf{b}$$

That is to say, the new **a'** basis vector replaces the old **b** basis vector; the new **b'** basis vector replaces the old **c** vector and the new **c'** basis vector replaces the old **a** basis vector (Figure 6.4b).

Pnam, (**a–cb**)

$$\mathbf{a}' \to \mathbf{a}, \mathbf{b}' \to -\mathbf{c}, \mathbf{c}' = \mathbf{b}$$

That is to say, the new **a'** basis vector replaces the old **a** basis vector; the new **b'** basis vector replaces the old –**c** vector and the new **c'** basis vector replaces the old **b** basis vector (Figure 6.4c).

Pmnb, (**bac**)

$$\mathbf{a}' \to \mathbf{b}, \mathbf{b}' \to \mathbf{a}, \mathbf{c}' = \mathbf{c}$$

That is to say, the new **a'** basis vector replaces the old **b** basis vector; the new **b'** basis vector

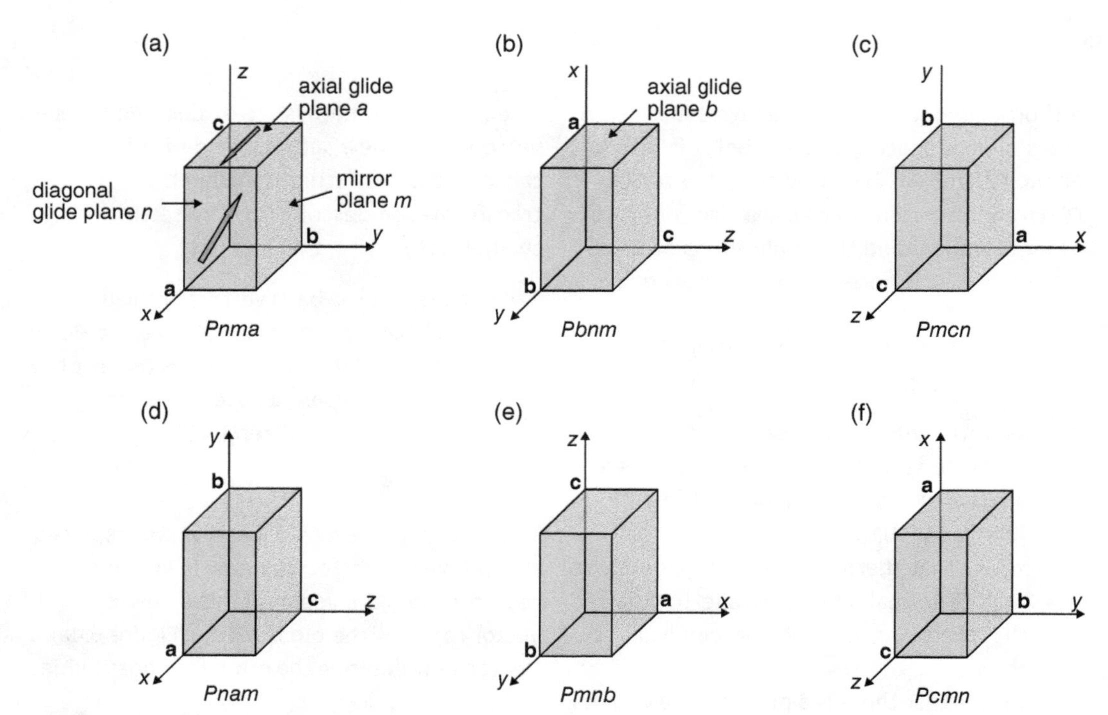

Figure 6.3 Alternative settings for space group No. 62, *Pnma*.

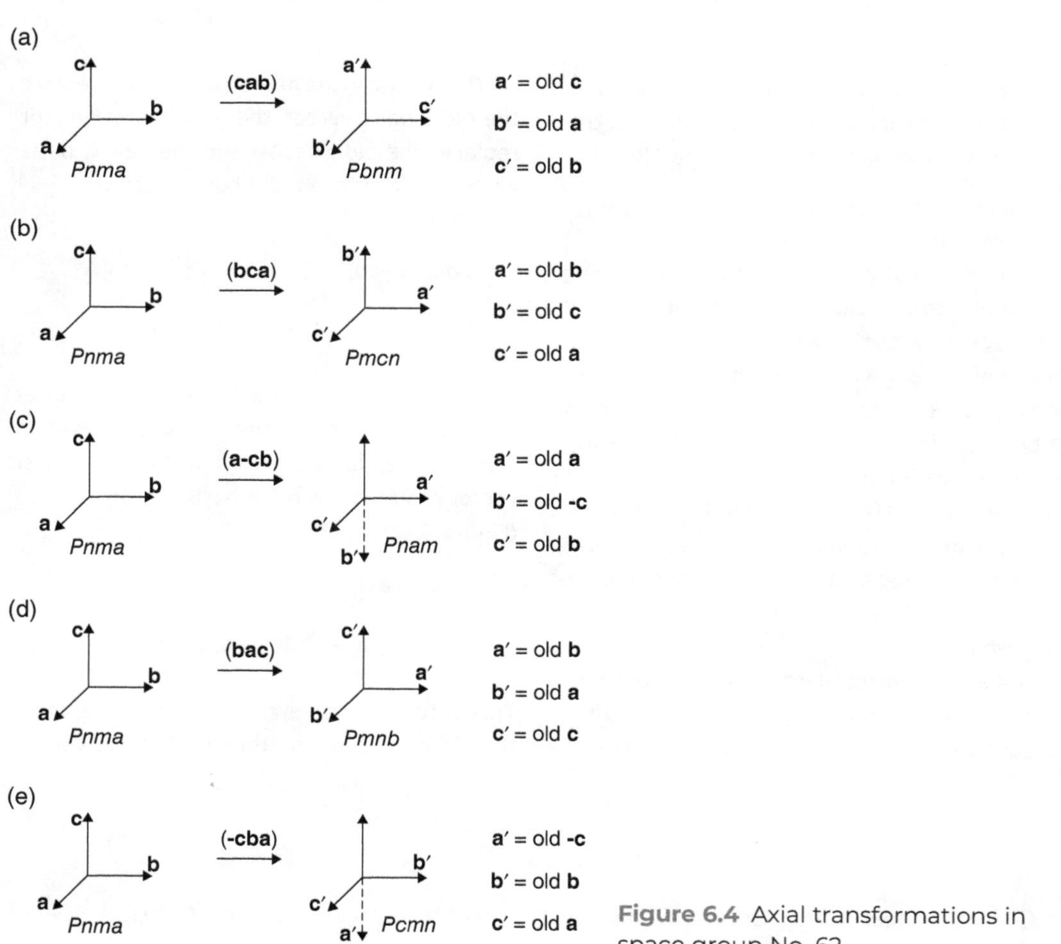

Figure 6.4 Axial transformations in space group No. 62.

replaces the old **a** vector and the new **c′** basis vector replaces the old **c** basis vector (Figure 6.4d).

Pcmn, (**−cba**)

$$\mathbf{a}' \rightarrow -\mathbf{c}, \mathbf{b}' \rightarrow \mathbf{b}, \mathbf{c}' = \mathbf{ca}$$

That is to say, the new **a′** basis vector replaces the old **−c** basis vector; the new **b′** basis vector replaces the old **b** vector and the new **c′** basis vector replaces the old **a** basis vector (Figure 6.4e).To summarise, the meaning of these new symbols is:

Pbnm: *b*: an axial glide plane normal parallel to **a**, glide vector parallel to **b**
n: a diagonal glide plane normal parallel to **b**
m: a mirror plane normal parallel to **c**
Pmcn: *m*: a mirror plane normal parallel to **a**
c: an axial glide plane normal parallel to **b**, glide vector parallel to **c**
n: a diagonal glide plane normal parallel to **c**
Pnam: *n*: a diagonal glide plane normal parallel to **a**
a: an axial glide plane normal parallel to **b**, glide vector parallel to **a**
m: a mirror plane normal parallel to **c**
Pmnb: *m*: a mirror plane normal parallel to **a**
n: a diagonal glide plane normal parallel to **b**
b: an axial glide plane normal parallel to **c**, glide vector parallel to **b**
Pcmn: *c*: an axial glide plane normal parallel to **a**, glide vector parallel to **c**
m: a mirror plane normal parallel to **b**
n: a diagonal glide plane normal parallel to **c**

The point group and space group are closely related. In order to determine the point group corresponding to a space group, the initial lattice letter of the space group is ignored, and the translation symmetry operators, *a*, *b*, *c*, *n*, *d*, are replaced by the mirror operator, *m*. Table 6.6 gives some examples.

6.4 THE GRAPHICAL REPRESENTATION OF THE SPACE GROUPS

A space group is generally represented by two figures, one that shows the effect of the operation of the symmetry elements present and one that shows the location of the various symmetry elements. In some space groups, it is possible to choose between several different origins for the diagrams, in which case both alternatives are presented.

Each diagram is a projection, down a unit cell axis, of a unit cell of the structure. The position

Table 6.6 Point groups and space groups

Crystallographic space group number	Short symbol	Full symbol	Crystallographic point group
3	$P2$	$P121$	2
6	Pm	$P1m1$	m
10	$P2/m$	$P1\ 2/m\ 1$	$2/m$
200	$Pm\bar{3}$	$P\ 2/m\ \bar{3}$	$m\bar{3}$
203	$Fd\bar{3}$	$F\ 2/d\ \bar{3}$	$m\bar{3}$
216	$F\bar{4}3m$	$F\bar{4}3m$	$\bar{4}3m$
219	$F\bar{4}3c$	$F\bar{4}3c$	$\bar{4}3m$

Table 6.7 Graphical symbols used in space group diagrams

Symmetry axes perpendicular to plane of diagram

o	Centre of symmetry
	Twofold rotation axis
	Twofold screw axis: 2_1
	Twofold rotation axis with centre of symmetry
	Twofold screw axis with centre of symmetry
▲	Threefold rotation axis
△	Threefold inversion axis: $\bar{3}$
	Threefold screw axis: 3_1
	Threefold screw axis: 3_2
◆	Fourfold rotation axis
	Fourfold inversion axis: $\bar{4}$
	Fourfold rotation axis with centre of symmetry
	Fourfold screw axis: 4_1
	Fourfold screw axis: 4_2
	Fourfold screw axis: 4_2 with centre of symmetry
	Fourfold screw axis: 4_3
	Sixfold rotation axis
	Sixfold inversion axis $\bar{6}$
	Sixfold rotation axis with centre of symmetry
	Sixfold screw axis: 6_1
	Sixfold screw axis: 6_2
	Sixfold screw axis: 6_3
	Sixfold screw axis: 6_3 with centre of symmetry
	Sixfold screw axis: 6_4
	Sixfold screw axis: 6_5

Symmetry planes normal to plane of diagram

————	Mirror plane
– – – – ·	Axial glide plane: vector ½ a, b or c parallel to plane
. . . .	Axial glide plane: vector ½ a, b or c normal to plane
– · – · – ·	Diagonal glide plane: vector ½ a, b or c parallel to plane plus ½ a, b or c normal to plane

Symmetry planes parallel to plane of diagram

	Mirror plane
	Axial glide plane, vector ½ a, b or c in direction of the arrow
1/4	Diagonal glide plane, vector ½ (a + b), (b + c) or (a + c) in direction of the arrow

Equivalent position diagrams

O	Motif in a general position in the plane of the diagram
O+	Motif at arbitrary height x, y or z above the plane of the diagram
⊙	Enantiomorph of motif in a general position in the plane of the diagram
⊙+	Enantiomorph of motif at an arbitrary height x, y or z above the plane of the diagram

of an atom or a motif is represented by a **circle**, and the origin of the cell is chosen with respect to the symmetry elements present. Symmetry elements are placed at the corners or along the edges of the unit cell whenever possible. The enantiomorphic form of a motif, generated by a mirror reflection from the original, is represented by a **circle containing a comma**. There is a set of standard graphical symbols used in these figures, the most important of which are set out in Table 6.7.

As an example, consider the diagrams appropriate to the space group No. 75, *P*4 (Figure 6.5). The space group symbol reveals that the unit cell is primitive, and so contains only one lattice point. The principal symmetry element present is a tetrad axis parallel to the **c**-axis. The tetrad axis is always chosen to be parallel to the **c**-axis, and there is no need to specify the space group symbol as *P*114. The origin of the unit cell coincides with the tetrad axis, and the projection is down this axis, and conventionally, the *x*-direction points from top to bottom and the *y*-direction from left to right.

The disposition of the symmetry elements is shown in Figure 6.5a. The diagrams reveal all of the symmetry elements present, including 'extra ones' generated by the principal

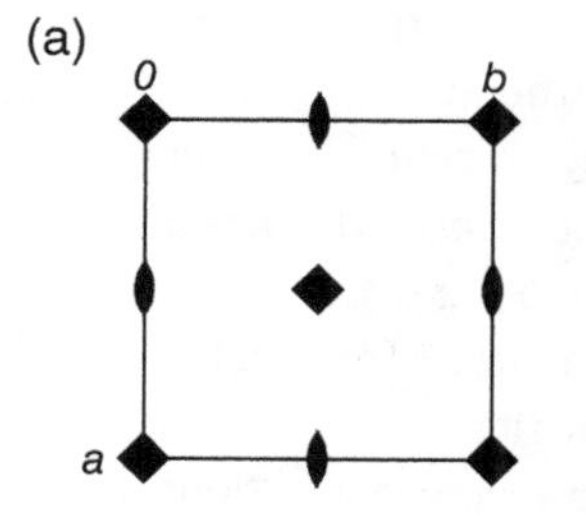

Space group No. 75: *P* 4

Fourfold rotation axis parallel to c (perpendicular to the plane of the diagram)

Twofold rotation axis parallel to c (perpendicular to the plane of the diagram)

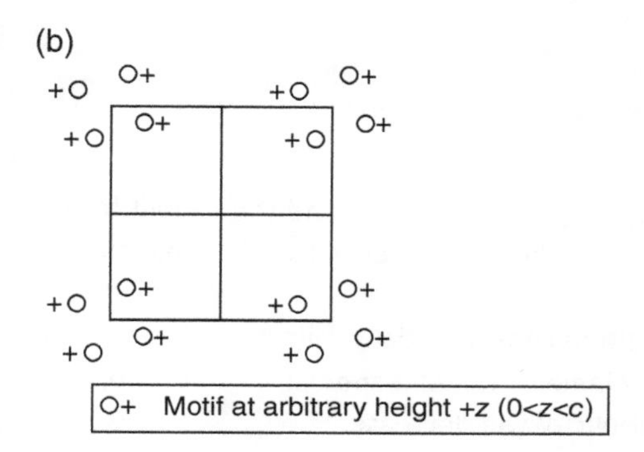

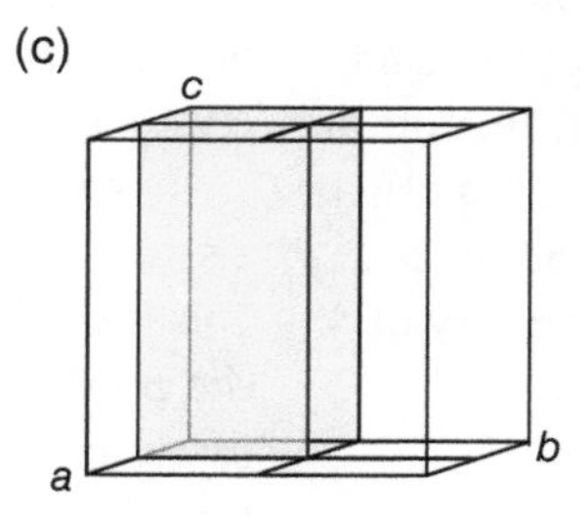

Figure 6.5 Space group diagrams for space group No. 75, *P*4: (a) symmetry elements; (b) positions of motif generated by symmetry operations; (c) asymmetric unit.

symmetry elements described in the space group symbol. For example, the position of the tetrad axis at the unit cell corner also generates another at the cell centre, and diad axes at the centre of each cell edge (Figure 6.5a). The operation of the tetrad axis on a motif (represented by a circle) generates four other copies in the unit cell (Figure 6.5b). The + sign by the motif indicates that there is no change in the height of this unit due to operation of the tetrad axis.

The smallest part of the unit cell that will reproduce the whole cell when the symmetry operations are applied is called the **asymmetric unit**. This is not unique, and several alternative asymmetric units can be found for a unit cell, but it is clear that rotation or inversion axes must always lie at points on the boundary of the asymmetric unit. The conventional asymmetric unit for the space group *P*4 is indicated in Figure 6.5c. The extension of the asymmetric unit is one unit cell in height, so that it is defined by the relations $0 \leq x \leq ½$, $0 \leq y \leq ½$, $0 \leq z \leq 1$.

Wyckoff letters, as described in Chapter 3 for the plane groups, label the different site symmetries that can be occupied by an atom in a unit cell. In space group 75, *P*4, the position chosen for the atom, (*x*, *y*, *z*), is a **general position** in the unit cell, and the positions generated by the symmetry operators present are called **general equivalent positions**. These are:

$$(1)\ x, y, z \quad (2)\ \bar{x}, \bar{y}, z \quad (3)\ \bar{y}, x, z \quad (4)\ y, \bar{x}, z$$

These are given the Wyckoff symbol *d*, and the multiplicity of the position is 4, as set out in Table 6.8.

If the atom position chosen initially falls upon a symmetry element, called a **special position**, the multiplicity will decrease. For example, an atom placed on the diad axis will not be repeated by the operation of the diad. However, as there are two diad axes, an atom at one will necessarily imply an atom at the other. This position will then have a multiplicity of 2, whereas a general point has a multiplicity of 4. These special positions are found at:

$$0, ½, z \quad ½, 0, z$$

The multiplicity is two, and the Wyckoff symbol is *c* (Table 6.8).

In a unit cell conforming to this group, there are also special positions associated with the tetrad axes, at the cell origin (0, 0), and at the cell centre (½, ½), (Figure 6.5a). The multiplicity of an atom located on a tetrad axis will be 1. These two positions are ½, ½, *z* and 0, 0, *z*.

The total number of positions in the unit cell is then as set out in Table 6.8.

6.5 BUILDING A STRUCTURE FROM A SPACE GROUP: Cs_3P_7

As in two dimensions, a crystal structure can be built from a motif plus a lattice. The motif is the minimal collection of atoms needed to generate the whole unit cell contents by application of the symmetry operators specified by the space group, or the **generator**. The positions of the atoms in a structure can be organised in a compact way by using the symmetry properties of the space group that is appropriate to the crystal structure.

In this example, we will consider how to build a structure from the space group $P4_1$, (No. 76). This is similar to the space group described in the previous section, *P*4, but here a four-fold screw axis along the **c**-axis replaces the ordinary tetrad. The space group diagrams are drawn in

Table 6.8 Positions in the space group *P*4

Multiplicity	Wyckoff letter	Site symmetry	Coordinates of equivalent positions
4	*d*	1	(1) x, y, z (2) $\bar{x}, \bar{y}, z$ (3) $\bar{y}, x, z$ (4) $y, \bar{x}, z$
2	*c*	2	0, ½, *z* ½, 0, *z*
1	*b*	4	½, ½, *z*
1	*a*	4	0, 0, *z*

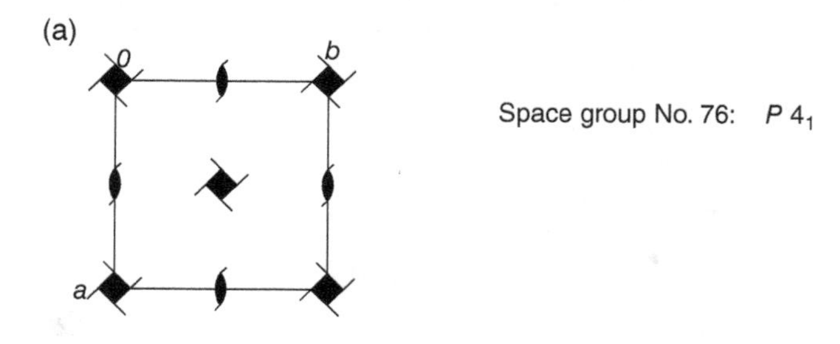

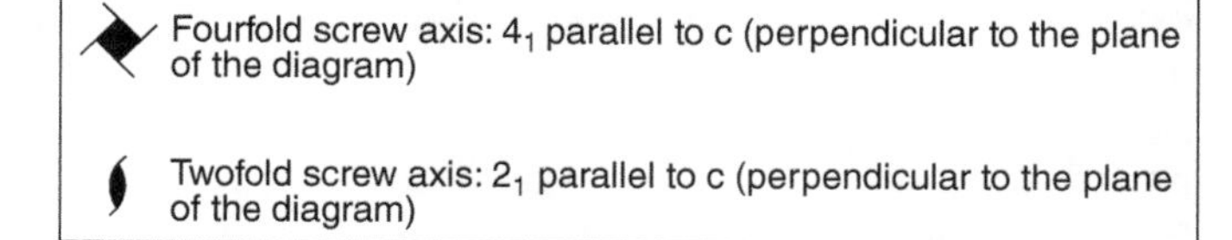

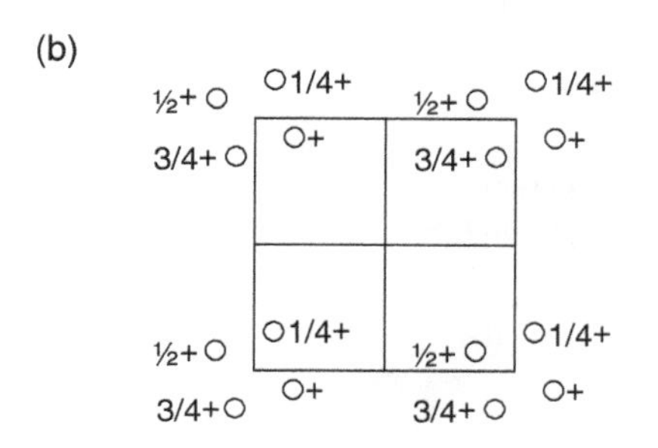

O+	Motif at arbitrary height +*z* (0<*z*<*c*)
O1/4+	Motif at arbitrary height *z* + *c*/4
O½+	Motif at arbitrary height *z* + *c*/2
O3/4+	Motif at arbitrary height *z* + 3*c*/4

Figure 6.6 Space group diagrams for space group No. 76, *P*4$_1$: (a) symmetry elements; (b) positions of motif generated by symmetry operations.

Figure 6.6. Note that the height of the motif is set out beside the symbol in Figure 6.6b. These heights can be obtained following the methods described for the 4_2 screw axis. Figure 6.6 is superficially very similar to the diagrams applicable to the space group *P*4 (Figure 6.5). However, the screw axis does not generate diads, but twofold screw axes, 2_1. The only positions available are the equivalent general positions, set out in Table 6.9.

The structure of caesium phosphide, Cs_3P_7, belongs in space group $P4_1$, and has lattice parameters $a = b = 0.9046$ nm, $c = 1.6714$ nm. The structure is tetragonal, and the screw axis runs parallel to the **c**-axis. All of the atoms must be placed in equivalent general positions, the coordinates of which are set out in Table 6.10, making *Z*, the number of Cs_3P_7 units in the unit cell, equal to four.

Note that there are three different Cs atoms specified, numbered Cs1, Cs2, and Cs3. The unit cell with just these three atoms present is drawn, projected down the **c**-axis (Figure 6.7a), down the **a**-axis (Figure 6.7b), and in perspective (Figure 6.7c). Note that the atom at height $z = 0$ is repeated at the top and bottom of the unit

Table 6.9 Positions in the space group $P4_1$

Multiplicity	Wyckoff letter	Site symmetry	Coordinates of equivalent positions
4	*a*	1	(1) x, y, z (2) $\bar{x}, \bar{y}, z + ½$ (3) $\bar{y}, x, z + ¼$ (4) $y, \bar{x}, z + ¾$

Table 6.10 Crystallographic data for Cs_3P_7 in the space group $P4_1$

Atom	Multiplicity and Wyckoff letter	Coordinates of atoms, *x*, *y*, *z*		
Cs1	4*a*	−0.2565	0.3852	0
Cs2	4*a*	0.4169	0.7330	0.8359
Cs3	4*a*	0.0260	0.8404	0.9914
P1	4*a*	0.790	0.600	0.8105
P2	4*a*	0.443	0.095	0.947
P3	4*a*	0.106	0.473	0.893
P4	4*a*	0.357	0.024	0.061
P5	4*a*	0.629	0.794	0.0318
P6	4*a*	0.998	0.341	0.705
P7	4*a*	0.011	0.290	0.840

Crystal system: tetragonal; lattice parameters: $a = 0.9046$ nm, $b = 0.9046$ nm, $c = 1.6714$ nm, $\alpha = \beta = \gamma = 90°$, $Z = 4$.
Source: Data taken from Meyer, T., Hoenle, W. and von Schnering, H.G. (1987). *Z. Anorg. Allg. Chem.* **552**: 69–80.

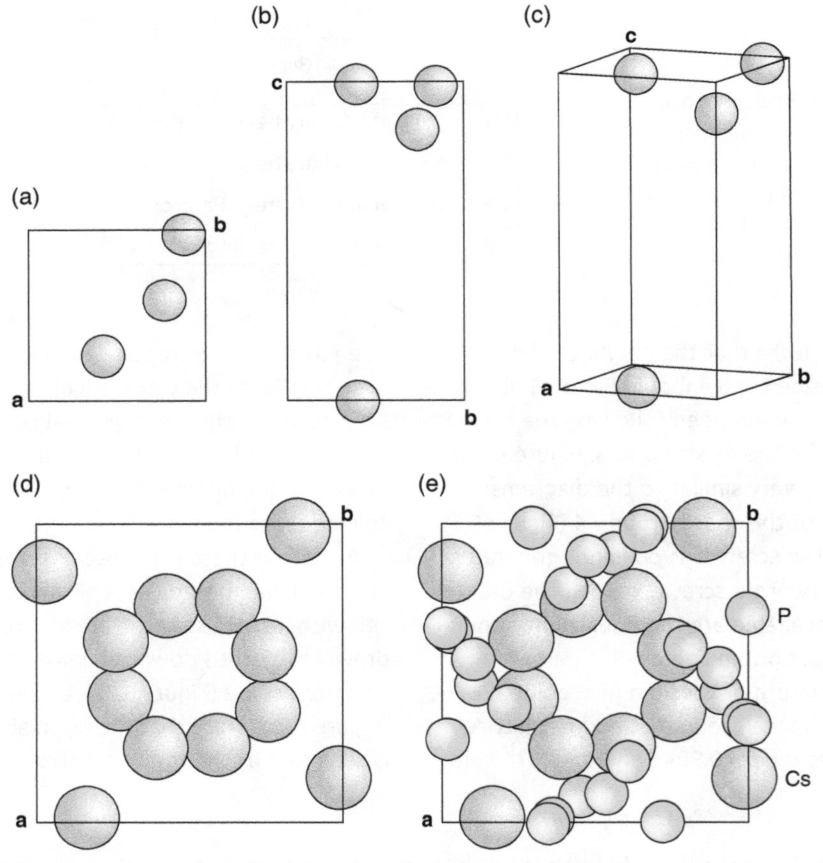

Figure 6.7 The structure of Cs_3P_7: (a) unit cell projected down the **c**-axis, atoms Cs1, Cs2, Cs3 only; (b) unit cell projected down the **a**-axis, atoms Cs1, Cs2, Cs3 only; (c) perspective view of unit cell, atoms Cs1, Cs2, Cs3 only; (d) unit cell projected down the **c**-axis, with all 12 Cs atoms present; (e) unit cell projected down the **c**-axis, with all atoms present.

cell. Each of these atoms will be replicated four times, by the action of the symmetry axis, so that the unit cell will contain 12 Cs atoms. The positions of these atoms are given in Table 6.11, listed as Cs1, Cs1_2, Cs1_3, etc., to show which atoms are derived from those in the original specification. The structure, with all the Cs atoms shown, projected down the **c**-axis, is drawn in Figure 6.7d.

Similarly, there are seven different P atoms listed, giving a total of 28 P in the unit cell when all the symmetry operations are completed. The positions of these atoms are obtained in the same way as those for the Cs atoms. The projection of the complete structure down the **c**-axis is drawn in Figure 6.7e.

Note that the use of the cell symmetry allows one to specify just 10 atoms instead of the total cell contents of 40 atoms. However, even in a relatively simple structure such as this one, it is difficult to obtain an idea of the disposition of the atoms in the unit cell, or how the crystal is built up. For this, computer graphics are necessary. In addition, a description in terms of polyhedra, as discussed in Chapter 8, often makes structures and relations between series of related phases clear.

Table 6.11 The positions of the Cs atoms in a unit cell of Cs_3P_7

Atom	Coordinates of atoms, *x, y, z*		
Cs1	0.7435	0.3852	0
Cs1_2	0.6148	0.7435	¼
Cs1_3	0.2565	0.6148	½
Cs1_4	0.3852	0.2565	¾
Cs2	0.4169	0.7330	0.8359
Cs2_2	0.2670	0.4169	0.0859
Cs2_3	0.5831	0.2670	0.3359
Cs2_4	0.7330	0.5831	0.5859
Cs3	0.0260	0.8404	0.9914
Cs3_2	0.1596	0.0260	0.2414
Cs3_3	0.9740	0.1569	0.4914
Cs3_4	0.8404	0.9740	0.7414

6.6 THE STRUCTURE OF DIOPSIDE, $MgCaSi_2O_6$

A more complex example of structure building, with four different atom types in the unit cell, is provided by the mineral diopside, $MgCaSi_2O_6$, which was one of the first crystal structures to be determined, during the early years of X-ray crystallography. The crystallographic data are given in Table 6.12.

The space group diagrams are given in Figure 6.8a and b. The convention for monoclinic crystals is to project the cell down the unique **b**-axis. The origin of the unit cell is chosen at a centre of symmetry, the **a**-axis points to the right and the **c**-axis down the page. The **b**-axis is then perpendicular to the plane of the figure.

Table 6.12 Crystallographic data for diopside, $CaMgSi_2O_6$

Atom	Multiplicity and Wyckoff letter	Coordinates of atoms, *x, y, z*		
Ca1	4*e*	0	0.3069	¼
Mg2	4*e*	0	0.9065	0.464
Si1	8*f*	0.284	0.0983	0.2317
O1	8*f*	0.1135	0.0962	0.1426
O2	8*f*	0.3594	0.2558	0.3297
O3	8*f*	0.3571	0.0175	0.9982

Crystal system: monoclinic; lattice parameters: a = 0.95848 nm, b = 0.86365 nm, c = 0.51355 nm, α = 90°, β = 103.98°, γ = 90°, Z = 4. *Source*: Data taken from Dove, M.T. (1989). *American Mineralogist* **74**: 774–779.

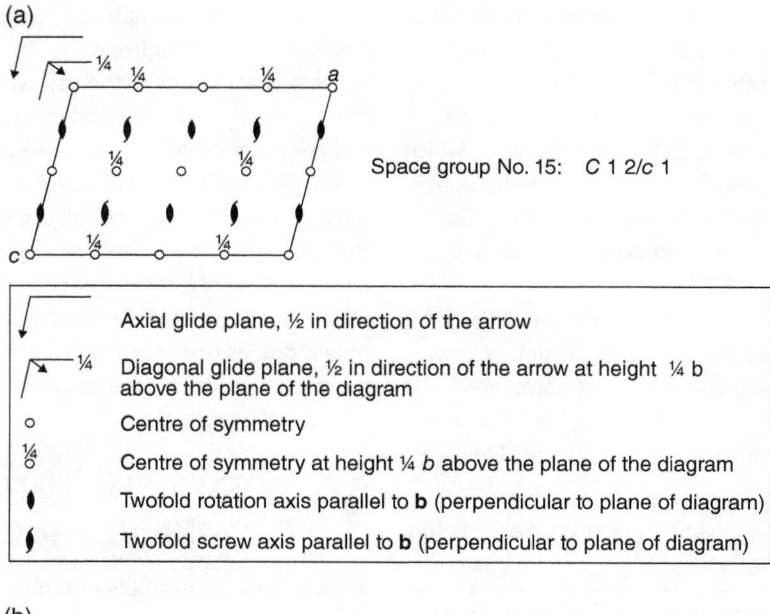

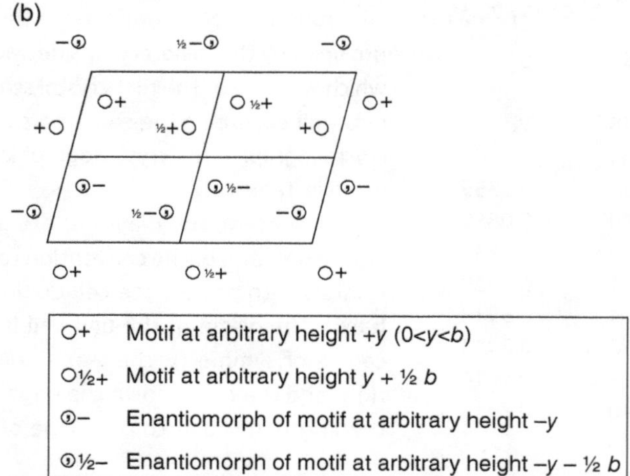

Figure 6.8 Space group diagrams for space group No. 15, *C* 1 2/*c* 1: (a) symmetry elements; (b) positions of motif generated by symmetry operations.

The space group symbol indicates that the most important symmetry element is a diad, which runs parallel to the **b**-axis. Perpendicular to this is a glide plane, the glide vector of which is parallel to the **c**-axis, and of value $c/2$, which generate a set of 2_1 screw axes. Heights by some symbols represent the height of the operator above the plane of the figure as fractions of the cell parameter b (Figure 6.8a). The action of the symmetry operators on a motif at a general position shows that this is sometimes transformed into its mirror image,

Table 6.13 Positions in the space group $C\,1\,2/c\,1$

Multiplicity	Wyckoff letter	Site symmetry	Coordinates of equivalent positions
8	*f*	1	(1) x, y, z (2) $\bar{x}, y, \bar{z} + ½$ (3) $\bar{x}, \bar{y}, \bar{z}$ (4) $x, \bar{y}, z + ½$ (5) $x + ½, y + ½, z$ (6) $\bar{x} + ½, y + ½, \bar{z} + ½$ (7) $\bar{x} + ½, \bar{y} + ½, \bar{z}$ (8) $x + ½, \bar{y} + ½, z$
4	*e*	2	$0, y, ¼; 0, \bar{y}, ¾$
4	*d*	$\bar{1}$	¼, ¼, ½; ¾, ¼, 0
4	*c*	$\bar{1}$	¼, ¼, 0; ¾, ¼, ½
4	*b*	$\bar{1}$	0, ½, 0; 0, ½, ½
4	*a*	$\bar{1}$	0, 0, 0; 0, 0, ½

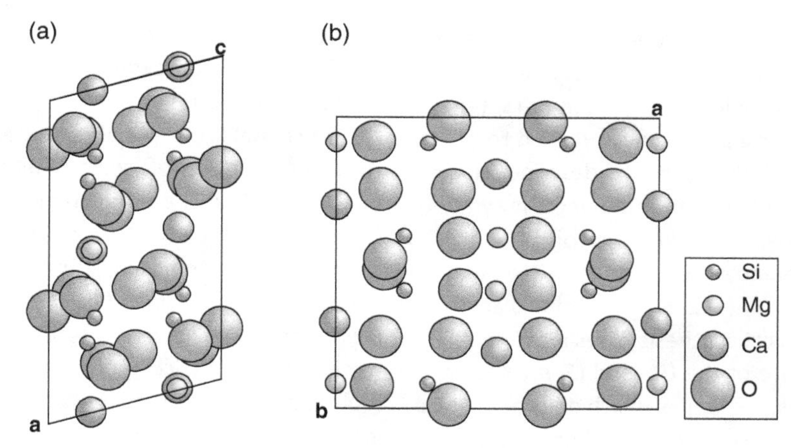

Figure 6.9 The structure of diopside, $CaMgSi_2O_6$: (a) unit cell projected down the **b**-axis; (b) unit cell projected down the **c**-axis.

shown by the presence of a comma within the circle that represents the motif. A motif above the plane of the figure is denoted by a + sign, and those below the plane by a – sign. Those designated ½+ and ½– are translated by a distance of ½ of the cell parameter *b* with respect to those unlabelled (Figure 6.8b). The equivalent positions in the cell are given in Table 6.13. The dots in the column listing the site symmetry indicate that the positions do not contribute to the site symmetry. When these are applied to the atoms listed in Table 6.12, the value of the space group approach to crystal structure building is apparent, because only six atoms are listed in Table 6.12, although the unit cell contains 40 atoms. The structure is projected down the unique **b**-axis (Figure 6.9a), and down the **c**-axis (Figure 6.9b). It is also clear that the structure is not easily visualised in this atomic representation. (See Chapter 8, Figure 8.17 for an alternative representation of the structure in terms of polyhedra.)

6.7 THE STRUCTURE OF ALANINE, $C_3H_7NO_2$

Alanine, 2-aminopropionic acid, $C_3H_7NO_2$, is an amino acid, and one of the naturally occurring building blocks of proteins. Naturally occurring amino acid molecules occur in only one configuration, the 'left-handed' form of the optically active enantiomers (see Section 5.5). This modification of alanine was described in older literature as L-alanine, or, L-α-aminopropionic acid, where the letter L describes the configuration of the molecule. (This nomenclature is confusing, as the prefix, L- is similar to the prefix *l*-, meaning laevorotatory and now superseded.) In fact, the natural molecule is dextrorotatory, and is written as *d*-alanine or (+)-alanine or (*S*)-(+)-2-aminopropionic acid, often shortened to *S*-alanine. The letter symbol *S* describes the configuration of the molecule and the symbol (+) records that it is dextrorotatory. The laevorotatory molecular form is written as (*R*)-alanine or (*R*)-(−)-2-aminopropionic acid. Normal chemical synthesis gives a racemic (equal) mixture of the (*R*)- and (*S*)-molecules, designated (*R*, *S*)-alanine, which does not rotate the plane of linearly polarised light. (In older literature, this is also called DL-alanine.)

Of central interest is the arrangement in space of the four groups surrounding the chiral carbon atom, and in particular, which of the two configurations causes light to rotate to the left as against to the right. To determine the absolute configuration of the molecule it is simplest to start with crystals of the optically active (*S*)-alanine. The presence of a single enantiomer in a crystal has implications for the possible symmetry of the crystal, and all enantiomorphous molecules crystallise in one of the space groups that lack inversion centres or mirror planes. There are only 65 appropriate space groups. The crystallographic data for naturally occurring (*S*)-alanine, which adopts the orthorhombic space group $P\,2_12_12_1$, No. 19, is given in Table 6.14.

The space group diagrams are given in Figure 6.10. The space group symbol indicates that the most important symmetry elements are two-fold screw axes running throughout the unit cell parallel to the three axes. These symmetry operators turn a motif above the plane of the diagram into one below the plane, shown by the presence of +, ½ +, − and ½ −, where the fractions represent half of the appropriate unit cell edge. The equivalent positions in the cell are given in Table 6.15. It is seen that only one general position occurs, with multiplicity of four, generating four molecules of alanine in the unit cell.

The crystal structure of the optically active (*S*)-alanine consists of layers of alanine molecules, all of the same enantiomer (Figure 6.11). The structure displayed as a distribution of atoms in the unit cell is not helpful (see, e.g. urea in Chapter 1), and it is preferable to delineate molecules using the familiar ball and stick approach. This allows the molecular geometry and packing to be visualised. Even so, the structure of a molecule is not easy to see from a two-dimensional figure, but in fact, the hydrogen atom is below the central carbon atom, the NH_2 group lies in front and to the right of it, the CH_3 group lies to the left, and the COOH group lies to the rear. The configuration is identical to that given in Figure 4.1.

A comparison of the optically inactive crystals of synthetic (*R*, *S*)-alanine with the optically active natural crystals of is of interest. The crystallographic data for (*R*, *S*)-alanine are given in Tables 6.16 and 6.17. Because mirror image forms of the alanine molecule pack into the unit cell, the space group should contain

Table 6.14 Crystallographic data for (*S*)-alanine [L-alanine, (*S*)-(+)-2-aminopropionic acid]

Atom	Multiplicity and Wyckoff letter	Coordinates of atoms, *x*, *y*, *z*		
C1	4*a*	0.55409	0.14081	0.59983
C2	4*a*	0.46633	0.16105	0.35424
C3	4*a*	0.25989	0.09069	0.30332
N1	4*a*	0.64709	0.08377	0.62438
O1	4*a*	0.72683	0.2558	0.3297
O2	4*a*	0.44086	0.18403	0.76120
H1	4*a*	0.70360	0.05720	0.19790
H2	4*a*	0.77830	0.18970	0.20860
H3	4*a*	0.58070	0.14920	0.01100
H4	4*a*	0.42130	0.24980	0.33590
H5	4*a*	0.19360	0.11090	0.13400
H6	4*a*	0.12610	0.10560	0.43570
H7	4*a*	0.30410	0.00510	0.30560

Crystal system: orthorhombic; lattice parameters: $a = 0.5928$ nm, $b = 1.2260$ nm, $c = 0.5794$ nm, $Z = 4$. Space group, $P2_12_12_1$, No. 19. *Source*: Data taken from Destro, R., Marsh, R.E., and Bianchi, R. (1989). *Journal of Physical Chemistry* **92**: 966–973.

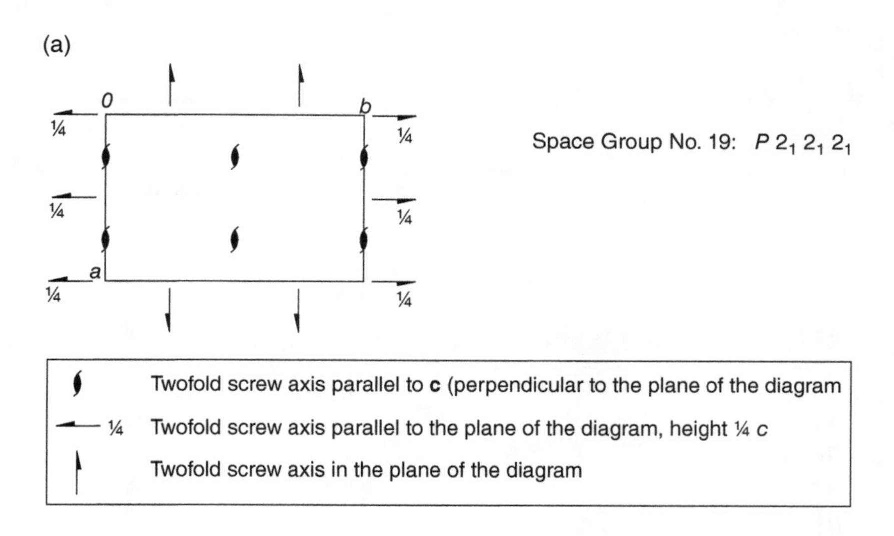

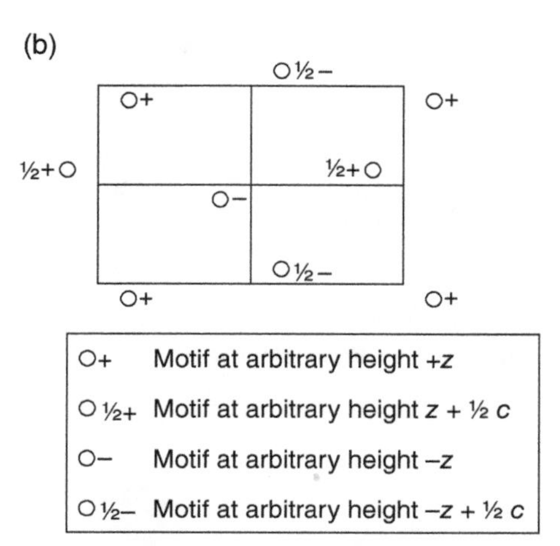

Figure 6.10 Space group diagrams for space group No. 19, $P\,2_1\,2_1\,2_1$: (a) symmetry elements; (b) positions of motif generated by symmetry operations.

Table 6.15 Positions in the space group $P\,2_1\,2_1\,2_1$

Multiplicity	Wyckoff letter	Site symmetry	Coordinates of equivalent positions
4	*a*	1	(1) x, y, z (2) $\bar{x} + ½, \bar{y}, z + ½$ (3) $\bar{x}, y + ½, \bar{z} + ½$ (4) $x + ½, \bar{y} + ½, \bar{z}$

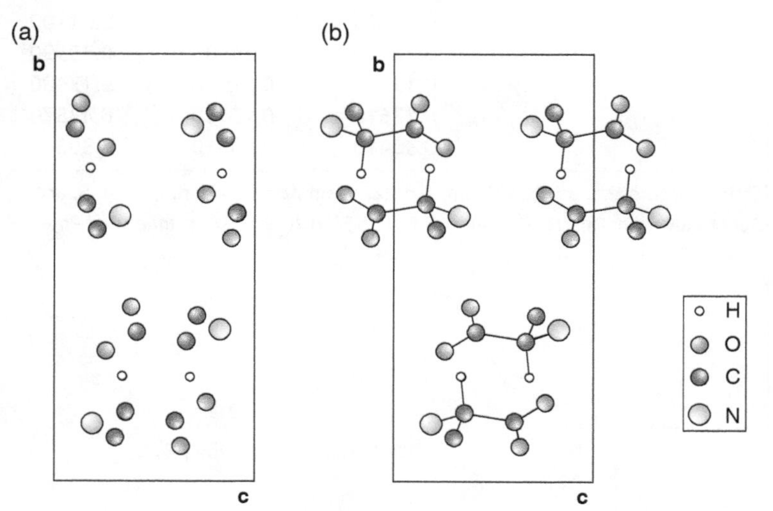

Figure 6.11 The structure of optically active (*S*)-alanine projected down the **a**-axis: (a) atom distribution; (b) conventional chemical bonds shown. Most of the hydrogen atoms have been omitted for clarity, and only that attached to the chiral carbon atom is included.

Table 6.16 Crystallographic data for (*R*, *S*)-alanine [DL-alanine, (*R*, *S*)-2-aminopropionic acid]

Atom	Multiplicity and Wyckoff letter	Coordinates of atoms, *x*, *y*, *z*		
C1	4*a*	0.66363	0.22310	0.33800
C2	4*a*	0.64429	0.31330	0.58180
C3	4*a*	0.59233	0.02160	0.29160
N1	4*a*	0.63948	0.39760	0.16590
O1	4*a*	0.58960	0.48620	0.60450
O2	4*a*	0.68614	0.20090	0.74090
H1	4*a*	0.74190	0.18040	0.32370
H2	4*a*	0.60600	−0.03080	0.13870
H3	4*a*	0.51530	0.06050	0.30690
H4	4*a*	0.61040	−0.09300	0.29970
H5	4*a*	0.68150	0.51600	0.19420
H6	4*a*	0.56800	0.43510	0.17410
H7	4*a*	0.65440	0.34630	0.02610

Crystal system: orthorhombic; lattice parameters: $a = 1.2026$ nm, $b = 0.6032$ nm, $c = 0.5829$ nm, $Z = 4$. Space group, $P\,n\,a\,2_1$, No. 33. *Source*: Data taken from Subha Nandhini, M., Krishnakumar, R.V., and Natarajan, S. (2001). *Acta Crystallographica* **C57**: 614–615.

Table 6.17 Positions in the space group $P\,n\,a\,2_1$, No. 33

Multiplicity	Wyckoff letter	Site symmetry	Coordinates of equivalent positions
4	*a*	1	(1) x, y, z (2) $\bar{x}, \bar{y}, z + ½$ (3) $x + ½, \bar{y} + ½, z$ (4) $\bar{x} + ½, y + ½, z + ½$

(a)

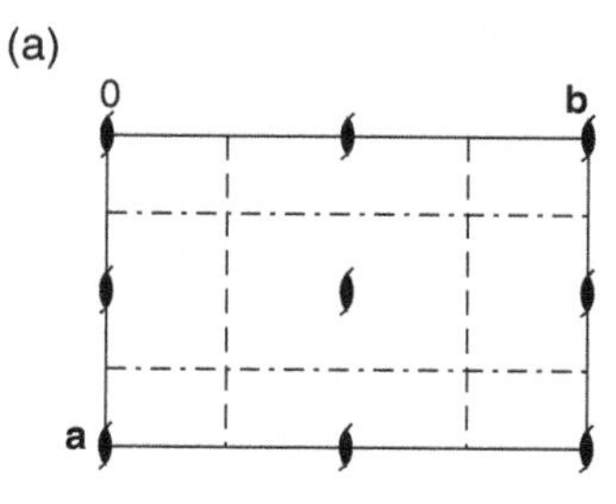

Space Group No. 33: $P\,n\,a\,2_1$

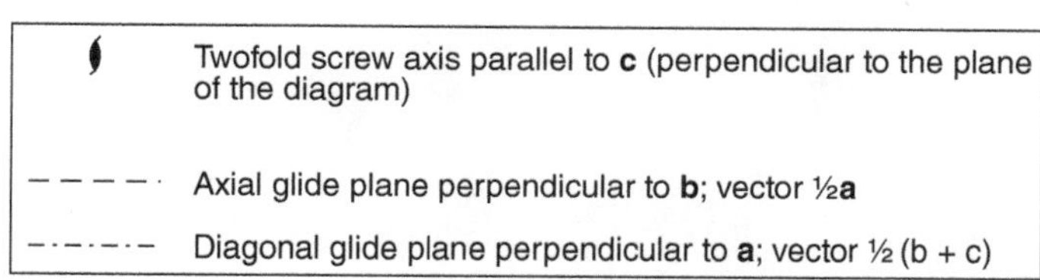

(b)

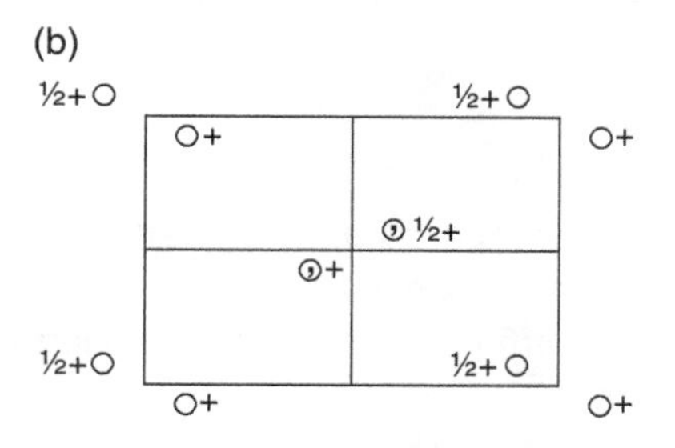

O+	Motif at arbitrary height +*z*
O ½+	Motif at arbitrary height *z* + ½ **c**
⊙+	Enantiomorph of motif at arbitrary height **z**
⊙ ½+	Enantiomorph of motif at arbitrary height *z* + ½ **c**

Figure 6.12 Space group diagrams for space group No. 33, $P\,n\,a\,2_1$: (a) symmetry elements; (b) positions of motif generated by symmetry operations.

mirror or glide planes. The space group diagrams (Figure 6.12) confirm this supposition. The symmetry elements consist of a twofold screw axis along the **c**-axis, an axial glide plane perpendicular to the **b**-axis, and a diagonal glide plane perpendicular to the **a**-axis (Figure 6.12a). The axial glide plane has the effect of moving the motif by a vector ½ **a**, parallel to the **a**-axis,

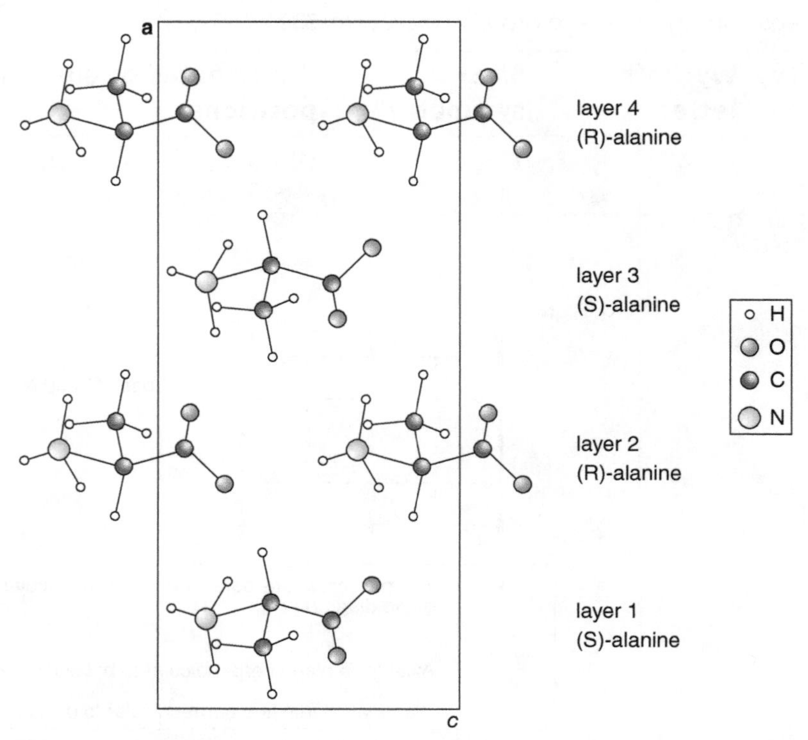

Figure 6.13 The structure of (*R*, *S*)-alanine projected down the **b**-axis. All hydrogen atoms are included.

and the diagonal glide moves the motif by a vector ½ **b** parallel to the **b**-axis and ½ **c** parallel to the **c**-axis (Figure 6.12b). The effect of these symmetry operations is to create enantiomorphic copies of any motif in the unit cell. If the motif is taken as an alanine molecule, the symmetry elements present in the space group will generate a structure that contains layers of alanine molecules of one type interleaved with layers of the other enantiomorph.

The crystal structure confirms this arrangement (Figure 6.13). As with (*S*)-alanine, the structure contains layers of alanine molecules. Each layer contains just one of the enantiomers. The molecules in layer 1 are identical to those in layer 3, and both consist of the naturally occurring configuration, (*S*)-alanine. Enantiomers with the opposite configuration, (*R*)-alanine, form layers 2 and 4. (To show this difference, more hydrogen atoms have been included compared with Figure 6.11.) The optical activity of the crystal is internally suppressed by the presence of the two forms of the molecule.

The comparison of these two structures thus provides a good insight into the relationship between structure, symmetry, and physical properties of crystals. Be careful to note, though, that the unit cell and symmetry properties of the crystal are a consequence of the way in which the molecules pack into a minimum energy arrangement.

ANSWERS TO INTRODUCTORY QUESTIONS

What symmetry operation is associated with a screw axis?

Screw axes are rototranslational symmetry elements. A screw axis of order *n* operates on an object by (a) a rotation of $2\pi/n$ anticlockwise and then a translation by a distance **t** parallel to the axis, in a positive direction. The value of *n* is the order of the screw axis. For example, a screw axis running parallel to the **c**-axis in an orthorhombic crystal would entail an anticlockwise rotation in the **a**-**b**-plane, (001), followed by a translation parallel to +**c**. This is a right-handed screw rotation. If the rotation component of the operator is applied *n* times, the total rotation is equal to 2π. The value of *n* is the principal symbol used for a screw axis, together with a subscript giving the fraction of the repeat distance moved in the translation. Thus, a three-fold screw axis 3_1 involves an anticlockwise rotation of $2\pi/3$, followed by a translation of a distance equal to 1/3 the lattice repeat in a direction parallel to the axis. The three-fold screw axis 3_2 involves an anticlockwise rotation of $2\pi/3$, followed by a translation of a distance equal to 2/3 the lattice repeat in a direction parallel to the axis.

What is a crystallographic space group?

A crystallographic space group describes the symmetry of a three-dimensional repeating pattern, such as found in a crystal. Each space group is thus a collection of symmetry operators. There are 230 space groups, which can be derived by combining the 32 crystallographic point groups with the 14 Bravais lattices. Any three-dimensional crystal unit cell must have an internal symmetry that corresponds to one of the 230 crystallographic space groups. By varying the motif, an infinite number of crystal structures can be created, but they will all possess a combination of symmetry elements corresponding to one of the crystallographic space groups.

What are Wyckoff letters?

If an atom is placed in the unit cell of a crystal at a position, or site, (x, y, z), identical copies of the atom must, by operation of the symmetry elements present, lie at other sites, (x_1, y_1, z_1), (x_2, y_2, z_2), and so on. A Wyckoff letter is a site label, and each different type of site in the unit cell is given a different Wyckoff letter.

The maximum number of copies of a site is generated when the initial atom is at a general position in the unit cell, and the resulting set are called general equivalent positions. If, however, position of the atom coincides with a symmetry element, a special position, the symmetry operators will generate fewer copies. The number of copies of an atom that are generated for each distinguishable site in the unit cell is called the site multiplicity. The sites are listed in order of decreasing multiplicity, and Wyckoff letters are added, starting at the lowest position in the list, with letter *a* and then proceeding alphabetically to the equivalent general positions, which are labelled last, as *b*, *c*, *d*, and so on, depending upon the number of different sites in the unit cell.

PROBLEMS AND EXERCISES

Quick Quiz

1. The translation vector associated with a glide plane is:
   a. A quarter of a lattice translation vector
   b. A half of a lattice translation vector
   c. A full lattice translation vector
2. The translation vector associated with a 3_2 axis is:
   a. Two-thirds of a lattice translation vector

b. Half of a lattice translation vector
c. Two lattice translation vectors

3. The number of crystallographic space groups is:
   a. 210
   b. 230
   c. 250
4. The first place in the Hermann-Mauguin symbol for a space group refers to:
   a. The crystal system (cubic, etc.)
   b. The point group
   c. The Bravais lattice
5. The last three places in the Hermann-Mauguin symbol for a space group refer to:
   a. Axes
   b. Planes
   c. Directions
6. The letter *d* in a space group symbol means:
   a. A diamond glide plane
   b. A diagonal glide plane
   c. A double glide plane
7. A rototranslation axis in a space group symbol is represented by:
   a. n ($n = 2, 3, 4, 6$)
   b. $\bar{n}$ ($n = 3, 4, 6$)
   c. n_p ($n = 2, 3, 4, 6$, $p = 1, 2, 3, 4, 5$)
8. The symbol $2/m$ in a space group symbol means:
   a. A mirror runs parallel to a diad axis
   b. The normal to a mirror plane runs parallel to a diad axis
   c. A mirror plane contains a diad axis
9. The smallest part of the unit cell that will reproduce the whole cell when the symmetry operations are applied is:
   a. The motif
   b. The asymmetric unit
   c. The point group
10. The Wyckoff letter *a* is given to:
    a. The equivalent general sites
    b. The equivalent sites with the highest multiplicity
    c. The equivalent sites with the lowest multiplicity

Calculations and Questions

6.1. Draw diagrams similar to Figure 6.1f for the axes 3, $\bar{3}$, 3_1, 3_2.

6.2. Write out the full Hermann-Mauguin space group symbols for the short symbols: (a) monoclinic, No. 7, *Pc*; (b) orthorhombic, No. 47, *Pmmm*; (c) tetragonal, No. 133, $P\,4_2/n\,b\,c$; (d) hexagonal, No. 193, $P\,6_3/mcm$; (e) cubic, No. 225, $F\,m\,\bar{3}\,m$.

6.3. Tetragonal space group No. 79 has space group symbol *I* 4.
   a. What symmetry elements are specified by the space group symbol?
   b. Draw diagrams equivalent to Figure 6.5a and b for the space group.
   c. What new symmetry elements appear that are not specified in the space group symbol?

6.4. Determine the general equivalent positions for a motif at *x, y, z* in the tetragonal space group No. 79, *I* 4, (see Question 6.3).

6.5. Structural data for PuS_2 are given in the following table:

Atom	Multiplicity and Wyckoff letter	Coordinates of atoms, *x, y, z*		
Pu1	1*a*	0	0	0
Pu2	1*b*	½	½	0.464
S1	1*a*	0	0	0.367
S2	1*b*	½	½	0.097
S3	2*c*	½	0	0.732

Crystal system: tetragonal; lattice parameters: $a = b = 0.3943$ nm, $c = 0.7962$ nm. Space group, *P*4*mm*, No. 99.
Source: Data taken from Marcon, J.P. and Pascard, R. (1966). *C. R. Acad. Sci.* **262**: 1679–1681.

The equivalent positions in this space group are:

Multiplicity	Wyckoff letter	Site symmetry	Coordinates of equivalent positions
8	*g*	1	(1) x, y, z (2) $\bar{x}, \bar{y}, z$ (3) $\bar{y}, x, z$ (4) $y, \bar{x}, z$ (5) $x, \bar{y}, z$ (6) $\bar{x}, y, z$ (7) $\bar{y}, \bar{x}, z$ (8) y, x, z
4	*f*	.*m*.	x, ½, z; $\bar{x}$, ½, z; ½, x, z; ½, $\bar{x}$, z
4	*e*	.*m*.	x, 0, z; $\bar{x}$, 0, z; 0, x, z; 0, $\bar{x}$, z
4	*d*	..*m*	x, x, z; $\bar{x}, \bar{x}, z$; $\bar{x}, x, z$; $x, \bar{x}, z$
2	*c*	2*mm*	½, 0, z; 0, ½, z
1	*b*	4*mm*	½, ½, z
1	*a*	4*mm*	0, 0, z

The space group diagrams are:

(a)

(b)

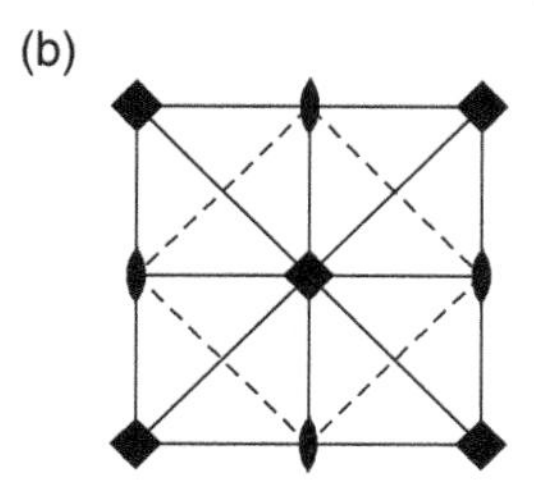

a. What symmetry elements are specified in the space group symbol?
b. What additional symmetry elements arise due to the operation of those in (a)?

c. How many Pu atoms are there in the unit cell?
d. What are the positions of the Pu atoms?
e. How many S atoms are there in the unit cell?
f. What are the positions of the S atoms?

6.6. The space group for perovskite, $CaTiO_3$, is No. 62. In the standard setting this is *Pnma*, but the space group quoted is usually *Pbmn*. Use the transformation (**cab**) to obtain the atom positions in the alternative space group setting.

6.7. The (abbreviated) crystallographic data for Sr_2TiO_4 are:

Atom	Multiplicity and Wyckoff letter	Coordinates of atoms, *x*, *y*, *z* (0, 0, 0) + (½, ½, ½) +		
Sr1	4*e*	0	0	0.3532
Ti1	2*a*	½	½	½
O1	4*c*	½	0	½
O2	4*e*	½	½	0.6558

Crystal system: tetragonal; lattice parameters: a = 0.3933 nm, c = 1.27535 nm, Z = 4.

Space group, *I* 4/*mmm*, No. 139.

Data taken from K. Kamamura et al. (2015). *Inorganic Chemistry*, **54**, 3896.

a. Sketch the atom positions in the unit cell.
b. What is the relationship of the structure to that of $SrTiO_3$? (see Appendix B for the details for the ideal perovskite structure of $SrTiO_3$, and Section 9.7).

Chapter 7
Diffraction and Crystal Structure Determination

What crystallographic information does Bragg's law give?
What is an atomic scattering factor?
What are the advantages of neutron diffraction over X-ray diffraction?

Radiation incident upon a crystal is scattered in a variety of ways. When the wavelength of the radiation is similar to that of the atom spacing in a crystal, the scattering, which is termed **diffraction**, gives rise to a set of well-defined beams arranged with a characteristic geometry, to form a **diffraction pattern**. The positions and intensities of the diffracted beams are a function of the arrangements of the atoms in space and some other atomic properties, especially, in the case of X-rays, the atomic number of the atoms. Thus, if the positions and the intensities of the diffracted beams[1] are recorded, it is possible to deduce the arrangement of the atoms in the crystal, that is, the crystal structure that gave rise to the pattern.

[1] Diffracted beams are frequently referred to as 'reflections', 'spots' or 'lines'. The use of the word 'reflection' stems from the geometry of diffraction (see Section 7.1, Bragg's Law). The use of the terms 'spots' and 'lines' arose because X-ray diffraction patterns were once recorded on photographic film. Single crystals gave rise to a set of spots, whilst polycrystalline samples gave a series of rings or lines on the film. The terms 'reflections', 'lines' or 'spots' will be used here as synonyms for 'beams'.

Crystals and Crystal Structures, Second Edition. Richard J. D. Tilley.

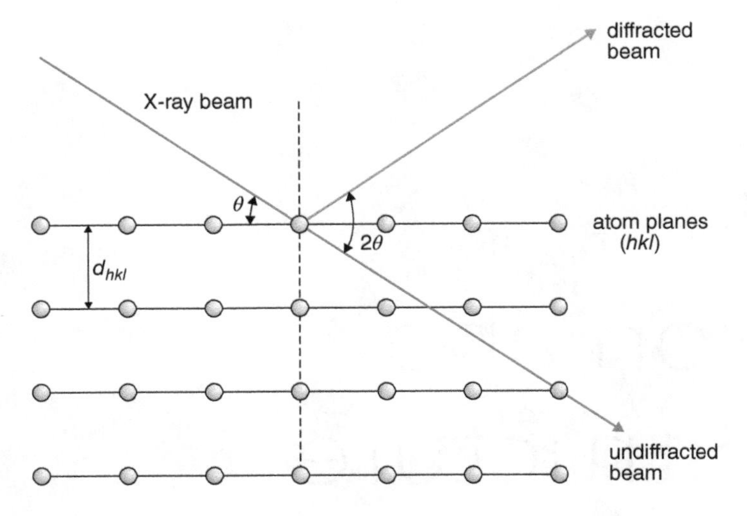

Figure 7.1 The geometry of Bragg's law for the diffraction of X-rays from a set of crystal planes, (*hkl*), with interplanar spacing d_{hkl}.

Historically, crystal structures have been mainly determined using the diffraction of X-rays, supplemented by neutron and electron diffraction, which give information that X-ray diffraction cannot supply.[2] The first crystal structure determined was that of NaCl, by W.H and W.L. Bragg, in 1913. In 1957 the structure of penicillin was published, and in 1958 the first three-dimensional structure of a protein, myoglobin, was determined. Currently the structures of many complex proteins and related biomolecules have been published, most recently using the technique of cryo-electron microscopy (see Section 7.16), allowing for great advances in the understanding of the biological function of these vital molecules.

[2] Note that it is crystallographic convention to use the Ångström unit, Å, as the unit of length. The conversion factor is 1 Å = 10^{-10} m; 1 nm = 10 Å. Here the SI unit, nm, will mainly be used, but Å may be employed from time to time, to avoid unnecessary confusion with crystallographic literature.

7.1 THE OCCURRENCE OF DIFFRACTED BEAMS: BRAGG'S LAW

A beam of radiation will only be diffracted when it impinges upon a set of planes in a crystal, defined by the Miller indices (*hkl*), if the geometry of the situation fulfils quite specific conditions, defined by Bragg's law:

$$n\lambda = 2d_{hkl} \sin\theta \qquad (7.1a)$$

where n is an integer, λ is the wavelength of the radiation, d_{hkl} is the interplanar spacing (the perpendicular separation) of the (*hkl*) planes, and θ is the diffraction angle or **Bragg angle** (Figure 7.1). The angle between the direction of the incident and diffracted beam is equal to 2θ. Bragg's law defines the conditions under which diffraction occurs, and gives the **position** of a diffracted beam, without any reference to its intensity.

There are a number of features of Bragg's law that need to be emphasised. Although the geometry of Figure 7.1 is identical to that of reflection, the physical process occurring is diffraction, and the angle of incidence conventionally used in optics to describe reflection is the complementary angle to that used in the Bragg equation. Moreover, there is no constraint on the angle at which a mirror will reflect light. However, planes of atoms will only diffract radiation when illuminated at the angle $\sin^{-1}(n\lambda/2d_{hkl})$.

Equation 7.1a includes an integer n, which is the **order** of the diffracted beam. Crystallographers take account of the different orders of diffraction by changing the values of (hkl) to fit. For example, the first-order reflection from (111) planes occurs at an angle given by:

$$\sin\theta\,(\text{1st}-\text{order } 111) = 1\lambda/2d_{111}$$

The second-order reflection from the same set of planes then occurs at an angle:

$$\sin\theta\,(\text{2nd}-\text{order } 111) = 2\lambda/2d_{111} = \lambda/d_{111}$$

However, $d_{111} = 2d_{222}$ so this can be written more simply as:

$$\sin\theta\,(222) = 1\lambda/2d_{222}$$

Similarly, the third-order reflection from (111) planes is at an angle:

$$\sin\theta\,(\text{3rd}-\text{order } 111) = 3\lambda/2d_{111}$$

However, $d_{111} = 3d_{333}$, so this can be written more simply as:

$$\sin\theta\,(333) = 1\lambda/2d_{333}$$

The same is true for all planes, so that the Bragg equation used in crystallography is simply:

$$\lambda = 2d_{hkl}\sin\theta \qquad (7.1b)$$

where d_{hkl} is the spacing of the appropriate hkl planes.

Within a crystal there are an infinite number of sets of atom planes, and Bragg's law applies to all these. Thus, if a crystal in a beam of radiation is rotated, each set of planes will, in its turn, diffract the radiation, when the value of $\sin\theta$ becomes appropriate. This is the principle by which diffraction data are collected for the whole of the crystal. The arrangement of the diffracted beams, when taken together, is the **diffraction pattern** of the crystal.

Measurement of the diffraction pattern of a crystal will thus provide a list of d_{hkl} values. Allocating a value hkl to each diffracted beam, a process called **indexing** the diffraction pattern, allows the unit cell of the structure to be obtained.

7.2 THE GEOMETRY OF THE DIFFRACTION PATTERN

In order to obtain a picture of the diffraction pattern which will arise when a crystal is irradiated with an incident beam of X-rays, electrons or neutrons, a graphical construction first described by Ewald, can be employed.

(i) Draw the reciprocal lattice of the crystal, in an orientation equivalent to that of the crystal being irradiated (Figure 7.2a).
(ii) Draw a vector of length $1/\lambda$ from the origin of the reciprocal lattice in a direction parallel, and in an opposite direction, to the beam of radiation, where λ is the wavelength of the radiation (Figure 7.2b).
(iii) From the end of the vector, construct a sphere, called the **Ewald sphere** (shown as a circle in cross-section), of radius $1/\lambda$. Each reciprocal lattice point that the sphere (circle) touches (or nearly touches) will give a diffracted beam (Figure 7.2c and d).

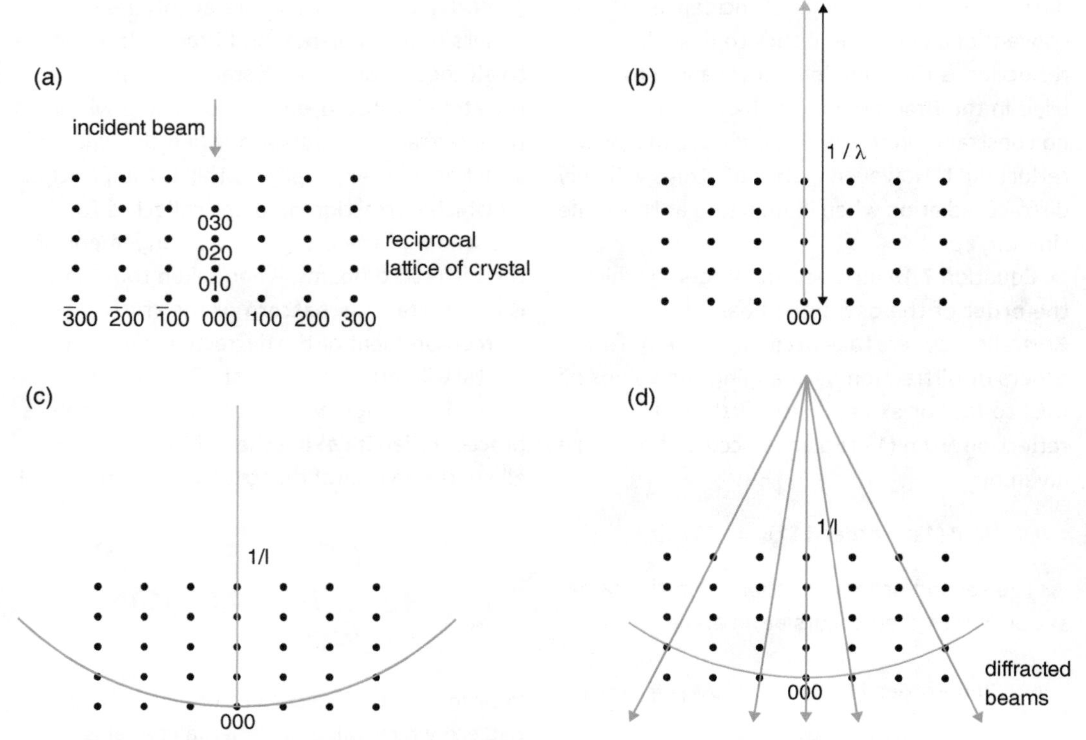

Figure 7.2 The Ewald construction: (a) the reciprocal lattice; (b) a vector of length 1/λ, drawn parallel to the beam direction; (c) a sphere passing through the 000 reflection, drawn using the vector in (b) as radius; (d) the positions of the diffracted beams.

Note that this construction is equivalent to Bragg's law. The distance of the *hkl* reciprocal lattice point from the origin of the reciprocal lattice is given by $1/d_{hkl}$, where d_{hkl} is the interplanar spacing of the (*hkl*) planes (Figure 7.3). The diffraction angle, θ, then conforms to the geometry:

$$\sin\theta = (1/2d_{hkl})/(1/\lambda) = \lambda/2d_{hkl}$$

Rearrangement leads to the Bragg equation:

$$2d_{hkl}\sin\theta = \lambda$$

The utility of the Ewald method of visualising a diffraction pattern is best understood via the formation of an electron diffraction pattern. This is because the Ewald sphere for electron diffraction is much larger than the spacing between the reciprocal lattice points. For electrons accelerated through 100 kV, which is typical for an electron microscope, the electron wavelength is 0.00370 nm, giving the Ewald sphere a radius of 270.27 nm^{-1}. The lattice parameter of copper (see Chapter 1), which can be taken as illustrative of many inorganic compounds, is 0.360 nm, giving a reciprocal lattice spacing $a*$ of 2.78 nm^{-1}. Because of this disparity in size, the surface of the Ewald sphere is nearly flat and lattice points in the plane, which is perpendicular to the electron beam and which passes through the origin, will touch the

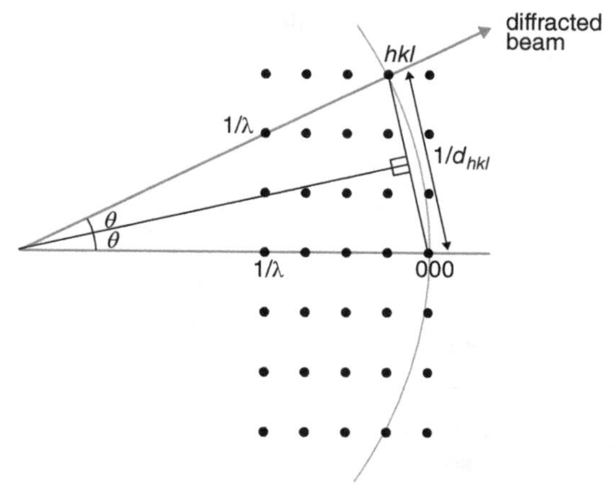

Figure 7.3 The geometry of the Ewald construction.

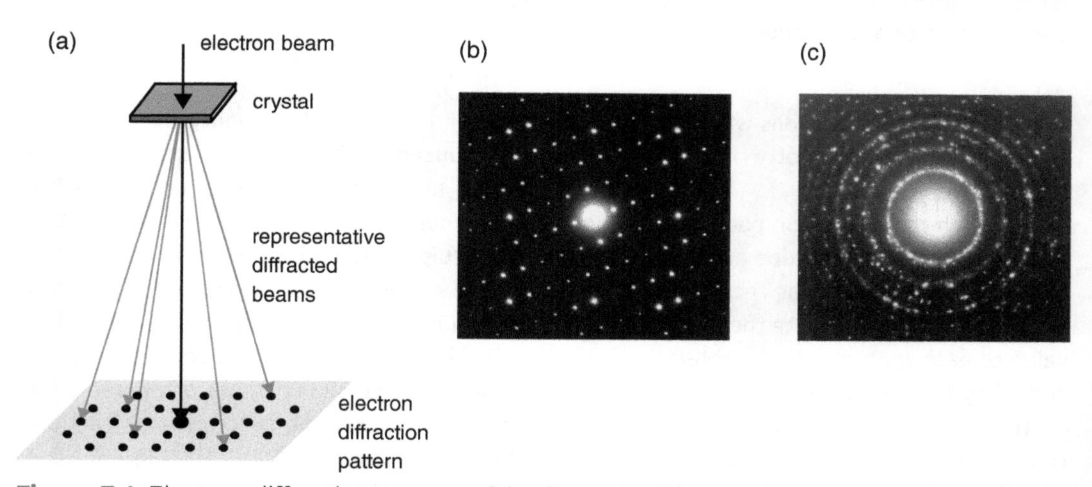

Figure 7.4 Electron diffraction patterns: (a) schematic diagram showing the formation of an electron diffraction pattern in an electron microscope; (b) an electron diffraction pattern from single crystal of $WNb_{12}O_{33}$; (c) electron diffraction pattern from polycrystalline TiO_2.

Ewald sphere. Hence all of the diffracted beams corresponding to these points will be produced simultaneously. If the diffracted beams are intercepted by a screen and displayed visually, the arrangement of diffraction spots will resemble a plane section through the reciprocal lattice, at a scale dependent upon the geometry of the arrangement (Figure 7.4a and b). Tilting the crystal will allow other sections of the reciprocal lattice to be visualised, and so the whole reciprocal lattice can be built up. (The mechanics of producing an observable diffraction pattern via the lenses in the microscope are described in Section 7.15.)

If the material examined consists of a powder of randomly arranged small crystallites, each will produce its own pattern of reflections. When large numbers of crystallites are present,

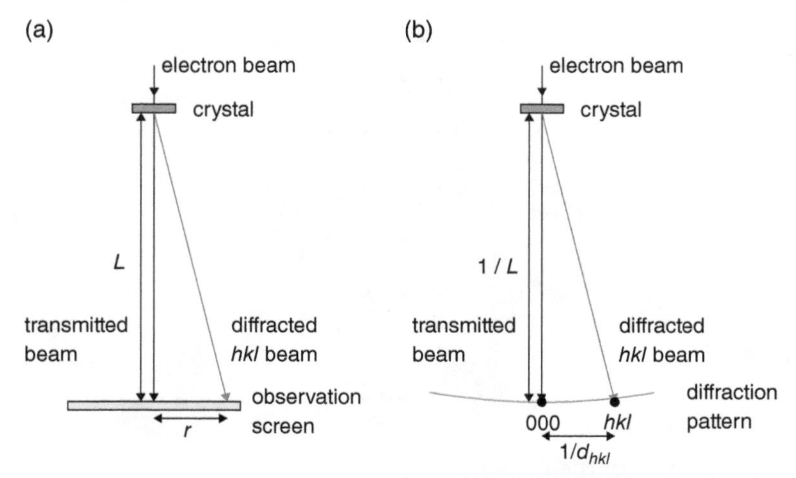

Figure 7.5 Comparison of real space and reciprocal space formation of diffraction patterns: (a) schematic formation of a diffraction pattern in an electron microscope; (b) the Ewald construction of a diffraction pattern.

all reciprocal lattice sections will be present. In this case, a pattern of spotty rings will be seen (Figure 7.4c).

Although the diffraction patterns obtained from an electron microscope are easy to understand in terms of the reciprocal lattice, it is still necessary to allocate the appropriate *hkl* value to each spot in order to obtain crystallographic information; that is, the pattern must be indexed. A comparison of the **real space** situation, in an electron microscope, with the **reciprocal space** equivalent, the Ewald sphere construction (Figure 7.5), shows that the relationship between the d_{hkl} values of the spots on the recorded diffraction pattern is given by the simple relationship:

$$r/L \approx (1/d_{hkl})/(1/\lambda)$$

$$d_{hkl} \approx \lambda L/r$$

where r is the distance from the origin of the diffraction pattern, 000, to the *hkl* spot and L is the distance over which the diffraction pattern is projected. In reality, as the pattern is produced by the action of lenses, the value of L is not readily accessible, and the pair of constants λL, called the **camera constant** for the equipment, is determined experimentally by the observation of a ring pattern from a known material. A polycrystalline evaporated thin film of TlCl, which adopts the CsCl structure, $a = 0.3834$ nm, has often been used for this purpose. The successive diffraction rings are indexed as 100, 110, 111, 200, 210, 211..., with the smallest first. The *d*-values of each plane are easily calculated using the formula (Table 2.4):

$$1/(d_{hkl})^2 = \left(h^2 + k^2 + l^2\right)/a^2 = (r/\lambda L)^2$$

A measurement of the ring diameters allows a value of λL to be calculated.

To index an electron diffraction pattern, that is, to give each spot appropriate *hkl* values, is straightforward when a value of the camera constant λL is known. The distance from the centre of the pattern to the reflection in question is measured, to give a value of r, and this is converted into a value of d using

$$d = \lambda L / r$$

A list of the d_{hkl}-values for the material under investigation is calculated using the equations given in Chapter 2, and the measured value of d is identified as d_{hkl} using the list. When two or more diffraction spots have been satisfactorily identified, the whole pattern can be indexed. Naturally, ambiguities arise because of errors in the assessment of the d-values from the diffraction pattern, but they can generally be removed by calculating the angles between the possible planes.

In the case of X-ray diffraction, the radius of the Ewald sphere is similar to that of the reciprocal lattice spacing. For example, the wavelength of CuKα radiation, typically used for X-ray investigations, is 0.15418 nm, giving an Ewald sphere radius of 6.486 nm^{-1}, which is comparable to the reciprocal lattice spacing of copper (2.78 nm^{-1}) and other inorganic materials. This means that far fewer spots are intercepted, and fewer diffracted beams are produced. The appearance of an X-ray diffraction pattern is then not so easy to visualise as that of an electron diffraction pattern, and rather complex recording geometry must be used to obtain diffraction data from the whole of the reciprocal lattice.

7.3 PARTICLE SIZE

Although Bragg's law gives a precise value for the diffraction angle θ, diffraction actually occurs over a small range of angles close to the ideal value because of crystal imperfections. More perfect crystals diffract over a very narrow range of angles whilst very disordered crystals diffract over a wide range. The *approximate* range of angles over which diffraction occurs, $\delta\theta$, centred upon the exact Bragg angle, θ, is given by:

$$\delta\theta \approx \lambda / (D_{hkl} \cos\theta)$$

where D_{hkl} is the crystal thickness in the direction normal to the (hkl) planes diffracting the radiation, and $\delta\theta$ is measured in radians. As D_{hkl} is a multiple of the interplanar spacing of the (hkl) set:

$$D_{hkl} = m\, d_{hkl}$$

where m is an integer. The apparent crystal size in a direction perpendicular to the (hkl) planes is then given by:

$$m d_{hkl} \approx \lambda / (\delta\theta \ \ \cos\theta)$$

Thus, the size and sharpness of the diffracted beams provide a measure of the perfection of the crystal: a sharp spot indicating a well-ordered crystal and a fuzzy spot an extremely disordered or very small crystallite. Amorphous materials give only a few broad ill-defined rings, if anything at all.

The influence of crystal size is displayed clearly on the reciprocal lattice. Ideally, each reciprocal lattice point is sharp, and diffraction will only occur when the Ewald sphere precisely intersects a particular point. The effect of finite crystal dimensions is to modify the shape of each reciprocal lattice point by drawing the point out in reciprocal space in a direction normal to the small dimension in real space. To a first approximation, a square crystal of side w will have each reciprocal lattice point modified to a cross with each arm of approximate length $1/w$, extended along a direction perpendicular to the square faces; a cubic crystal will be similar, with each arm pulled out along directions normal to the cube faces. A spherical crystal of diameter w will have each reciprocal lattice point broadened into a sphere of approximately $1/w$ diameter; a needle of diameter w will have each reciprocal lattice point broadened into a disc of approximate diameter $1/w$ normal to the needle axis, and a thin crystal, of thickness w, will have the reciprocal lattice point drawn out into a spike of approximate length $1/w$ in a direction normal to the sheet (Figure 7.6a–d). The effect of this broadening is to greatly increase the areas of reciprocal lattice sections sampled in an electron microscope. For example, a thin film has reciprocal lattice points drawn out into spikes normal to the film, allowing the Ewald sphere to intersect considerably more points than in the case of a thick film (Figure 7.7).

7.4 THE INTENSITIES OF DIFFRACTED BEAMS

The experimental techniques outlined in the previous sections allow the lattice parameters of a crystal structure to be determined. However, the determination of the appropriate crystal lattice, face-centred cubic as against body-centred, for example, requires information on the intensities of the diffracted beams. More importantly, in order to proceed with a determination of the complete crystal structure, it is vital to understand the relationship between the intensity of a beam diffracted from a set of (*hkl*) planes and the atoms that make up the planes themselves.

The intensities of diffracted beams vary from one radiation type to another, and are found to depend upon the following components:

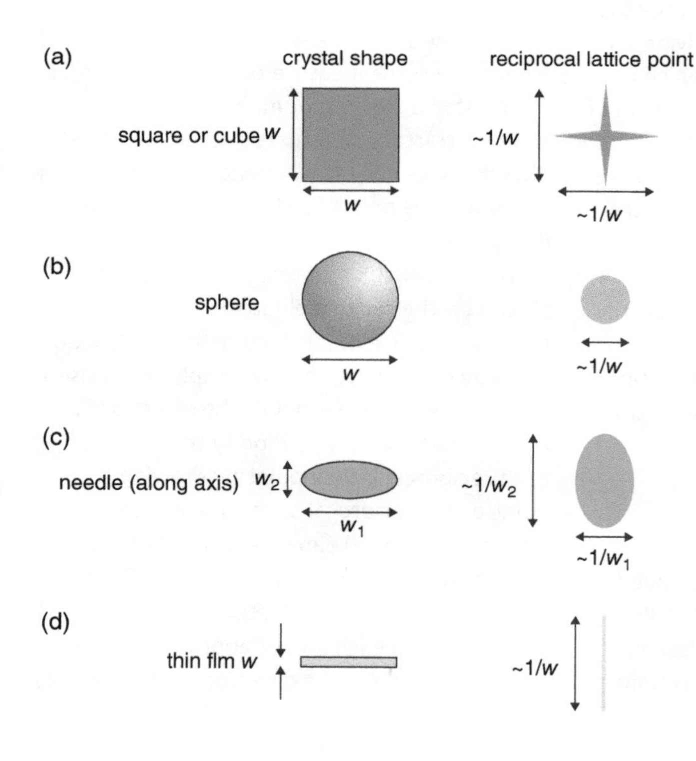

Figure 7.6 The effect of crystallite shape on the form of a diffraction spot: (a) square or cube; (b) sphere; (c) elliptical needle; (d) thin film.

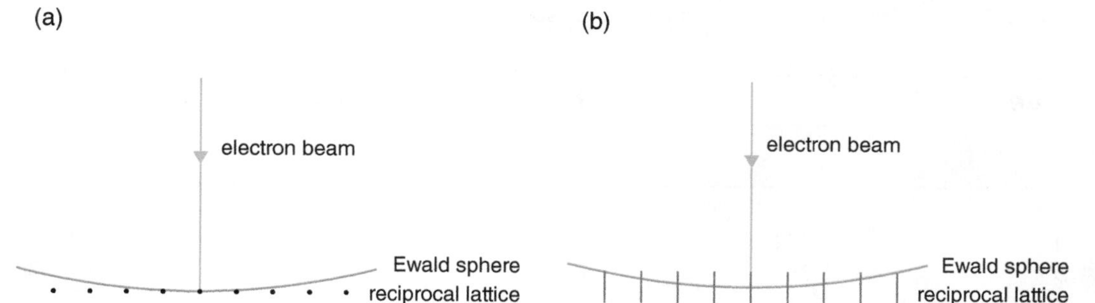

Figure 7.7 The Ewald sphere intersects fewer reciprocal lattice points in a thick film, (a), than a thin film, (b).

(i) The nature of the radiation.
(ii) The angle that the crystal makes with the diffracted beam.
(iii) The diffracting power of the atoms present, the **atomic scattering factor**.
(iv) The arrangement of the atoms in the crystal, the **structure factor**.
(v) The thermal vibrations of the atoms, the **temperature factor**.
(vi) The polarisation of the beam of radiation, the **polarisation factor**.
(vii) The thickness, shape and perfection of the crystal, the **form factor**, as touched upon in Section 7.3.
(viii) The time over which each plane is able to diffract radiation as the crystal rotates, the **Lorentz factor**.
(ix) For diffraction from a powder rather than a single crystal, the number of equivalent (*hkl*) planes present, the **multiplicity**.

To explain these factors, it is convenient to describe the determination of intensities with respect to X-ray diffraction. The differences that arise with other types of radiation are outlined below.

In the following four sections the important contribution made by the arrangement and types of atoms in crystals is considered. The other factors, which can be regarded as correction terms, are described later in this chapter. In the case of X-rays, the initial beam is almost undiminished on passing through a small crystal, and the intensities of the diffracted beams are a very small percentage of the incident beam intensity. For this reason, it is reasonable to assume that each diffracted X-ray photon is scattered only once. Scattering of diffracted beams back into the incident beam direction is ignored. This approximation is the basis of the **kinematical theory** of diffraction.

7.5 THE ATOMIC SCATTERING FACTOR

The scattering of the X-ray beam increases as the number of electrons, or equally, the atomic number (proton number), of the atom, Z, increases. Thus heavy metals such as lead, $Z = 82$, scatter X-rays far more strongly than light atoms such as carbon, $Z = 6$. Neighbouring atoms such as cobalt, Co, $Z = 27$, and nickel, Ni, $Z = 28$, scatter X-rays almost identically. The scattering power of an atom for a beam of X-rays is called the **atomic scattering factor**, or **atomic form factor**, f_a.

Table 7.1 The Cromer-Mann coefficients for sodium, firstly with λ in Å, and secondly with λ in nm

Index	**Cromer-Mann coefficients**			
	1	2	3	4
λ in Å				
a	4.763	3.174	1.267	1.113
b	3.285	8.842	0.314	129.424
c	0.676			
λ in nm				
a	4.763	3.174	1.267	1.113
b	0.03285	0.08842	0.00314	1.29424
c	0.676			

These and other values used in this chapter are taken from http://www-structure.llnl.gov.

Atomic scattering factors were originally determined experimentally, but now can be calculated using quantum mechanics. The atomic scattering factors, derived from quantum mechanical calculations of the electron density around an atom, are (approximately) given by the equation:

$$f_a = \sum_{i=1}^{4} a_i \cdot \exp\left[-b_i\left(\frac{\sin\theta}{\lambda}\right)^2\right] + c \qquad (7.2)$$

The constants, a_i and b_i, called the **Cromer-Mann coefficients**, vary for each atom or ion. The Cromer-Mann coefficients for sodium, Na, with an atomic number $Z = 11$, for use in Eq. (7.2) are given in Table 7.1. The units of f_a are electrons, and the wavelength of the radiation is in Å. To use the SI unit of nm (10 Å = 1 nm) for the wavelength, use the same values of a_i and c, but divide the values of b_i by 100 (Table 7.1).

It is found that the scattering is strongly angle-dependent, this being expressed as a function of $(\sin\theta)/\lambda$, which, by using the Bragg equation, is equal to $1/2d_{hkl}$. The scattering factor curves for all atoms have a similar form (Figure 7.8). At $(\sin\theta)/\lambda = 0$, the scattering factor is equal to the atomic number, Z, of the element in question. For example, the element sodium, with $Z = 11$, has a value of the scattering factor at $(\sin\theta)/\lambda = 0$ of 11. For ions, the scattering factor is similarly dependent upon the number of electrons present. For example, the scattering factor for the sodium ion Na^+ at $(\sin\theta)/\lambda = 0$ is equal to $Z - 1$, i.e. 10, rather than 11. Similarly, the atomic scattering factors of K^+ ($Z = 17$) and Cl^- ($Z = 19$) are identical.

Electrons also scatter radiation by another effect, the Compton effect. Compton scattering adds to the general background of scattered X-rays, and is usually included in the instrumental and other correction factors that are applied when the intensity of a diffracted spot is measured.

7.6 THE STRUCTURE FACTOR

To obtain the total intensity of radiation scattered by a unit cell, the scattering of all of the atoms in the unit cell must be combined. This is carried out by adding together the waves scattered from each set of (*hkl*) planes independently, to obtain a value called the

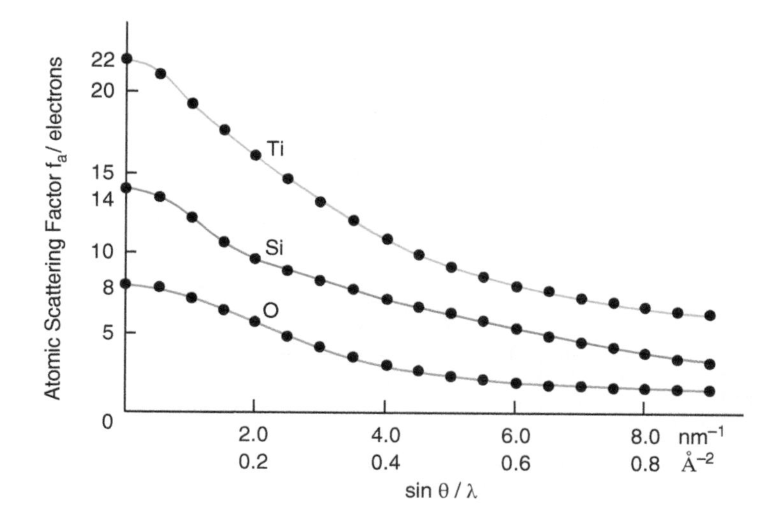

Figure 7.8 Atomic scattering factors for titanium (Ti), silicon (Si) and oxygen (O), as a function of $\sin\theta/\lambda$.

structure factor, **F**(*hkl*), for each *hkl* plane. It is calculated in the following way.

The *amplitude* (i.e. the 'strength') of the diffracted radiation scattered by an atom in a plane (*hkl*) will be given by the value of f_a appropriate to the correct value of $\sin\theta/\lambda$ (= $1/2d_{hkl}$) for the (*hkl*) plane. However, because the scattering atoms are at various locations in the unit cell, the waves scattered by each atom are out of step with each other as they leave the unit cell. The difference by which the waves are out of step is called the **phase difference** between the waves.

The way in which the scattered radiation adds together is governed by the relative phases of each scattered wave. The phase difference between two scattered waves, the amount by which they are out of step, is represented by a 'phase angle' measured in radians. Waves that are in step have a phase angle of 0 (or an even multiple of 2π). Waves that are completely out of step have a phase difference of π (or an odd multiple of π). If two waves with a phase difference of 2π are added, the result is a wave with *double the amplitude* of the original, whilst if the phase difference is π, the result is *zero*. Intermediate values of the relative phases give intermediate values for the amplitude of the resultant wave (Figure 7.9).

The phase difference between the waves scattered by two atoms will depend upon their relative positions in the unit cell and the directions along which the waves are superimposed. The directions of importance are those specified by the Bragg equation, which, for simplicity, are denoted by the indices of the (*hkl*) planes involved in the scattering, rather than the angle itself. Although X-rays are scattered by the electron cloud of an atom, it is adequate, for the present purposes, to regard the atoms in the unit cell as occupying point-like positions, *x*, *y*, *z*. The phase of the wave (in radians) scattered from an atom A at a position x_A, y_A, z_A, into the (*hkl*) reflected beam is:

$$2\pi(hx_A + ky_A + lz_A) = \phi_A$$

where x_A, y_A, z_A are the fractional coordinates with respect to the unit cell edges, as normally used.

To illustrate this, suppose that the unit cell contains just two identical atoms, M1 at the cell

(a)

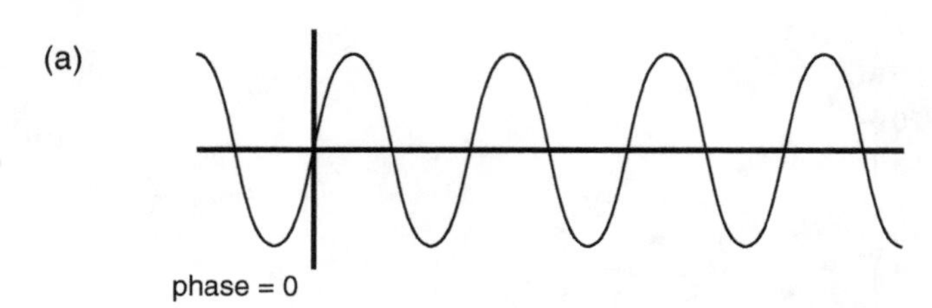

(b)

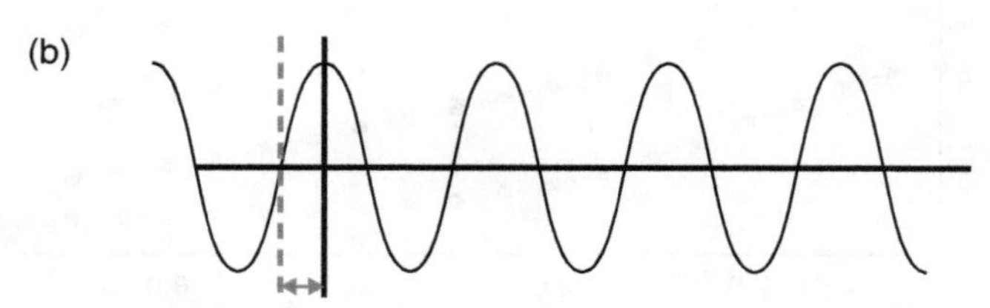

(c)

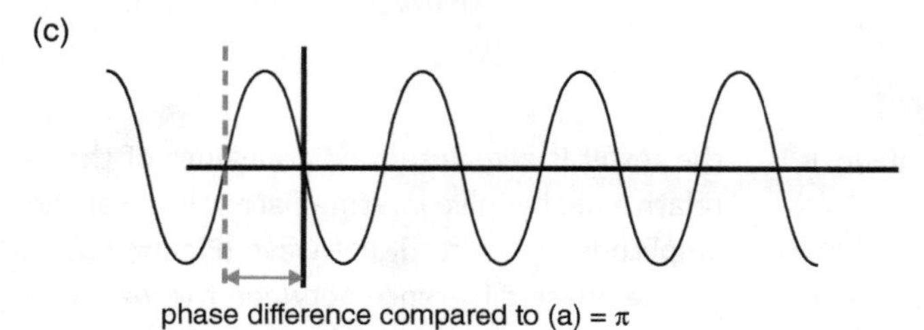

(d)

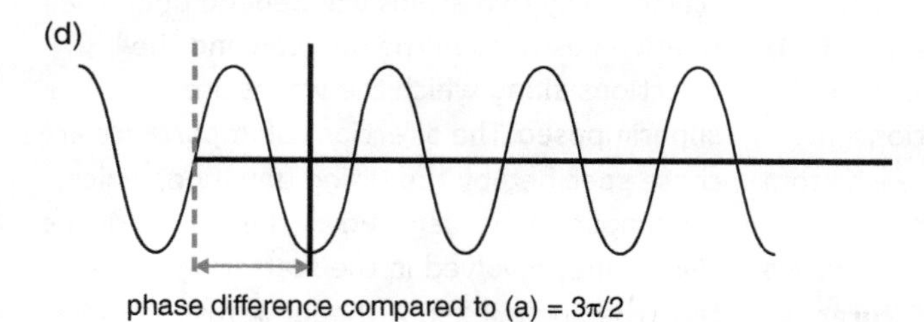

(e)

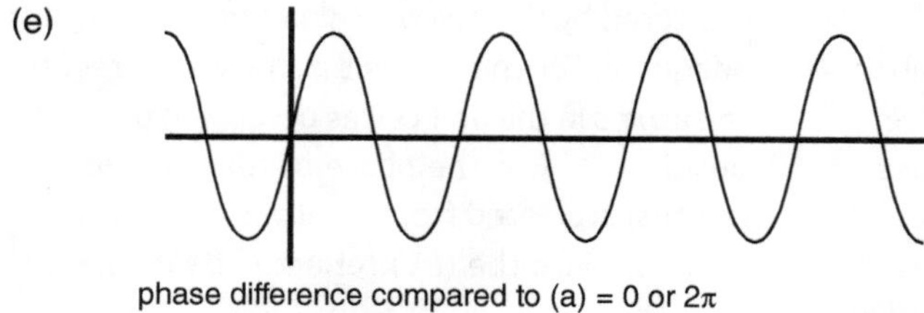

Figure 7.9 The phase difference between waves: (a) 0; (b) $\pi/2$; (c) π; (d) $3\pi/2$; (e) 2π.

origin (0, 0, 0), and M2 at the cell centre (½, ½, ½) (Figure 7.10a). The reflection from (100) planes will be non-existent, because the waves scattered by the M1 atoms at the cell origin, which have a phase:

$$\text{M1} \quad 2\pi\,(1 \times 0) = 0$$

are combined with a similar wave, arising from the M2 atoms at the cell centre, with phase:

$$\text{M2} \quad 2\pi\,(1 \times \tfrac{1}{2}) = \pi$$

to give a zero resultant amplitude (Figure 7.10b). In contrast, the amplitude of the

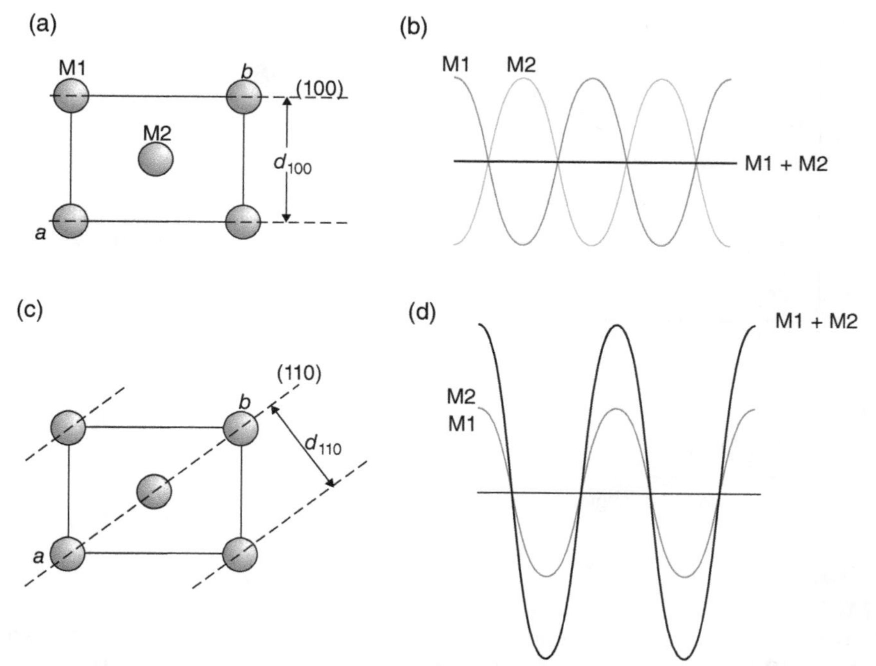

Figure 7.10 The scattering of waves from a unit cell: (a) (100) planes; (b) the waves from M1 and M2 are completely out of step and result in zero amplitude; (c) (110) planes; (d) the waves from M1 and M2 are exactly in step, resulting in doubled amplitude.

reflection from (110) (Figure 7.10c) will be high, because this time the waves from both M1 and M2 are in step (Figure 7.10d), with relative phases:

$$\text{M1} \quad 2\pi(1 \times 0 + 1 \times 0) = 0$$

$$\text{M2} \quad 2\pi(1 \times \tfrac{1}{2} + 1 \times \tfrac{1}{2}) = 2\pi$$

Thus, even in this simple case, the amplitude of the resultant scattered wave can vary between the limits of zero to $2 \times f_M$, depending upon the planes that are scattering the waves.

When the scattered waves from all of the atoms in a unit cell are added, the form of the resultant will depend, therefore, upon both the scattering power of the atoms involved and the individual phases of all of the separate waves. A clear picture of the relative importance of the scattering from each of the atoms in a unit cell, including the phase information, and how they add together to give a final value of $\mathbf{F}(hkl)$, can be obtained by adding the scattering from each atom using vector addition (Appendix A). The scattering from an atom A is represented by a vector $\mathbf{f}_A$ that has a length equal to the scattered amplitude (i.e. the strength of the scattering) of f_A and a phase angle ϕ_A. To draw the diagram, a vector $\mathbf{f}_A$ of length f_A is drawn at an angle ϕ_A in an anticlockwise direction to the horizontal (x-) axis (Figure 7.11a). In order to picture the scattering from a complete unit cell, the scattering from each of the N atoms in the cell is represented by vectors $\mathbf{f}_n$ of length f_n drawn at an angle ϕ_n to the horizontal (Figure 7.11b), where n runs from 1 to N. Note that the scattering from an atom at the origin of the unit cell, position 000, will be represented by a vector along the horizontal, as

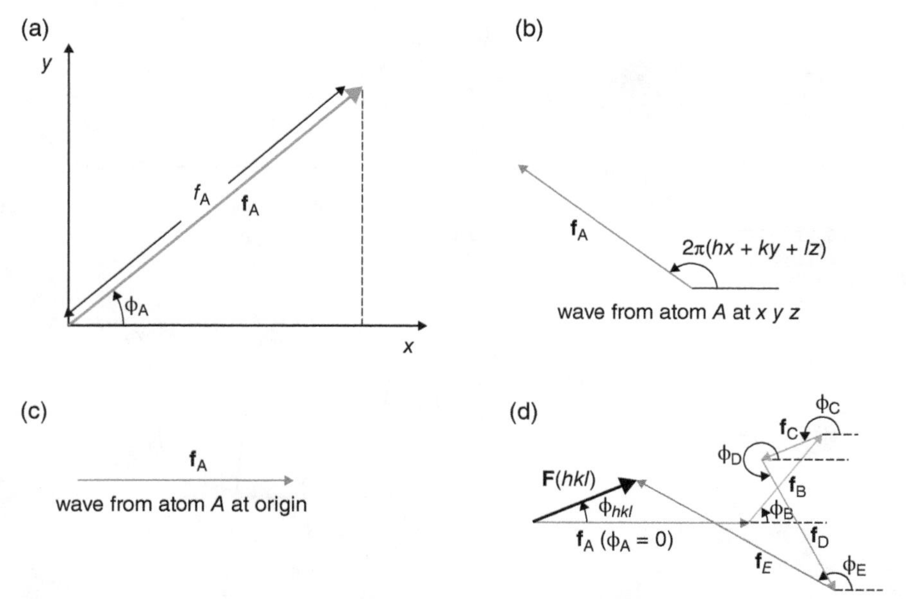

Figure 7.11 The representation of scattered waves as vectors: (a) a scattered wave vector in the x, y plane; (b) the wave scattered by an atom A at (x, y, z); (c) the wave scattered by atom A at the origin (000); (d) the addition of waves scattered by five atoms, A, B, C, D and E.

the phase angle will be zero (Figure 7.11c). Each successive vector is added, head to tail, to the preceding one, with ϕ_n always being drawn in an anticlockwise direction with respect to the horizontal. The resultant of the addition of all of the $\mathbf{f}_n$ vectors, using the 'head to tail' method of graphical vector addition, (see Appendix A), gives a vector $\mathbf{F}(hkl)$ with a phase angle ϕ_{hkl} (Figure 7.11d). The numerical length of the vector $\mathbf{F}(hkl)$ is written $F(hkl)$. It is seen that the value of ϕ_{hkl} is equal to the sum of all of the phase angles, ϕ_n, of the scattered waves from the n atoms.

7.7 STRUCTURE FACTORS AND INTENSITIES

The graphical method gives a lucid picture of scattering from a unit cell, but is impractical as a method for calculation of the intensities of diffracted beams. The pictorial summation must be expressed algebraically for this purpose. The simplest way of carrying this out is to express the scattered wave as a **complex amplitude** (see Appendices E and F):

$$\mathbf{f}_A = f_A \exp\,[2\pi i(hx_A + ky_A + lz_A)] = f_A e^{i\phi_A} \qquad (7.3)$$

The amplitude (i.e. the numerical size), f_A, is the *modulus* of $\mathbf{f}_A$, written $|\mathbf{f}_A|$.

Equation (7.3) can be written in an equivalent form as a complex number:

$$\begin{aligned}\mathbf{f}_A &= a_{hkl} + i b_{hkl} \\ &= f_A\{\cos\phi_A + i\sin\phi_A\} \\ &= f_A\{\cos\,[2\pi(hx_A + ky_A + lz_A)] \\ &\quad + i\sin\,[2\pi(hx_A + ky_A + lz_A)]\}\end{aligned}$$

This representation of the scattering can be drawn on an **Argand diagram** used to display complex numbers (Figure 7.12). The depiction is

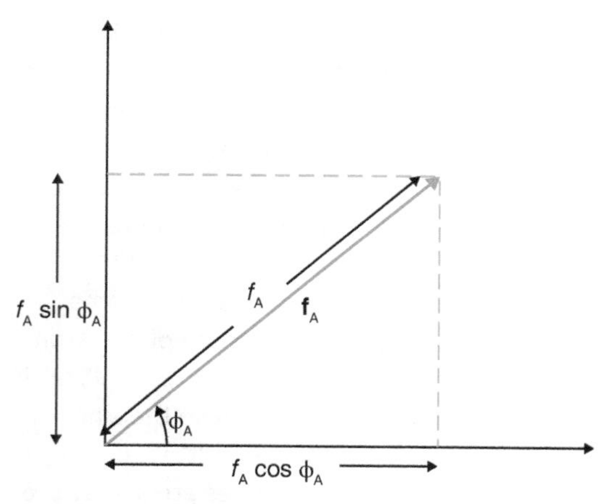

Figure 7.12 Representation of $\mathbf{f}_A$ as a complex number on an Argand diagram.

often said to be in the **Gaussian plane** or in the **complex plane**. When $\mathbf{f}_A$ is plotted in the Gaussian plane, the projection of $\mathbf{f}_A$ along the horizontal axis, $f_A \cos \phi_A$, represents the real part of the complex number, a_{hkl}. Similarly, the projection of $\mathbf{f}_A$ along the vertical axis, $f_A \sin \phi_A$, represents the imaginary part of the complex number, b_{hkl}. It is now possible to add the scattered amplitudes in the complex plane graphically, as in Figure 7.11d, or better still, algebraically (Appendix F).

In this form, the total scattering from a unit cell containing N different atoms is simply:

$$\begin{aligned}\mathbf{F}(hkl) &= \sum_{n=1}^{N} f_n \exp\left[2\pi i(hx_n + ky_n + lz_n)\right] \\ &= F(hkl)e^{i\phi_{hkl}} \\ &= \sum_{n=1}^{N} f_n \cos 2\pi(hx_n + ky_n + lz_n) \\ &\quad + i\sum_{n=1}^{N} f_n \sin 2\pi(hx_n + ky_n + lz_n) \\ &= A_{hkl} + i\,B_{hkl}\end{aligned}$$

where hkl are the Miller indices of the diffracting plane, and the summation is carried out over all N atoms in the unit cell, each of which has an atomic scattering factor, f_n, appropriate to the (hkl) plane being considered. The magnitude of the scattering is $F(hkl)$, the modulus of $\mathbf{F}(hkl)$, written $|\mathbf{F}(hkl)|$, and ϕ_{hkl} is the phase of the scattering factor, which is the sum of all of the phase contributions from the various atoms present in the unit cell.

The **intensity** scattered into the hkl beam by all of the atoms in the unit cell, $I_0(hkl)$, is given by $|\mathbf{F}(hkl)|^2$, the modulus of $\mathbf{F}_{hkl}$ squared (see Appendix F):

$$\begin{aligned}I_0(hkl) &= |\mathbf{F}(hkl)|^2 = F(hkl)^2 \\ &= \left\{\sum_{n=1}^{N} f_n \exp\left[2\pi i(hx_n + ky_n + lz_n)\right]\right\}^2 \\ &= \left\{\sum_{n=1}^{N} f_n \cos 2\pi(hx_n + ky_n + lz_n) + i\sum_{n=1}^{N} f_n \sin 2\pi(hx_n + ky_n + lz_n)\right\}^2 \\ &= A_{hkl}^2 + B_{hkl}^2\end{aligned}$$

If the unit cell has a centre of symmetry, the sine terms, B_{hkl}^2, can be omitted, as they sum to zero.

The calculation of the intensity reflected from a plane of atoms in a known structure is straightforward, although tedious without a computer. Because of the instrumental and other factors that affect measured intensities, calculated intensities, which ignore these factors, are generally listed as a fraction or percentage of the strongest reflection.

7.8 NUMERICAL EVALUATION OF STRUCTURE FACTORS

Although structure factor calculations are carried out by computer, it is valuable to evaluate a few examples by hand, to gain an appreciation of the steps involved. As an example, the calculation of the structure factor and intensity for the 200 reflection from a crystal of the rutile form of TiO_2 (see Chapter 1) is reproduced here. The details of the tetragonal unit cell are:

Lattice parameters: $a = b = 0.4594$ nm, $c = 0.2959$ nm

Atom positions: Ti: 0, 0, 0; ½, ½, ½

O: $^3/_{10}$, $^3/_{10}$, 0; ⅘, ⅕, ½; $^7/_{10}$, $^7/_{10}$, 0; ⅕, ⅘, ½

There are two Ti atoms and four oxygen atoms in the unit cell.

The calculation proceeds in the following way.

(i) Estimate the value of $\sin\theta/\lambda$ appropriate to (200):

$$\sin\theta/\lambda = 1/(2\,d_{200}) = 2.177\ \text{nm}^{-1}\ \left(0.2177\ \text{Å}^{-1}\right)$$

(ii) Determine the atomic scattering factors for Ti and O at this value from tables of data or Eq. (7.2), using the appropriate Cromer-Mann coefficients. (Approximate values can be found from Figure 7.8.) Precise values are: f_{Ti}, 15.513; f_O, 5.326.

(iii) Calculate the phase angles of the waves scattered by each of the atoms in the unit cell. Note that the unit cell is centrosymmetric, and so the sine terms (B_{hkl}) can be omitted. In this case, the phase angles are simply the cosine (A_{hkl}) terms, [$\cos 2\pi(hx_n + ky_n + lz_n)$], which in this case simplifies to [$\cos 2\pi(2x_n)$]. The results are given in Table 7.2.

(iv) $F(200)$ is given by adding the values in the following column of Table 7.2.

$$F(200) = 2(15.513) - 4(4.3097) = 13.79$$

(v) The phase angle ϕ_{200}, associated with $F(200)$, is the sum of all of the individual phase angles in Table 7.2. It is seen that $\phi_{200} = 10\pi\ (1800°) = 2\pi\ (0°)$.

(vi) Calculate the intensity, $I_0(200)$, of the diffracted beam, equal to $|\mathbf{F}(hkl)|^2$:

$$I_0(200) = |\mathbf{F}(200)|^2 = F(200)^2 = 190.16$$

This procedure is repeated for all of the other *hkl* reflections in the unit cell.

Table 7.2 Calculation of $F(200)$ for TiO_2, rutile

Atom	Phase angle, ϕ, radians	Phase angle, ϕ, °	$\cos\phi$	$f_a \cos\phi = A_{200}$	$\sin\phi$	$f_a \sin\phi = B_{200}$
Ti(1)	0	0	1.0	15.513	0	0
Ti(2)	2π	360	1.0	15.513	0	0
O(1)	1.2π	216	−0.8090	−4.3087	−0.5878	−3.1305
O(2)	3.2π	576 (= 216)	−0.8090	−4.3087	−0.5878	−3.1305
O(3)	2.8π	504 (= 144)	−0.8090	−4.3087	0.5878	3.1305
O(4)	0.8π	144	−0.8090	−4.3087	0.5878	3.1305

Table 7.3 Calculation of $F(100)$ for TiO_2, rutile

Atom	Phase angle, ϕ, radians	Phase angle, ϕ, °	$f_a \cos\phi = A_{100}$
Ti(1)	$2\pi\ (1 \times 0)$	0	F(Ti) × 1
Ti(2)	$2\pi\ (1 \times ½)$	180	F(Ti) × −1
O(1)	$2\pi\ (1 \times 3/10)$	108	F(O) × −0.3090
O(2)	$2\pi\ (1 \times 4/5)$	288	F(O) × 0.3090
O(3)	$2\pi\ (1 \times 7/10)$	252	F(O) × −0.3090
O(4)	$2\pi\ (1 \times 1/5)$	72	F(O) × 0.3090

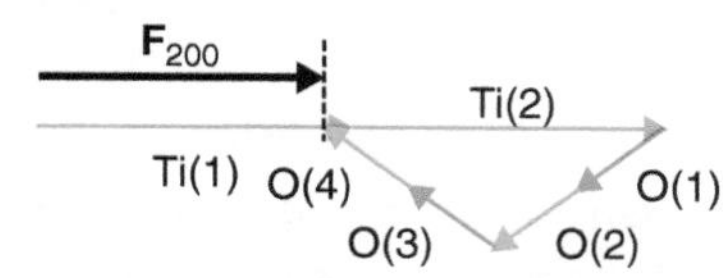

Figure 7.13 The vector addition of scattered waves contributing to the 200 reflection of rutile, TiO_2.

(vii) In order to check that the sine terms do add to zero, the calculation can be repeated to find $\sum f_n \sin 2\pi\ (2x)$. These terms do indeed add to zero (Table 7.3).

To relate this numerical assessment to the vector addition described above, the values given in Table 7.2 are displayed graphically in Figure 7.13. The value obtained for F(200) from this figure is 13.8, compared with the arithmetical value of 13.6. The phase angle ϕ_{hkl} is zero (2π).

An examination of Table 7.2 and Figure 7.13 makes clear the importance of the phases of the waves. To a large extent these control the intensity. If the rutile calculation is repeated for the (100) plane, the phase angles for the two Ti atoms are:

$$\text{Ti(1)},\quad \cos 2\pi\,(1 \times 0) = 1$$

$$\text{Ti(2)},\quad \cos 2\pi\,(1 \times ½) = -1$$

That is, the waves scattered by the two Ti atoms are exactly out of step and will cancel. It is immediately apparent that the (100) reflection will be weak, as it will depend only upon the oxygen atoms. In fact, these cancel too, so that the intensity turns out to be zero (Table 7.3). The phase angle, ϕ_{100}, equal to the sum of all of the separate phase angles, is equal to $5\pi(900°) = \pi(180°)$, and the sum of the terms $f_a \cos\phi$, A_{100}, is zero, so that there is no diffracted beam (Table 7.3).

7.9 SYMMETRY AND REFLECTION INTENSITIES

The atoms in the unit cell will determine the intensity of a diffracted beam of radiation via the structure factor, as described above. This means that there will be relationships between the various **F**(hkl) values that arise because of the symmetry of the unit cell. The symmetry of the structure will therefore play an important part in determining the intensity diffracted. One consequence of this is **Friedel's law**, which states that the structure factors, **F**, of the pair of reflections hkl and $\bar{h}\bar{k}\bar{l}$ are **equal in magnitude but have opposite phases**. That is:

$$F(hkl) = F\left(\bar{h}\bar{k}\bar{l}\right);\quad \phi_{hkl} = -\phi\bar{h}\bar{k}\bar{l}$$

Such pairs of reflections are called **Friedel pairs**. Because of this, the intensities can be expressed as:

$$F(hkl)^2 = I_0(hkl) = F(\bar{h}\bar{k}\bar{l})^2 = I_0(\bar{h}\bar{k}\bar{l})$$

The intensities of Friedel pairs will be equal. This will cause the diffraction pattern from a crystal to appear centrosymmetric even for crystals that lack a centre of symmetry. Diffraction is thus a centrosymmetric physical property, which means that the point symmetry of any diffraction pattern will belong to one of the 11 Laue classes (Section 5.2).

The most important consequence of the symmetry elements present in a crystal is that some (hkl) planes have $F(hkl) = 0$, and so will never give rise to a diffracted beam, irrespective of the atoms present. Such 'missing' diffracted beams are called **systematic absences.** This can most easily be described in terms of the vector representation described above. Suppose that a crystal is derived from a body-centred lattice. In the simplest case, the motif is one atom per lattice point, and the unit cell contains atoms at 0 0 0 and ½ ½ ½ (Figure 7.14a). (Note that the cell can have any symmetry). The structure factor for each (hkl) set is given by:

$$\mathbf{F}(hkl) = f_a \exp 2\pi i\,(h \times 0 + k \times 0 + l \times 0) + f_a \exp 2\pi i\,(h \times \tfrac{1}{2} + k \times \tfrac{1}{2} + l \times \tfrac{1}{2})$$

If $h + k + l$ is *even*, $\mathbf{F}(hkl)$ becomes

$$\mathbf{F}(hkl) = f_a \exp 2\pi i\,(0) + f_a \exp 2\pi i\,(2n/2)$$

where n is an integer. The Gaussian plane representation (Figure 7.14b) shows that the vectors are parallel and add together to give a positive value for $\mathbf{F}(hkl)$ for any value of n.

If $h + k + l$ is *odd*, $\mathbf{F}(hkl)$ becomes:

$$\mathbf{F}(hkl) = f_a \exp 2\pi i\,(0) + f_a \exp 2\pi i\,(n/2)$$

where n takes odd integer values, 1, 3, etc. The Gaussian plane diagram (Figure 7.14c) now shows that the vectors are opposed, and $\mathbf{F}(hkl)$ is 0, irrespective of the value of n. Thus all planes with $h + k + l$ even will give rise to a reflection, whilst all planes with $h + k + l$ odd will not.

All crystallographic unit cells derived from a body-centred lattice give rise to the same systematic absences. Similar considerations apply to the other Bravais lattices. The conditions that apply for diffraction *to occur* from (hkl) planes in the Bravais lattices, called **reflection conditions**, are listed in Table 7.4.

Apart from the Bravais lattice, other crystallographic features give rise to systematic

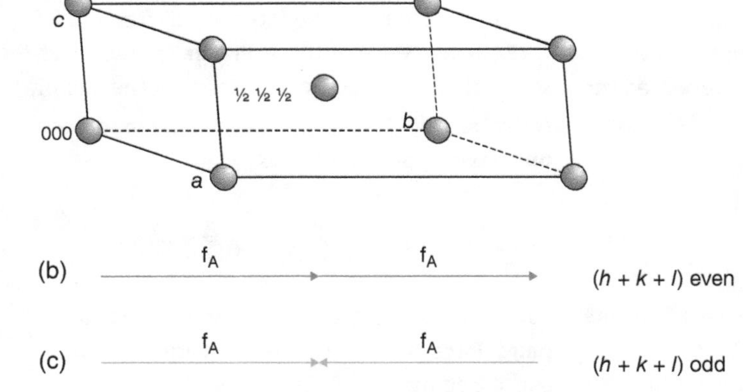

Figure 7.14 Diffraction from a body-centred unit cell: (a) a body-centred unit cell; (b) vector addition of waves for reflections $h + k + l$ even; (c) vector addition of waves for reflections $h + k + l$ odd.

Table 7.4 Reflection conditions for Bravais lattice cells

Lattice type	Reflection conditions[a]
Primitive (*P*)	None
A-face-centred (*A*)	$k + l = 2n$
B-face-centred (*B*)	$h + l = 2n$
C-face-centred (*C*)	$h + k = 2n$
Body centred (*I*)	$h + k + l = 2n$
All-face-centred (*F*)	$h + k, h + l, k + l = 2n$ (h, k, l, all odd or all even)

[a] *n* is an integer.

absences. The symmetry elements that are responsible for systematic absences are (i) a centre of symmetry, (ii) screw axes, and (iii) glide planes. The systematic absences that occur on a diffraction pattern thus give detailed information about the symmetry elements present in a unit cell and the lattice type upon which it is based.

Systematic absences arise from symmetry considerations and always have *F*(*hkl*) equal to zero. They are quite different from **structural absences**, which arise because the scattering factors of the atoms combine so as to give a value of *F*(*hkl*) = 0 for other reasons. For example, the (100) diffraction spots in NaCl and KCl are systematically absent, as the crystals adopt the *halite* structure, which is derived from an all-face-centred (*F*) lattice (see Table 7.4). On the other hand, the (111) reflection is present in NaCl, but is (virtually) absent in KCl for structural reasons – the atomic scattering factor for K^+ is almost equal to that of Cl^-, as the number of electrons on both ions is 18.

The reciprocal lattice described in Chapter 2 consists of an array of points. Because a diffraction pattern is a direct representation of the reciprocal lattice, it is often useful to draw it as a **weighted reciprocal lattice**, in which the area allocated to each node is proportional to the structure factor *F*(*hkl*) of each reflection. The weighted reciprocal lattice omits all

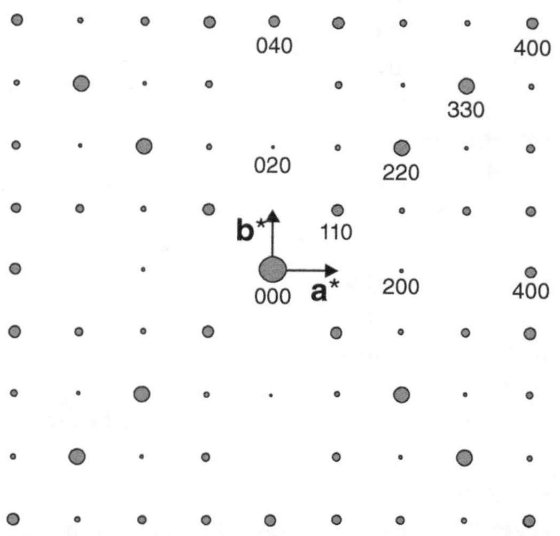

Figure 7.15 The *hk*0 section of the weighted reciprocal lattice of rutile, TiO_2.

reflections that are systematically absent, and so gives a clearer impression of the appearance of a diffraction pattern from a crystal (Figure 7.15).

7.10 THE TEMPERATURE FACTOR

The most important correction term applied to intensity calculations is the **temperature factor**. The calculations described above assume that the atoms in a crystal are stationary. This is not so, and in molecular crystals and some inorganic crystals the vibrations can be considerable, even at room temperature. This vibration has a considerable effect upon the intensity of a diffracted beam. The vibration frequency of an atom in a crystal is often taken to be approximately 10^{13} Hz at room temperature. The frequency of an X-ray beam of wavelength 0.154 nm (the wavelength of copper Kα radiation, which is frequently used for crystallographic work) is 1.95×10^{18} Hz. Consequently, atoms nominally in any (hkl) plane will, in fact, be continually displaced out of the plane by varying amounts on the timescale of the radiation. This has the effect of smearing out the electron density of each atom, and diminishing each atomic scattering factor. For planes with a large interplanar spacing, d_{hkl}, which diffract at small angles, the displacements are only a fraction of d_{hkl}, and the effect is small. However, the atomic displacements in planes with a small interplanar spacing, d_{hkl}, may be equal to or greater than d_{hkl} itself, causing considerable loss in diffracted intensity. It is for this reason that X-ray diffraction patterns from organic molecular crystals, which usually have low melting points and large thermal vibrations, are very weak at high Bragg angles, whilst high melting point inorganic crystals usually give sharp diffraction spots at high Bragg angles. Diffraction patterns from crystals in which atoms display large degrees of thermal vibration are obtained from crystals maintained at low temperatures, mounted on special cryogenic holders, to offset this problem. (The data for (*S*)-alanine, Section 6.7, were obtained from a crystal maintained at 23 K.)

Although chemical bonds link the atomic vibrations throughout the crystal, the thermal motion of any atom in a crystal is generally assumed to be independent of the vibration of the others. Under this approximation, the atomic scattering factor for a thermally vibrating atom, f_{th}, is given by:

$$f_{th} = f_a \exp\left[-B\left(\frac{\sin\theta}{\lambda}\right)^2\right] \qquad (7.4)$$

where f_a is the atomic scattering factor defined for stationary atoms, B is the **atomic temperature factor** (also called the **Debye-Waller factor** or **B-factor**), θ is the Bragg angle and λ the wavelength of the radiation. The effect of the temperature factor is to reduce the value of the atomic scattering factor considerably at higher values of $\sin\theta/\lambda$ (Figure 7.16). The structure factor for a diffracted (hkl) beam is then:

$$\mathbf{F}(hkl) = \sum_{n=1}^{N} f_n \exp\left[2\pi i(hx_n + ky_n + lz_n)\right] \exp\left[-B\left(\frac{\sin\theta}{\lambda}\right)^2\right]$$

$$\mathbf{F}(hkl)^2 = \left\{\sum_{n=1}^{N} f_n \exp\left[2\pi i(hx_n + ky_n + lz_n)\right] \exp\left[-B\left(\frac{\sin\theta}{\lambda}\right)^2\right]\right\}^2 = I(hkl)$$

and the corrected diffracted intensity, $I(hkl)$ is given by:

$$I(hkl) = I_0(hkl) \exp\left[-2B\left(\frac{\sin\theta}{\lambda}\right)^2\right]$$

The atomic temperature factor is related to the magnitude of the vibration of the atom concerned by the equation:

$$B = 8\pi^2 U$$

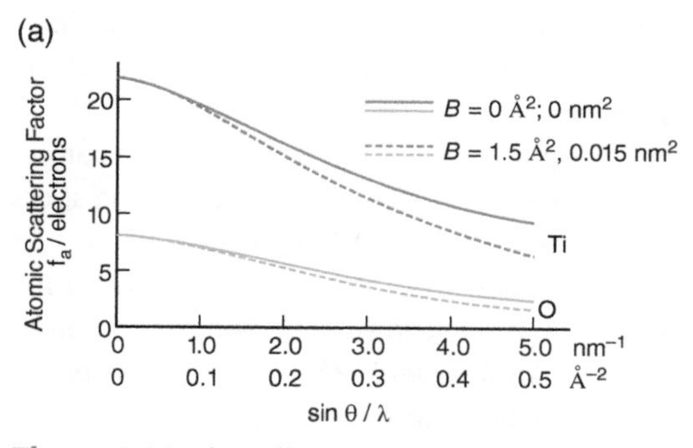

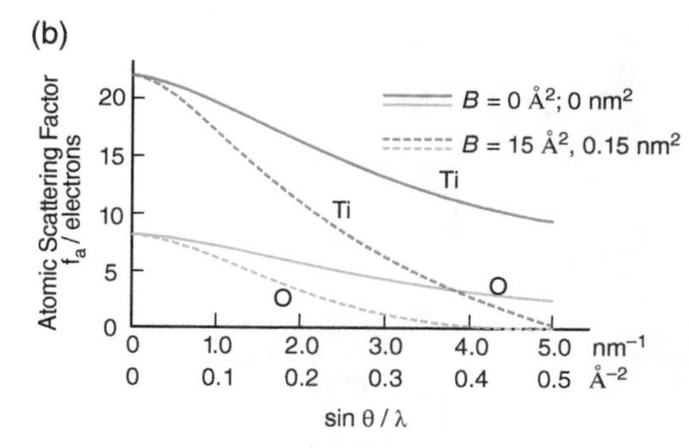

Figure 7.16 The effect of the temperature factor on the scattering power of titanium (Ti) and oxygen (O) atoms: (a) $B = 0$ and 0.015 nm^2 (1.5 Å^2); (b) $B = 0$ and 0.15 nm^2 (15 Å^2).

where U is the **isotropic temperature factor**, which is equal to the square of the mean displacement of the atom, $<\overline{r}^2>$, from the normal equilibrium position, r_0. Because crystallographers record the atomic dimensions in Å (10^{-10} m), B is quoted in units of Å^2. The value of B can range from 1 to 100 Å^2, (0.1–10 nm^2), although values of the order of 1–10 Å^2 (0.1–1 nm^2) are normal. To use the SI unit of nm for the wavelength of the radiation in Eq. (7.4), the value of B in Å^2 must be divided by 100. In crystal structure drawings, the atom positions are often drawn as spheres with a radius proportional to B.

Although the value of U gives a good idea of the overall magnitude of the thermal vibrations of an atom, these are not usually isotropic (the same in all directions). To display this, the atom positions in a crystal structure are not indicated by spheres of a radius proportional to B, but by ellipsoids with axes proportional to the mean displacement of the atoms in three principle directions. The isotropic temperature factor, U, is replaced by the **anisotropic temperature factors** U_{11}, U_{22}, U_{33}, U_{12}, U_{13}, U_{23}. These define the size and orientation of the thermal ellipsoid with respect to the crystallographic axes. Structures of molecules, in particular, are displayed in this way, using an ORTEP diagram, which is an acronym for Oak Ridge Thermal Ellipsoid Program. This pictorial presentation gives an idea of the three-dimensional molecular shape by drawing the bond distances and angles and representing the atoms by ellipsoids. Figure 7.17 shows an ORTEP diagram of the molecule $Ru_3(CO)_{10}(PMe_2napth)_2$.

7.11 POWDER X-RAY DIFFRACTION

Although powder diffraction may not be the first choice for structure determination, powder X-ray diffraction is used routinely for the identification of solids, especially in mixtures, in a wide range of sciences from geology to forensics.[3] Two main sample geometries are used in X-ray powder diffraction experiments. In the first, a finely powdered sample of the

[3] In the very first Sherlock Holmes story (*A Study in Scarlet*, first published in *Beeton's Christmas Annual*, 1887), emphasis is placed upon Holmes' ability to tell different soil types 'at a glance', and this important ability features in a number of stories. Of course, Holmes had to make do with optical means to characterise the differing soils. The routine use of powder X-ray diffraction to quantify different soil types has made this a much more powerful tool, which is widely in use today, both geologically and in forensic science.

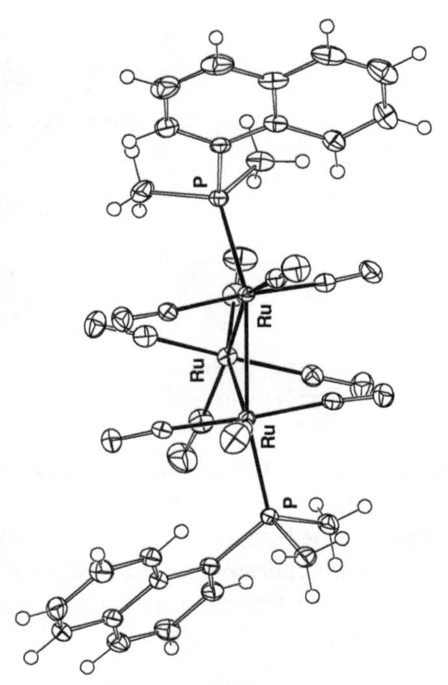

Figure 7.17 ORTEP diagram of the molecule $Ru_3(CO)_{10}(PMe_2napth)_2$. All atoms are represented as thermal ellipsoids except for hydrogen. Source: Reprinted from Bruce, M.I., Humphrey, P.A., Schmutzler, R. et al. (2005). Ruthenium carbonyl clusters containing PMe2(nap) and derived ligands (nap = 1-naphthyl): generation of naphthalyne derivatives. *Journal of Organometallic Chemistry* **690**: 784–791, with permission from Elsevier.

material to be investigated is formed into a cylinder by introducing it into a hollow glass fibre, or by binding it with a gum. In the second, the powder is formed into a flat plate, by applying it to sticky tape or a similar adhesive. In either case, the powder is irradiated with a beam of X-rays.

Each crystallite will have its own reciprocal lattice, and if the crystals are randomly orientated, each reciprocal lattice will be randomly orientated. Because of this, the overall reciprocal lattice appropriate to the powder will consist of a series of concentric spherical *hkl* shells rather than discrete spots. In such cases the diffraction pattern from a powder placed in the path of an X-ray beam gives rise to a series of cones rather than spots (Figures 7.4 and 7.18a). The positions and intensities of the diffracted beams are recorded along a narrow strip surrounding the sample (Figure 7.18b) to yield a characteristic pattern of diffracted 'lines' or peaks (Figure 7.18c). The position of a line (not the intensity) is found to depend only upon the spacing of the crystal planes involved in the diffraction and the wavelength of the X-rays used, via Bragg's law. The angle between the transmitted or 'straight through' beam and the diffracted beam is then 2θ, so the angle at the apex of a cone of diffracted rays is 4θ, where θ is the Bragg angle. In essence, the positions of the lines on a powder diffraction pattern simply depend upon the unit cell dimensions of the phase.

The intensities of the diffraction lines depend upon the factors already mentioned, but further corrections must be included if the calculated intensities are to correspond accurately with the observed intensities. An important factor is the **multiplicity** of the reflections in the powder pattern, *p*. This correction term can be easily understood by deriving the powder pattern from the diffraction 'spot' pattern of a single crystal (Figure 7.19a). The intensity of each spot is well defined, and the intensity of, say, the 200 spot, I (200), can be measured independently of all other reflections. If an X-ray beam is incident upon a random array of crystals, instead of a single crystal, each diffraction spot will lie at a random angle. With a small number of crystals, a 'spotty ring pattern' will form. Figure 7.19b shows the pattern generated from the single crystal pattern in Figure 7.19a if there are four other identical crystals present, rotated anticlockwise successively by (i) 25°, (ii) 39°,

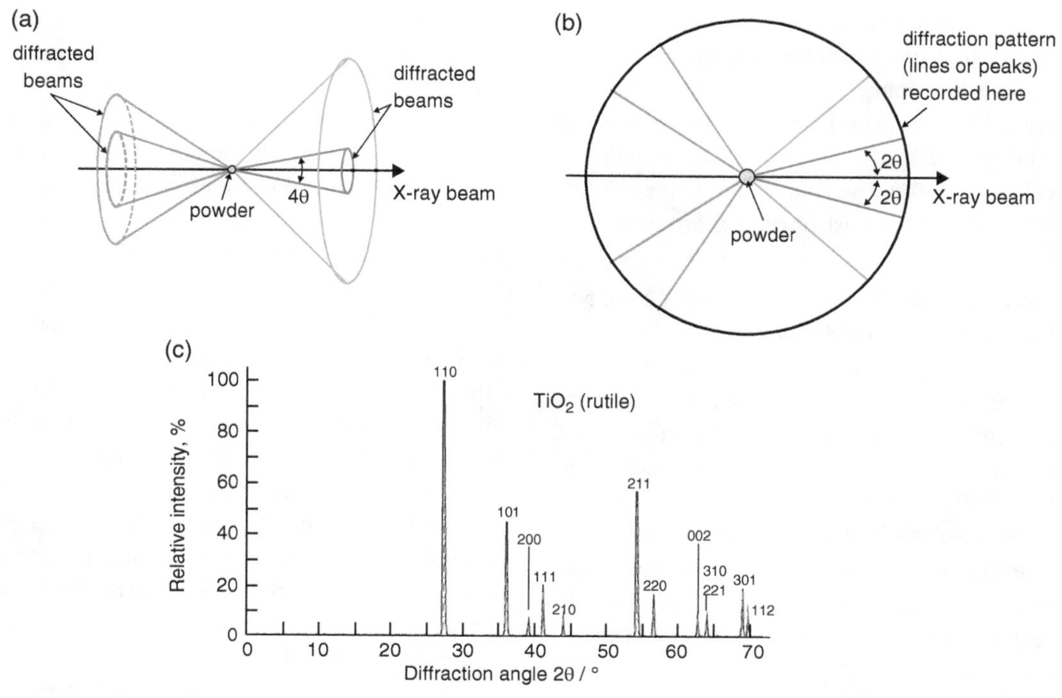

Figure 7.18 (a) X-ray diffraction from a powder sample; (b) the diffraction pattern is recorded along a circle; (c) diffraction pattern from powdered rutile, TiO_2.

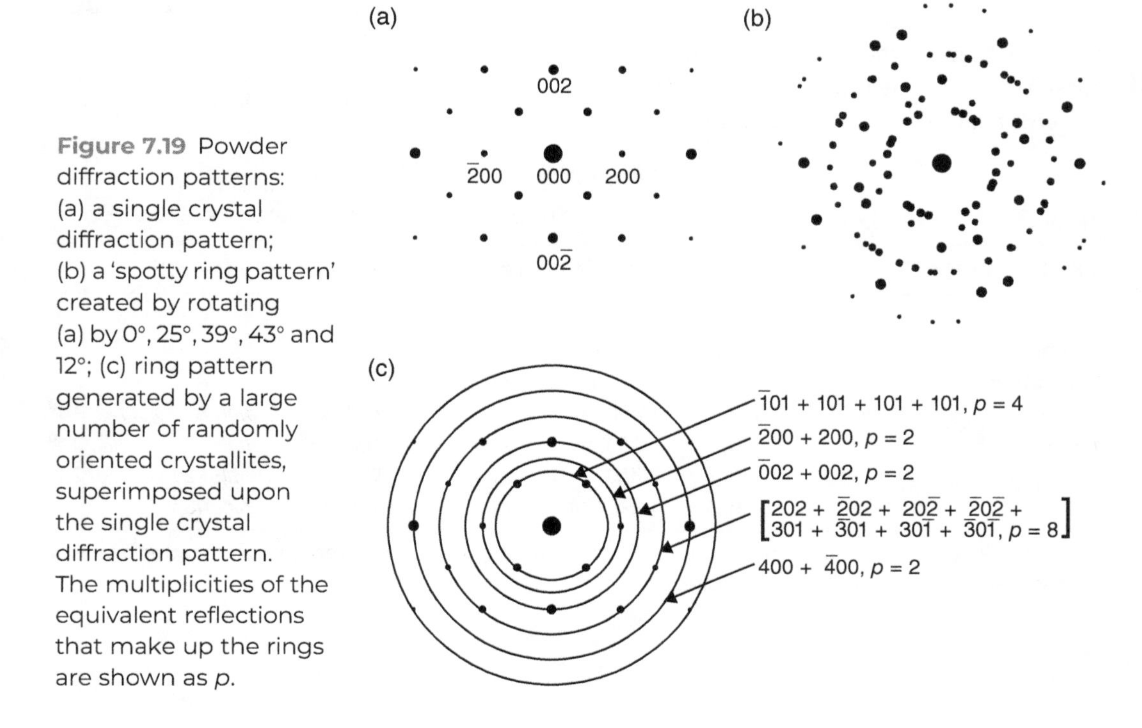

Figure 7.19 Powder diffraction patterns: (a) a single crystal diffraction pattern; (b) a 'spotty ring pattern' created by rotating (a) by 0°, 25°, 39°, 43° and 12°; (c) ring pattern generated by a large number of randomly oriented crystallites, superimposed upon the single crystal diffraction pattern. The multiplicities of the equivalent reflections that make up the rings are shown as p.

(iii) 43° and (iv) 12°. In principle it is still possible to measure unique intensities, such as *I*(200), but it is becoming difficult. If large numbers of crystallites are present, a 'ring pattern' will form (Figure 7.19c). It is now impossible to measure a value for *I*(200). The ring arising from a (200) reflection will consist of the superposition of both (200) and ($\bar{2}00$) intensities. The best that can now be obtained is *I*(200) plus *I*($\bar{2}00$), and the intensity is double that expected from a single (200) diffracted beam. Hence the intensities measured on a powder pattern are greater than that of a single reflection by a multiplicity. For the (200) reflection described, the multiplicity is 2. In a ring pattern, the intensity of each ring is the sum of the intensities of more than one single crystal spot. The multiplicities of the 101 set (101, $\bar{1}01$, $10\bar{1}$ and $\bar{1}0\bar{1}$) will be 4. The multiplicities of other rings are noted in Figure 7.19c.

In general, the **multiplicity of equivalent reflections**, *p*, will depend upon the unit cell type. For any (*hkl*) reflection the multiplicity will be at least two (Table 7.5).

There are two other important angle-dependent factors that also need to be considered when the intensities of powder patterns are evaluated: the polarisation correction and the Lorentz correction. Both of these depend upon the experimental setup employed. The first of these relates to the **polarisation** of the diffracted beam. The incident beam of X-rays is usually unpolarised, but the inclusion of focusing crystal monochromators in the equipment will change this. Either way, the polarisation of the beam produces an angle-dependent reduction in intensity. The second factor is related to the time that an (*hkl*) plane spends in the diffracting position. Bragg diffraction takes place over a range of angles, not just at the exact value of θ given by Bragg's law (Section 7.1). In terms of the reciprocal lattice, it can be likened to each reciprocal lattice point having a volume (Figure 7.6). The time that each diffraction spot spends near enough to the Ewald sphere to give rise to a diffracted beam is then found to be angle-dependent. This is termed the **Lorentz factor**.

The Lorentz and polarisation factors are usually combined, for powder X-ray diffraction, into a single *Lp* correction term. As this varies with the experimental arrangement, there is no single universal correction factor that can be applied. A representative formulation is:

$$Lp = (1 + \cos^2 2\theta)/\left(\sin^2\theta \cos\theta\right)$$

Table 7.5 Multiplicity of equivalent reflections for powder diffraction patterns

Crystal class	Diffracting plane and multiplicity[a]					
Triclinic	All, 2					
Monoclinic	0*k*0, 2	*h*0*l*, 2	*hkl*, 4			
Orthorhombic	*h*00, 2	0*k*0, 2	00*l*, 2	*hk*0, 4	0*kl*, 4	*h*0*l*, 4	*hkl*, 8
Tetragonal	00*l*, 2	*h*00, 4	*hh*0, 4	*hk*0, 8[a]	0*kl*, 8	*hhl*, 8	*hkl*, 16[a]
Trigonal	00*l*, 2	*h*00, 6	*hh*0, 6	hk0, 12[a]	0*kl*, 12[a]	*hhl*, 12[a]	*hkl*, 24[a]
Hexagonal	00*l*, 2	*h*00, 6	*hh*0, 6	hk0, 12[a]	0*kl*, 12[a]	*hhl*, 12[a]	*hkl*, 24[a]
Cubic	*h*00, 6	*hh*0, 12	*hk*0, 24[a]	*hhh*, 8	*hhl*, 24	*hkl*, 48[a]	

[a]In some point groups the multiplicity is made up from contributions of two different sets of diffracted beams at the same angle but with different intensities.

The intensity of a powder diffraction ring is then written as:

$$I(hkl) = I_0(hkl)\exp\left[-2B\left(\frac{\sin\theta}{\lambda}\right)^2\right]\left(\frac{1+\cos^2 2\theta}{\sin^2\theta\ \cos\theta}\right)p$$

Another correction factor arises because the X-ray beam is absorbed as it passes through the sample. The effect of this is to reduce the intensities of reflections that pass through large volumes of sample compared with those that pass through small volumes. Correction factors will depend upon sample shape and size, and a general formula applicable to all sample geometries cannot be given. However, this term must be included in careful work.

In essence, a powder diffractogram contains as much information as a single crystal experiment. When the intensity and the positions of the diffraction pattern are taken into account, the pattern is unique for a single substance. The X-ray diffraction pattern of a substance can be likened to a fingerprint. In effect, the pattern of lines on the powder diffraction pattern of a single phase is virtually unique, and mixtures of different crystals can be analysed if a reference set of patterns is consulted.

To illustrate this, the powder diffraction patterns of two materials with closely similar structures and unit cells are shown in Figure 7.20. These are the *rutile* form of PbO_2,

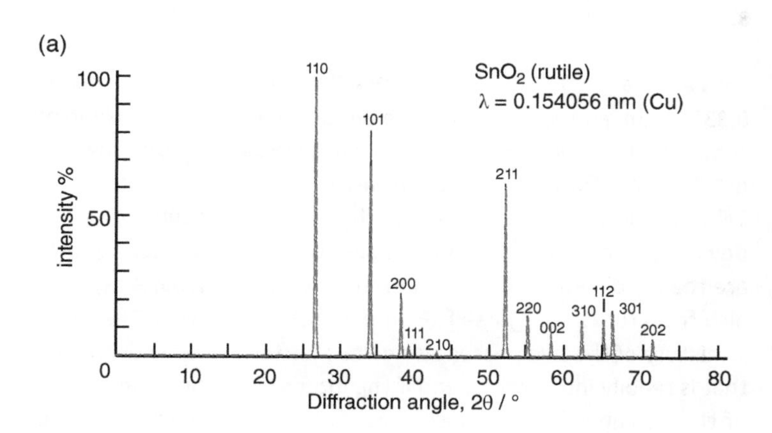

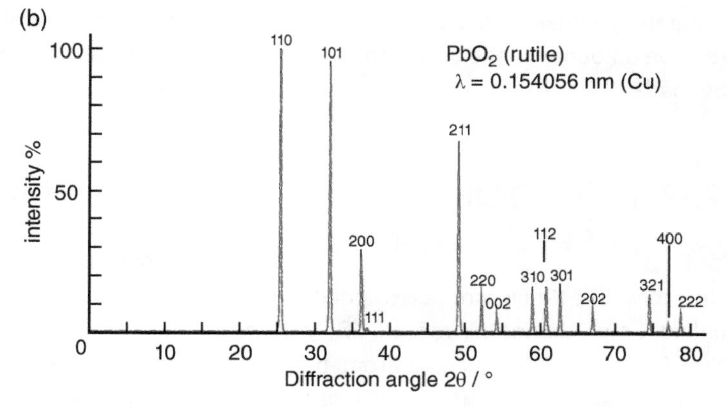

Figure 7.20 Powder diffraction patterns of *rutile* structure: (a) tin dioxide, cassiterite, SnO_2; (b) lead dioxide, PbO_2.

Table 7.6 Powder diffraction data for the rutile forms of SnO_2 and PbO_2, Cu radiation, $\lambda = 0.1540562$ nm

SnO_2 $a = 0.4737$ nm, $c = 0.3186$ nm			PbO_2 $a = 0.4946$ nm, $c = 0.3379$ nm		
hkl	***d*, nm**	**Relative intensity**	***hkl***	***d*, nm**	**Relative intensity**
110	0.3350	100	110	0.3497	100
101	0.2644	81	101	0.2790	95.5
200	0.2368	22.4	200	0.2473	29.5
111	0.2309	3.8	111	0.2430	1.5
210	0.2118	1.3	211	0.1851	67.4
211	0.1764	61.7	220	0.1749	15.1
220	0.1675	14.6	002	0.1689	7.4
002	0.1593	7.0	310	0.1564	16.1
310	0.1498	13.2	112	0.1521	16.1
112	0.1439	13.2	301	0.1482	17.4
301	0.1415	16.5	202	0.1395	9.4
202	0.1322	6.3	321	0.1271	13.7
321	0.1215	9.7	400	0.1237	3.8

with a tetragonal unit cell, $a = 0.49460$ nm, $c =$ 0.33790 nm, and SnO_2, which also adopts the *rutile* structure with $a = 0.47370$ nm, $c = 0.31860$ nm. The data for these structures are given in Table 7.6. It is seen that although the two powder patterns show a family similarity, they are readily differentiated from each other and also from rutile (TiO_2) itself (Figure 7.18c). A mixture of the two phases also gives a pattern that is readily interpreted, and if the intensities of the two lines from the two phases are compared, a quantitative assessment of the relative amounts of the two materials can be made.

7.12 NEUTRON DIFFRACTION

Neutrons, like X-rays, are not charged, but unlike X-rays, do not interact significantly with the electron cloud around an atom, but only with the massive nucleus. Because of this, neutrons penetrate considerable distances into a solid. Neutrons are diffracted following Bragg's law, and the intensities of diffracted beams can be calculated in a similar way to that used for X-ray beams. Naturally, neutron atomic scattering factors, which differ considerably from X-ray atomic scattering factors, need to be used. The technique is quite different from X-ray diffraction in practice, not least because neutrons need to be generated in a nuclear reactor and need to be slowed to an energy that is suitable for diffraction experiments. Despite this difficulty, neutron diffraction has a number of significant strong points, and is used routinely for structure determination. Neutron diffraction is usually carried out on powder samples, and structure refinements are carried out via the Rietveld method, often in conjunction with X-ray data for the same material.

An advantage of neutron diffraction is that it is often able to distinguish between atoms in a crystal that are difficult to separate with X-rays.

This is because X-ray scattering factors are a function of the atomic number of the elements, but this is not true for neutrons. Of particular importance is the fact that the neutron scattering factors of light atoms, such as hydrogen, H, carbon, C, nitrogen, N, and oxygen, O, are similar to, or even greater than, those of transition metals and heavy metal atoms. This makes it easier to determine their positions in the crystal compared with X-ray diffraction, and neutron diffraction is the method of choice for the precise determination of the positions of light atoms in a structure.

Another advantage of neutron diffraction is that neutrons have a spin and so interact with the magnetic structure of the solid, which arises from the alignment of unpaired electron spins in the structure. The magnetic ordering in a crystal is quite invisible to X-rays, but is revealed in neutron diffraction as extra diffracted beams, and hence a change in unit cell.

7.13 STRUCTURE DETERMINATION USING X-RAY DIFFRACTION

Crystal structures have most often been determined by diffraction of X-rays from single crystals of the sample. In order to generate a structure, it is necessary to convert a dataset consisting of diffraction spot position and intensity into an electron density map. The principle can be illustrated by imagining a one-dimensional crystal, composed of a line of atoms of various types. The contribution of the undiffracted beam is **F**(000), the structure factor for the 000 reflection. To this is added the contributions of, say, the 100 and $\bar{1}00$ reflections, **F**(100) and **F**($\bar{1}00$), to give a low resolution 'structure image'. More pairs of reflections are then added in order to improve the resolution and obtain a more realistic rendering of the atom chain. The structure can be represented by:

$$\text{structure} = \mathbf{F}(000) + \mathbf{F}(100) + \mathbf{F}(\bar{1}00) + \ldots$$

Unfortunately, values of **F**(hkl) are not available in the X-ray dataset. Recall that

$$\mathbf{F}(hkl) = F(hkl)e^{i\phi_{hkl}}$$

The value $F(hkl)$ is easily obtained as it is equal to the square root of the measured intensity. However, the phase angle, ϕ_{hkl}, for the reflection *cannot be recovered from the intensity*. This fact is known as the **phase problem**. The equation for the contrast must be rewritten in terms of the known experimentally determined $F(h00)$ values and the unknown phases, ϕ_{100}, ϕ_{200}, ϕ_{300}, etc. The equation for a one-dimensional chain is:

$$\begin{aligned}\text{contrast} &= F(000) + F(100)\cos 2\pi(x-\phi_{100}) \\ &\quad + F(200)\cos 2\pi(2x-\phi_{200}) \\ &\quad + F(300)\cos 2\pi(3x-\phi_{300}) + \ldots \\ &= F_{000} + \sum_{h=-\infty}^{+\infty} F_h \cos 2\pi(xh-\phi_h)\end{aligned}$$

where the summation is taken over all of the $h00$ and $\bar{h}00$ reflections. The units of the various scattering factors, $F(h00)$, are electrons. The electron density of the crystal in the x direction, given in electrons per unit distance, is related to the structure factor series by $\rho(x)$, where:

$$\rho(x) = (1/d_{100})\left[F_{000} + \sum_{h=-\infty}^{+\infty} F_h \cos 2\pi(xh-\phi_h)\right]$$

The one-dimensional equation can be generalised to three dimensions and the electron density of a crystal at any point in the unit cell x, y, z, is given by:

$$\rho(x,y,z) = \frac{1}{V}\sum_h\sum_k\sum_l F(hkl)$$

$$\exp\left[-2\pi i\left(hx + ky + lz - \phi_{(h,k,l)}\right)\right]$$

where V is the volume of the unit cell, and the indices h, k and l run from $-\infty$ to $+\infty$. Because values of $F(hkl)$ are available experimentally, the problem of computing the electron density or, equivalently, determining the crystal structure, reduces to how to obtain the phase, $\phi_{(h,k,l)}$ for each reflection, viz. how to solve the phase problem. The methods used are described in the following section.

The procedure of structure determination from a single crystal X-ray diffraction experiment can be summarised in the steps below.

(i) Obtain an accurate set of intensity values. The correction factors mentioned above, as well instrumental factors associated with the geometry of the diffractometer, must be taken into account. (In the early days of X-ray crystallography this in itself was no easy task.)
(ii) Determine the unit cell and index the diffracted reflections in terms of hkl. Determine the point group and space group of the crystal, making use of systematic absences.
(iii) Construct a possible model for the crystal, using physical or chemical intuition, or established techniques for solving the phase problem (see Section 7.14).
(iv) Compare the intensities, or more usually the structure factors, expected of the model structure with those obtained experimentally.
(v) Adjust the atom positions in the model repeatedly to obtain improved agreement between the observed and calculated values. This process is called **refinement**. The crystal structure is generally regarded as satisfactory when a low value of the **reliability factor**, **residual**, or **crystallographic R factor**, is obtained. There are a number of ways of assessing the R factor, each of which gives a different number, and often more than one value is cited. The most usual expression used is:

$$R = \frac{\sum_{hkl}\left||F_{obs}| - |F_{calc}|\right|}{\sum_{hkl}|F_{obs}|}$$

where F_{obs} and F_{calc} are the observed and calculated structure factors, and the modulus values, written $|F_{obs}|$, etc. mean that the absolute value of the structure factor is taken, and negative signs are ignored. Structure determinations for well crystallised inorganic or metallic compounds have R values of the order of 0.03. The values of R for organic compounds or complex molecules such as proteins are generally higher.

When a powder is examined, many diffracted beams overlap so that the procedure of structure determination is more difficult. In particular this makes space group determination less straightforward. Nevertheless, powder diffraction data are now used routinely to determine the structures of new materials. An important technique used to solve structures from powder diffraction data is that of **Rietveld refinement**. In this method, the exact shape of each diffraction line, called the **profile**, is calculated and matched with the experimental data. Difficulties arise not only because of overlapping reflections, but also because instrumental factors add significantly to the profile of a diffracted beam. Nevertheless, Rietveld refinement of powder diffraction patterns is routinely used to determine the structures of materials that cannot readily be prepared in a form suitable for single crystal X-ray study.

7.14 SOLVING THE PHASE PROBLEM

Initially, all crystal structures were solved by constructing a suitable model making use of any physical properties that reflect crystal symmetry, and chemical intuition regarding formulae and bonding, computing F_{hkl} values and comparing them with observed F_{hkl} values. This 'trial and error' method was used to solve the structures of fairly simple crystals, such as the metals and minerals described in Chapter 1. (Many of the structural relations described in the following chapter originate in attempts to arrive at good starting structures for subsequent X-ray analysis.) However, the method is very labour-intensive, and to a large extent success depends upon the intuition of the researcher.

One of the first mathematical tools to come to the aid of the crystallographer was the development of the **Patterson function**, described in 1934 and 1935. The Patterson function is:

$$P(u,v,w) = \frac{1}{V}\sum_h\sum_k\sum_l F(hkl)^2 \cos 2\pi(hu + kv + lw)$$

where the indices h, k and l run from $-\infty$ to $+\infty$. The Patterson function does not need phases, and the squares of the structure factors are available experimentally as the intensities of the hkl reflections. The Patterson function, when plotted on (u, v, w) axes, forms a map rather like the display of electron density. However, peaks in the Patterson map correspond to interatomic vectors and the peak heights are proportional to the product of the scattering factors of the two atoms at the vector head and tail. That is, suppose that there are atoms at A, B and C in a unit cell. The function $P(uvw)$ would show peaks at U, V, W, where $O - U$ is equal to $A - B$, $O - V$ is equal to $B - C$ and $O - W$ is equal to $A - C$, (Figure 7.21). The vectors can be related to the atom positions in the unit cell by a number of methods, including direct methods, described in the following paragraphs. The main drawback of the Patterson function is that large numbers of similar atoms give overlapping peaks that cannot be resolved. It is of greatest use in locating a small number of heavy atoms in a unit cell. In this way, the location of the positions of Co atoms in vitamin B_{12} led to the successful resolution of the complete structure by 1957. This technique is also of importance in the multiple isomorphous replacement (MIR) method for the solution of protein structures described below.

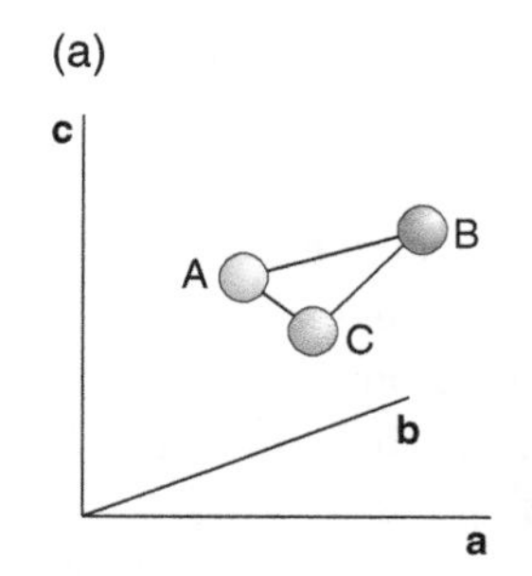

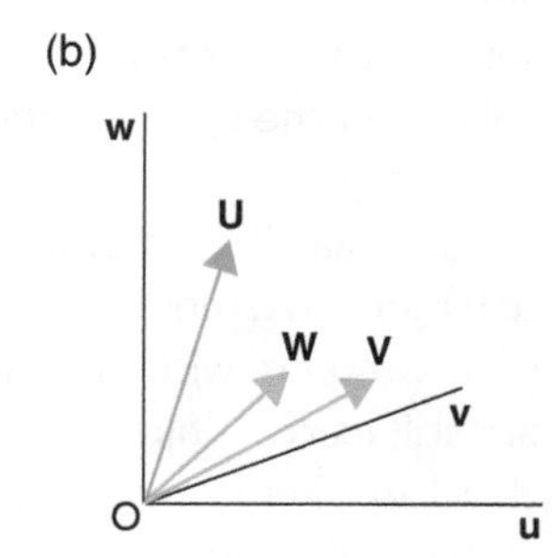

Figure 7.21 The Patterson function: (a) a molecule, containing atoms A, B, C, in a unit cell; (b) the corresponding Patterson function contains vectors **OU**, **OV**, and **OW** which correspond to the interatomic vectors A–B, B–C, A–C.

Direct methods replaced trial and error and other methods of deducing model structures in the middle of the twentieth century. In these techniques, statistical relationships between the amplitudes and phases of the strong reflections were established, and the mathematical methodology between these quantities was worked out, particularly by Hauptmann and Karle, in the 1940s and 1950s. (The Nobel Prize was awarded to these scientists in 1985 for these studies.) A number of algorithms, which exploited the growing power of electronic computers, used this mathematical framework to derive structures directly from the experimental dataset of position, intensity and *hkl* index. The use of these programmes allows the structures of molecular compounds with up to 100 or so atoms to be derived routinely, but as more atoms are added the computations become increasingly lengthy and subject to error. A rapid increase in computing power has led to an extension of direct methods to molecules with 500 atoms or more.

There are a number of computing approaches used in direct method techniques, one being the '*bake and shake*' method. In this procedure, a model structure is derived (the bake part). The phases are calculated and then perturbed, that is, changed slightly (the shake part), so as to arrive at a lower value, in accordance with the formulae implemented in the computer algorithms. When this is completed, a new set of atomic positions is calculated (a new bake) and the cycle is repeated for as long as necessary.

Although direct methods are being extended to larger and larger molecules, the structures of large proteins, with more than say 600 atoms, are still inaccessible to this technique. Such structures are mainly determined using the methods described below.

7.15 ELECTRON MICROSCOPY

7.15.1 Diffraction Patterns and Structure Images

Crystal diffraction patterns and images of the structure are observed using transmission electron microscopy.[4] Electrons are charged, and, as a consequence, they interact much more strongly with the outer electron cloud of atoms than X-rays. Because of this, electron scattering is far more intense than X-ray scattering, and an electron beam accelerated by 100 kV is only able to penetrate about 10 nm into a crystal before being either absorbed or entirely diffracted into other directions. Thus, although the condition for the diffraction of electrons is given by Bragg's law, a more complex theory is needed to calculate the intensity of the diffracted beams. In the case of X rays, it is reasonable to assume that each diffracted X-ray photon is scattered only once, and the kinematical theory of diffraction is adequate for X-ray diffraction (Section 7.4). Electrons are generally diffracted frequently on passing through a crystal, and each electron is scattered many times, even when traversing a slice of crystal only 1 or 2 nm in thickness. This implies that the intensity of the incident beam not only falls as electrons are diffracted, but also increases as electrons are re-diffracted back into the incident beam direction. The theory that is needed to account for this is called the **dynamical theory** of diffraction. The calculation of the diffraction pattern intensities is therefore not as easy as in the case of X-rays. However, the major advantage of electrons,

[4] Scanning electron microscopy, a technique which mainly gives information about surfaces, is not considered here.

which quite offsets this disadvantage, is that, because they are charged, they can be focused by magnetic lenses to yield an image of the structure. This reciprocal relationship between the formation of the diffraction pattern and the image is at the heart of structure determination, irrespective of the technique used. It is displayed most clearly, however, in high-resolution electron microscopy.

The steps by which an electron (or any) microscope forms an image is as follows. An electron beam is propagated along the microscope axis. (As electrons interact strongly with any molecules present, the electron microscope operates under a high vacuum.) An objective lens, which sits close to the sample, takes each of the separate sets of beams that have been diffracted by the crystal and focuses them to a spot in the back focal plane of the lens, to form the diffraction pattern. Lenses below the objective lens, called intermediate and projector lenses, are then focused so as to form a magnified image of the diffraction pattern on the viewing screen (Figure 7.22). The recorded electron diffraction pattern is a good approximation to a plane section through the weighted reciprocal lattice of the material (Section 7.2).

However, if the diffracted beams are allowed to continue, they create an image of the crystal, or more precisely, an image of the projected electron density in the crystal. This forms in the image plane of the objective lens, some way beyond the diffraction pattern Figure 7.23. This image comes about simply by the recombination of the diffracted beams. The focal length of the intermediate and projector lenses can be altered to focus either upon the diffraction pattern (in the back focal plane of the objective lens) or the image, formed in the image plane of the objective lens. Additionally, because the focal lengths can be varied continuously, the various intensity patterns that form on the viewing screen between, on the one hand, the diffraction pattern, and, on the other, the image, can be observed. The way in which the diffracted beams recombine to form an image is therefore lucidly demonstrated.

The image reconstructed in this way contains all of the information present in each diffracted beam, including the phase of each diffracted beam. This includes the structure factors, atomic scattering factors and phase angles for each reflection that contributes to the image. Now the detail in the image will depend critically upon how many diffracted beams contribute to it. If only the undiffracted 000 beam is used, almost no structural information is present in the image, although gross defects in the structure, such as precipitates and dislocations, can be imaged using a special technique called diffraction contrast.

If the undiffracted beam and a pair of other beams, say 100 and $\bar{1}00$, are combined, the image will consist of a set of 'lattice fringes' with a spacing of d_{100}, running parallel with the (100) planes in the crystal and an approximate amplitude $F(100)$ (Figure 7.24a). In electron microscope parlance, the microscope has 'resolved' the (100) 'lattice' planes. This fringe pattern is, in fact, a poor representation of the one-dimensional electron density (or even less precisely, the atoms) associated with the (100) planes in the crystal parallel to the **a**-axis of the unit cell. As other reflections, for example 200 and $\bar{2}00$, are added in, further sets of fringes are incorporated into the image, and these add or subtract from those already present to form a more complex fringe pattern, but still parallel to the original (100) set, and with a spacing of d_{200}, and approximate amplitude $F(200)$, so that the microscope has 'resolved' the (200) 'lattice' planes (Figure 7.24b). This is now a better picture of the electron density (or atoms) along

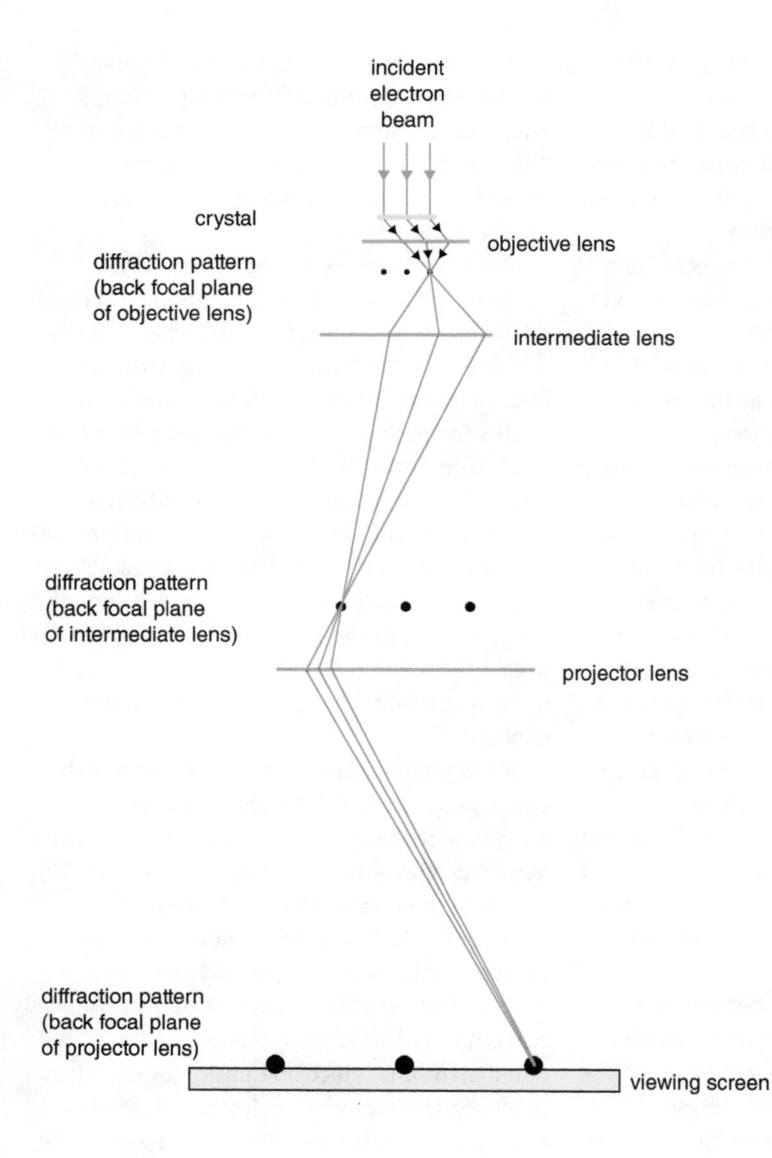

Figure 7.22 The formation of a magnified diffraction pattern in an electron microscope, (schematic).

the **a**-axis of the unit cell, dependent upon the values of **F**(h), but the overall fringe repeat remains equal to d_{200}.

As more and more pairs of $h00$ and $\bar{h}00$ beams are added in, the fringe profile becomes more and more complex, but the result can always be viewed in simple terms as a set of fringes parallel to (100), with spacing equal to d_{h00}. Should this degree of detail be imaged by the instrument, these planes, 300, 400, and higher, have been 'resolved'. The fringe pattern gives a better and better approximation to the electron density (or atom positions) in the crystal in this one direction.

If reflections corresponding to $0k0$ diffraction spots are considered, the fringes in

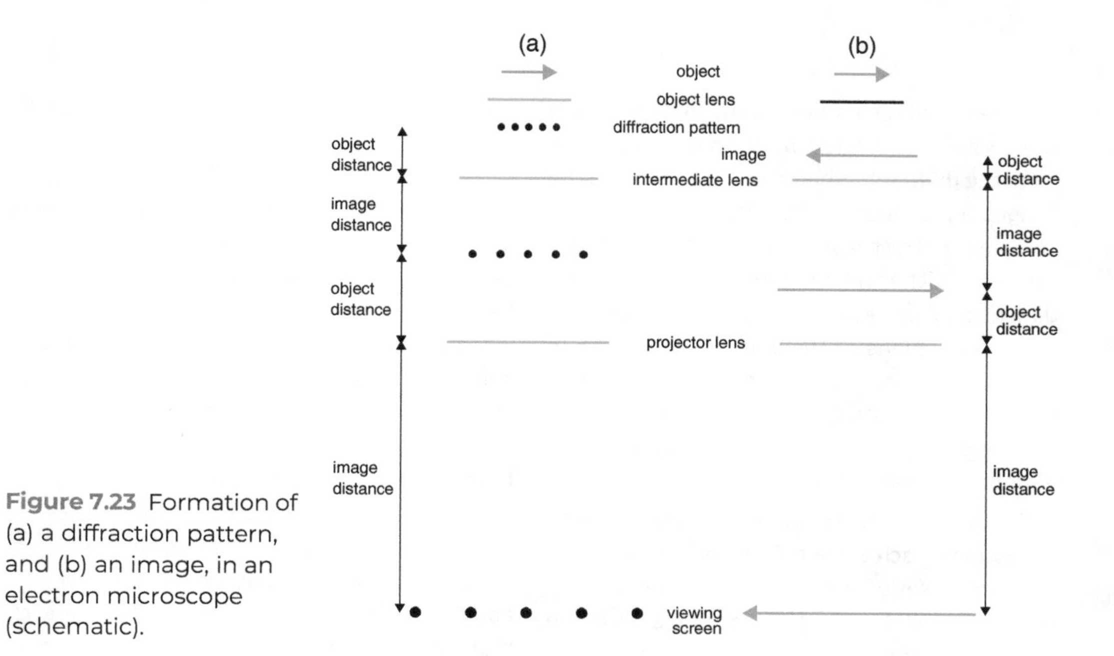

Figure 7.23 Formation of (a) a diffraction pattern, and (b) an image, in an electron microscope (schematic).

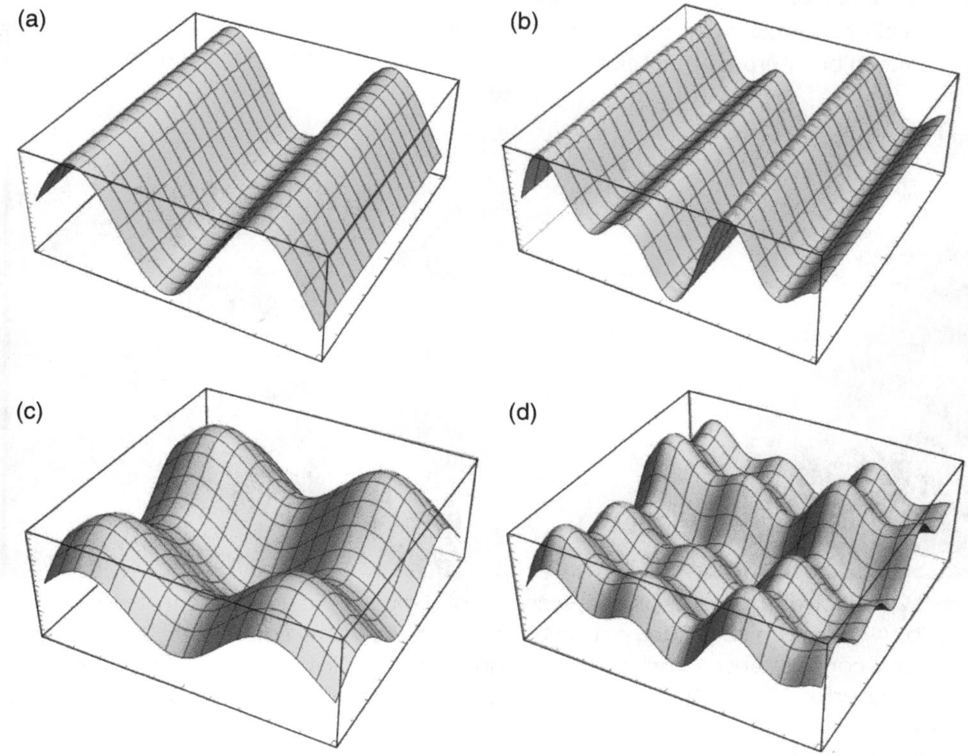

Figure 7.24 Schematic of the formation of an image in an electron microscope: (a) (000), (100) and $(\bar{1}00)$ beams; (b) (000), (100), $(\bar{1}00)$, (200) and $(\bar{2}00)$ beams; (c) (000), (100), $(\bar{1}00)$, (010) and $(0\bar{1}0)$ beams; (d) (000), (100), $(\bar{1}00)$, (200), $(\bar{2}00)$, (010), $(0\bar{1}0)$, (020) and $(0\bar{2}0)$ beams.

the image will run parallel to the (010) planes in the crystal, but will resemble, in outline, those from the (h00) set. When these are added to the fringes representing (*h*00) planes, a more complex contrast resembling a set of peaks is formed, which can be regarded as a poor two-dimensional projection of the electron density in the (001) plane of the unit cell (Figure 7.24c and d). More detail of the projected electron density (i.e. the structure) is obtained by including fringes at an angle, such as the members of the *hk*0 set.

The same is true for all series of *hkl* and $\overline{h}\,\overline{k}\,\overline{l}$ reflections. Each extra pair of reflections produces a set of fringes, of spacing d_{hkl}, running parallel to the (*hkl*) set of planes in the crystal, which are added into the resultant image contrast in proportion to the intensity of the reflections. With each step, the image contrast resembles the electron density in the object more and more. As greatest electron density usually occurs close to atomic nuclei, the image can then be interpreted in terms of projected atoms without significant loss of precision.

There are two important points to note. Firstly, the amount of information in the image, its **resolution**, is dependent upon the *number* of diffracted beams that are included in the image formation process. As more beams are incorporated into the image, the closer this approaches the real structure. Secondly, all of the information in each diffracted beam, including the phase of the structure factor, **F**(*hkl*), as well as any spurious information introduced by lens or instrumental defects, contributes to image formation. Thus, an electron micrograph contains information about non-periodic structures, such as defects, that may be present. As electron microscopes are able to image at atomic resolution, the defect structure of complex materials can be understood in a degree of detail not available to X-ray diffraction. For example, Figure 7.25a is a micrograph of a flake of niobium pentoxide, H-Nb_2O_5. The structure is composed of columns

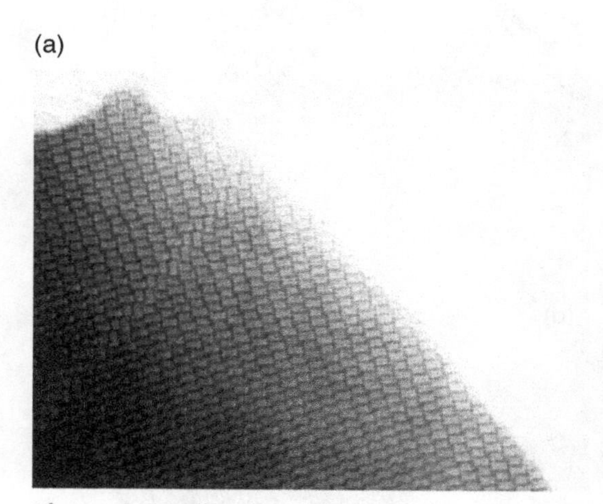

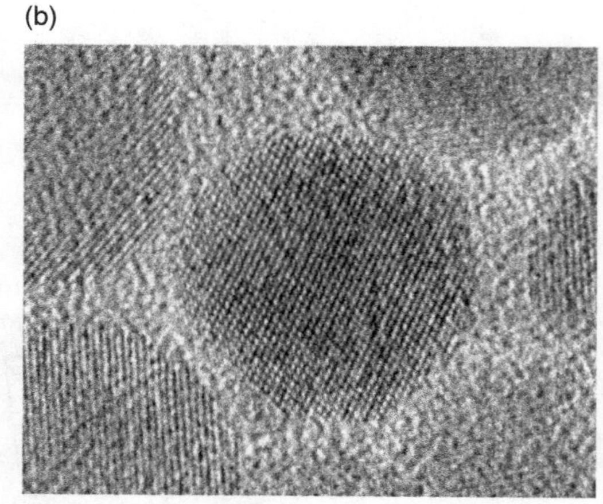

Figure 7.25 Electron micrographs: (a) a crystal of a niobium oxide close in composition to Nb_2O_5. (b) A cadmium sulphide particle of approximately 8 nm diameter. The uneven background contrast arises from the amorphous carbon film used to support the particles. *Source*: (b) courtesy Dr J.H. Warner and Dr R.D. Tilley, Victoria University of Wellington, New Zealand.

of corner-linked NbO_6 octahedra, which in projection, look like blocks or tiles. The blocks are of two sizes, approximately 1 × 1.4 and 1 × 1.8 nm. The regular way in which the blocks tile the surface is clearly revealed in Figure 7.25a, but towards the top of the image there is a lamella of structure in which the blocks, although of normal sizes, are arranged differently. This would not be revealed by X-ray diffraction. The technique is also especially valuable in the study of nanoparticles, which can be imaged in an electron microscope. For example, Figure 7.25b is the image of a cadmium sulphide nanoparticle approximately 8 nm in diameter. The particle is revealed by the two sets of overlapping 'lattice' fringes. Microstructural details for such particles via X-ray diffraction are difficult to obtain.

7.15.2 Diffraction and Fourier Transforms

The relationship between an electron diffraction pattern and the resulting electron micrograph can be treated mathematically by way of Fourier analysis. In 1822, Joseph Fourier published a theory that described heat conduction in terms of an infinite series of sine and cosine functions, now called a Fourier series. Since then, the method of using Fourier series to better understand the periodic properties of materials has been widely used. In the present context, both the crystal structure and the diffraction of a crystal are periodic structures that can be treated mathematically in terms of Fourier series.

Taking a one-dimensional example, any repetitive function $f(x)$ with a period 2π can be represented by an infinite series

$$f(x) = a_0/2 + a_1 \cos x + a_2 \cos 2x + a_3 \cos 3x + \ldots + b_1 \sin x + b_2 \sin 2x + b_3 \sin 3x \ldots$$

where a_0, a and b are constants, n is a positive integer, and x runs from $+\infty$ to $-\infty$. For a function which is even, $f(x) = f(-x)$ so the sine terms in the series reduce to zero, leaving a cosine series. For an odd function, in which $f(-x) = -f(x)$, the cosine terms reduce to zero, leaving a sine series.

A diffraction pattern is formed when the crystal structure is irradiated by a monochromatic beam of electrons. When the electron wave emerges from the crystal, the electric field has a space variation that mirrors the transmission profile of the crystal. (The atoms have imposed information upon the incident beam that is now being carried away from it.) Just at the exit of the crystal, the amplitude of the electromagnetic field in the outgoing wave can be represented by a collection of sinusoidal plane waves that are combined with the incident wave. Each of the plane sinusoidal waves moves away from the crystal at a definite angle θ that depends upon the wavelength of the initial beam and the interplanar spacing of the diffracting (hkl) planes in the crystal, d_{hkl}. The equation relating the two is the Bragg equation:

$$\sin \theta = m\lambda / 2d_{hkl}$$

Each of these new waves gives rise to an irradiance peak in the detector which makes up the Fraunhofer diffraction pattern. When a lens is used to bring the diverging waves to a focus at a fairly short distance from the crystal, the diffraction pattern is found in the back focal plane of the lens.

The diffraction pattern can be written as a Fourier series. Suppose we have diffraction from a single (one-dimensional) row of atoms. As the pattern is an even function, the Fourier series that represents the variation of the amplitude of the electric field component of the waves due to the crystal periodicity is given by:

$$E(x) = a_0/2 + \sum a_n \cos(2\pi n q_1 x) = a_0/2 + a_1 \cos 2\pi q_1 x + a_2 \cos 4\pi q_1 x + a_3 \cos 6\pi q_1 x + \ldots$$

where $q_1 = 1/d$ is the spatial frequency of the crystal. The constants $a_1, a_2, a_3, \ldots$ are the amplitudes of the harmonics with frequencies $q_1, 2q_1, 3q_1, \ldots$

However, this gives an unsymmetrical plot, and a diffraction pattern is symmetrical about the zero order. The series can be made symmetrical by using the fact:

$$\cos\theta = \cos -\theta$$

and so each term in the series can be split into two, to give:

$$E(x) = a_0/2 + (a_1/2)\cos(2\pi qx) + (a_1/2)\cos(-2\pi qx) + (a_2/2)\cos(4\pi qx) + (a_2/2)\cos(-4\pi qx) + (a_3/2)\cos(6\pi qx) + (a_3/2)\cos(-6\pi qx)\ldots$$

When this is plotted, a representation of the diffraction pattern is obtained. However, the diffraction pattern is, in fact, a plot of the time average of the Fourier transform.

A measurement of the position of a diffraction spot gives the value of mq and a measurement of the intensity of the spot gives the square of the amplitude of the component.

This last result holds broadly true universally, with one small change. Most diffracting objects are not infinitely long and so are not, strictly speaking, infinite periodic functions. To handle non-period functions, the Fourier summation is replaced by an integral and the term Fourier *series* is replaced by Fourier *transform*. An electron diffraction pattern is often called the Fourier transform of the crystal.

The reverse procedure, called the inverse Fourier transform, is also possible. The Fourier transform works both ways, so that we can say if B is the Fourier transform of A then A is the Fourier transform of B. Thus, the image formed in an electron microscope by allowing the diffracted beams to overlap can be treated as the inverse Fourier transform of the diffraction pattern and is often called the Fourier transform of the diffraction pattern.

7.16 PROTEIN CRYSTALLOGRAPHY

7.16.1 The Phase Problem

In principle, protein crystallography, or indeed the crystallography of any large organic molecule, is identical to that of the crystallography of inorganic solids. The most significant difference between proteins and the structures considered earlier is the enormous increase in the complexity between the two. For example, one of the smaller proteins, insulin, has a molar mass of approximately 5700 g mol^{-1}, and a run-of-the-mill (in size terms) protein, haemoglobin, has a molar mass of approximately 64 500 g mol^{-1}. When it is remembered that proteins are mostly made up of light atoms, H, C, N and O, the atomic complexity can be appreciated. This difficulty is reflected in the difference between the rate of development of inorganic crystallography compared with protein crystallography. The first inorganic crystal structures were solved by W.H. and W.L. Bragg in 1913, but it was not until the middle of the same century, a lapse of 50 years or so, before the structures of macromolecules and proteins began to be resolved.

The problems associated with the determination of protein structures were centred upon by two factors: the growth of perfect crystals of a size suitable for high-quality single crystal X-ray diffraction studies

(Section 7.16.2), and the difficulty of determining the phases of the many X-ray reflections obtained. The first protein structures were derived using a technique called **isomorphous replacement** (IR), developed in the late 1950s. The materials used are heavy metal derivatives of protein crystals. To obtain a heavy metal derivative of a protein, the protein crystal is soaked in a solution of a heavy metal salt. The metals most used are Pt, Hg, U, lanthanoids, Au, Pb, Ag and Ir. The heavy metal or a small molecule containing the heavy metal, depending upon the conditions used, diffuses into the crystal via channels created by the disordered solvent present. The aim is for the heavy metal to interact with some surface atoms on the protein, without altering the protein structure. This is never exactly achieved, but in suitable cases the changes in structure are slight.

The experimental situation is now that two sets of data are available: F_P, a structure factor magnitude list for the native protein, and F_{PH}, a structure factor magnitude list for the IR protein. The vector relationships between the three structure factors, $\mathbf{F}_P$, $\mathbf{F}_{PH}$, and that of the heavy metal atoms alone, $\mathbf{F}_H$ is:

$$\mathbf{F}_{PH} = \mathbf{F}_P + \mathbf{F}_H \tag{7.5}$$

The connection between these three pieces of data can be understood if the scattering factors are displayed in vector format (Figure 7.26), where $\mathbf{F}_P$ is the vector scattering factor of any *hkl* reflection from the protein, with magnitude F_P and phase angle ϕ_P, $\mathbf{F}_H$ is the vector scattering factor for the heavy atoms alone, with magnitude F_H and phase angle ϕ_H, and $\mathbf{F}_{PH}$ is the vector scattering factor of the heavy atom derivative of the protein, with magnitude F_{PH} and phase angle ϕ_{PH}.

The positions of the heavy metal atoms alone can be determined using techniques such as the Patterson function. Once the positions of the heavy metal atoms are known with reasonable precision, $\mathbf{F}_H$, F_H and ϕ_H can be calculated. Using Figure 7.26, it is possible to write:

$$F_{PH}^2 = F_P^2 + F_H^2 + 2F_P F_H \cos(\phi_P - \phi_H)$$

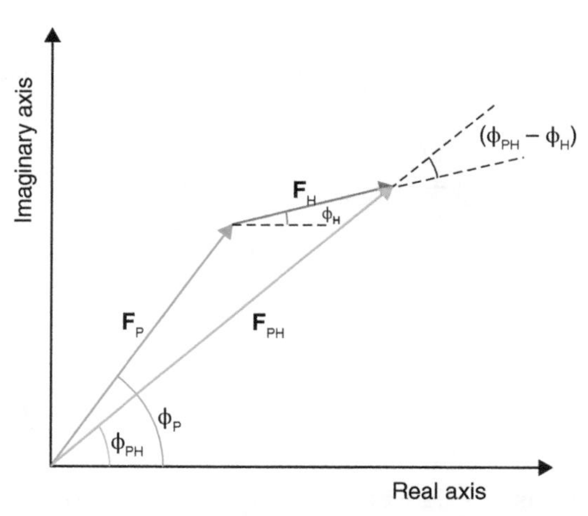

Figure 7.26 Representation in the Gaussian plane of the phase relationships between the structure factor of a pure protein, $\mathbf{F}_P$, a heavy atom, $\mathbf{F}_H$, and an isomorphic heavy atom derivative of the protein, $\mathbf{F}_{PH}$.

hence:

$$\cos(\phi_P - \phi_H) = (F_{PH}^2 - F_P^2 - F_H^2)/2F_P F_H$$

$$\phi_P = \phi_H + \cos^{-1}\left[(F_{PH}^2 - F_P^2 - F_H^2)/2F_P F_H\right] \tag{7.6}$$

Now this cannot be solved for ϕ_P unambiguously, as the $\cos^{-1}$ term has two roots (solutions), as displayed graphically in Figure 7.27. A circle of radius F_P delimits the possible positions of the vector $\mathbf{F}_P$. A circle of radius F_{PH}, drawn from the head of the vector $-\mathbf{F}_H$, represents all possible positions of vector $\mathbf{F}_{PH}$ with respect to vector $\mathbf{F}_H$. The positions where these two circles intersect represent the solutions of Eq. (7.6) above. Although this does not give a unique value for the phase angle, it is sometimes possible to augment the data from elsewhere to give a preferred value. Once sufficient phase angles have been estimated, structure refinement can proceed. This technique, the **single isomorphous replacement (SIR)** method, has been successfully used to determine a number of protein structures.

The value of ϕ_P can be found unambiguously if another (different) heavy atom derivative is made, so that now $\mathbf{F}_{H1}$, $\mathbf{F}_{H2}$, $\mathbf{F}_{HP1}$, $\mathbf{F}_{HP2}$ and $\mathbf{F}_P$ are to be determined. This is **multiple isomorphous replacement (MIR)** and is generally used rather than the SIR technique. The values for $\mathbf{F}_{H1}$ and $\mathbf{F}_{H2}$ can be determined using Patterson techniques. Two equations, similar to Eq. (7.6) above, now exist:

$$\phi_P = \phi_{H1} + \cos^{-1}\left[(F_{PH1}^2 - F_P^2 - F_{H1}^2)/2F_P F_{H1}\right]$$

$$\phi_P = \phi_{H2} + \cos^{-1}\left[(F_{PH2}^2 - F_P^2 - F_{H2}^2)/2F_P F_{H2}\right]$$

This pair of simultaneous equations can be solved to give a unique solution. If a diagram similar to Figure 7.27 is drawn, incorporating the data for both derivatives, three circles can be plotted, which intersect at one point, corresponding to the unique value of ϕ_P (Figure 7.28). The structure factor of the protein

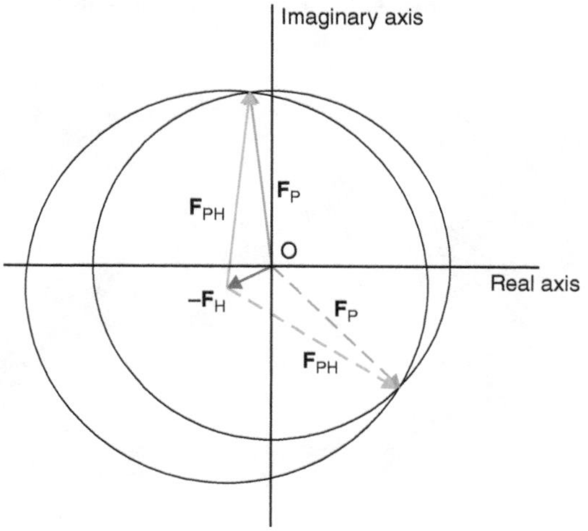

Figure 7.27 Representation in the Gaussian plane of the phase relationships derived by single isomorphous replacement (SIR) in a protein. The intersection of the two circles represents the two solutions to Eq. (7.6).

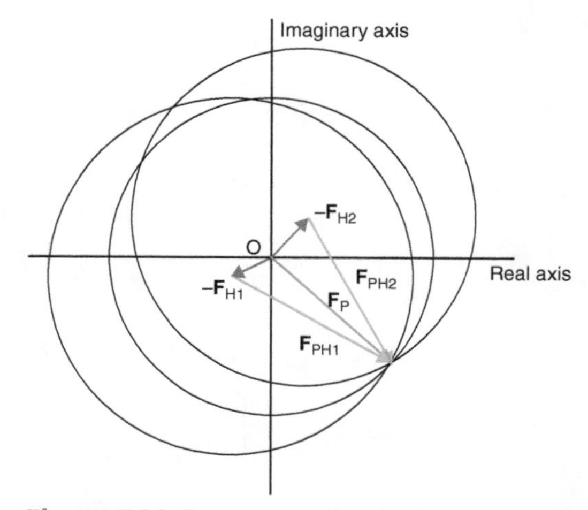

Figure 7.28 Representation in the Gaussian plane of the phase relationships derived by multiple isomorphous replacement (MIR) in a protein.

vector, $\mathbf{F}_P$, lies on a circle of radius F_P centred at O. The structure factors of the heavy metal derivatives, $\mathbf{F}_{PH1}$ and $\mathbf{F}_{PH2}$, lie on circles of radii F_{H1} and F_{H2} with centres at the tips of the vectors $-\mathbf{F}_{H1}$ and $-\mathbf{F}_{H2}$. The intersection of the three circles represents a unique solution to position of $\mathbf{F}_P$, corresponding to a single phase angle. When sufficient phase angles are known, refinement can proceed.

Protein structures are also determined by using **multiple anomalous dispersion (MAD)** techniques. In this method, scattering that does not obey the kinematic theory is used. The kinematic theory deals with single scattering of X-ray photons, which have low energy compared with the electron energy levels of the scattering atom. The scattered photon is supposed to suffer no loss of energy and no phase delay compared with the incident photon. This process is represented by the atomic scattering factor f_a (Section 7.5).

However, if the energy of the X-ray photon is enough to excite an electron in the atom from one energy level to another, supplementary processes occur. In addition to the photons scattered kinematically, others will be absorbed and then re-emitted. To account for this extra detail, the scattering factor f_a needs to be represented as a **complex scattering factor**

$$f_a = f'_a + \mathrm{i}\, f''_a$$

and the structure factor $\mathbf{F}(hkl)$ by a **complex structure factor**

$$\mathbf{F}(hkl) = \mathbf{F}'(hkl) + \mathbf{F}''(hkl)$$

The wavelength at which an electron is excited from one energy level to another for any atom is called the **absorption edge**. An atom will display several absorption edges, depending upon the energy levels available to the excited electron, and **anomalous absorption** occurs when the wavelength of the X-ray beam is close to the absorption edge of an atom in the sample. Anomalous scattering causes small changes in intensities, and in particular, Friedel's law (see Section 7.9) breaks down and $F(hkl)$ is no longer equal to its Friedel opposite, $F(\bar{h}\,\bar{k}\,\bar{l})$ at the same wavelength. The difference, ΔF_b, is called the **Bijvoet difference**, and the reflection pairs are called **Bijvoet pairs**.

$$\Delta F_b = F(hkl) - F\left(\bar{h}\,\bar{k}\,\bar{l}\right)$$

In addition, $F(hkl)$ is not constant but varies slightly with wavelength, to give an additional **dispersive difference**.

In general, the anomalous scattering from light atoms is negligible. However, anomalous scattering from heavy atoms, either native to the protein, or in isomorphous heavy atom derivative proteins, can be appreciable when the X-ay wavelength is chosen to be close to the absorption edge of the heavy atoms that have been used. The information provided by anomalous scattering can be used to solve the phase problem for proteins in the following way.

When anomalous scattering from Bijvoet pairs of a heavy metal atom derivative of a protein occurs, the vector relationship Eq. (7.5) now needs to be written:

$$\mathbf{F}_{PH}{}^{+} = \mathbf{F}_P + \mathbf{F}_H{}^{+} = \mathbf{F}_P{}^{+} + \mathbf{F}_H{}^{+\prime} + \mathbf{F}_H{}^{+\prime\prime}$$

$$\mathbf{F}_{PH}{}^{-} = \mathbf{F}_P + \mathbf{F}_H{}^{-} = \mathbf{F}_P{}^{-} + \mathbf{F}_H{}^{-\prime} + \mathbf{F}_H{}^{-\prime\prime}$$

where the superscripts + and – refer to the hkl and $\bar{h}\,\bar{k}\,\bar{l}$ Bijvoet pairs. If the protein contains only one type of anomalous scatterer, giving rise to **single anomalous dispersion**, the vectors $\mathbf{F}_H'$ and $\mathbf{F}_H''$ are perpendicular to each other, so if ϕ_H is the phase angle of $\mathbf{F}_H'$, the phase angle of $\mathbf{F}_H''$ is $\phi_H + \pi/2$. Figure 7.26 is now modified to Figure 7.29.

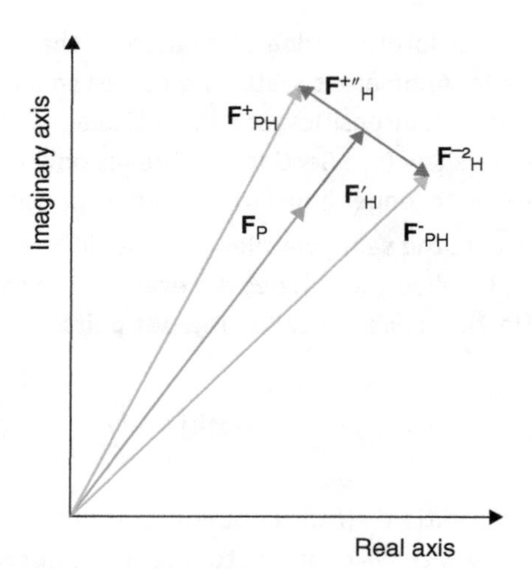

Figure 7.29 Representation in the Gaussian plane of the phase relationships between the structure factor of a protein, $\mathbf{F}_P$, those from a single anomalous scattering heavy atom, $\mathbf{F}_H$, and an isomorphic heavy atom derivative of the protein, $\mathbf{F}_{PH}$.

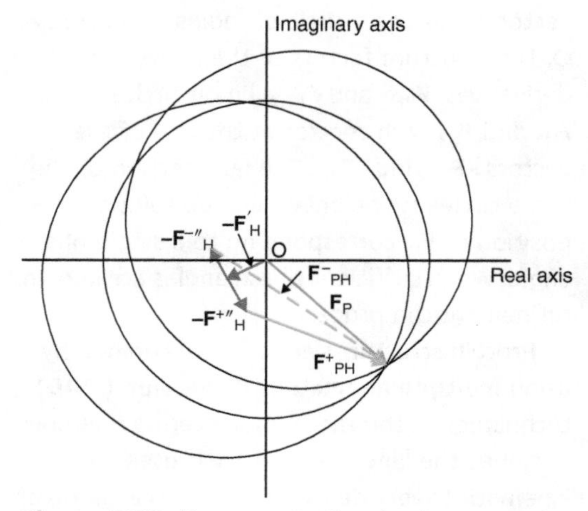

Figure 7.30 Representation in the Gaussian plane of the phase relationships derived by a single anomalous scattering atom in a heavy atom derivative protein. The intersection of the three circles represents a single phase angle.

This representation of the phase relationships between the structure factor of a protein, $\mathbf{F}_P$, and those from a single anomalous scattering heavy atom, $\mathbf{F}_H$ and an isomorphic heavy atom derivative of the protein, $\mathbf{F}_{PH}$, results in two scattering factors $\mathbf{F}_{PH}{}^+$ and $\mathbf{F}_{PH}{}^-$. Using this, it is possible to write equations analogous to Eq. (7.6):

$$F_{PH}{}^+ - F_{PH}{}^- = 2F_H'' \sin(\phi_{PH} - \phi_H)$$

hence:

$$\phi_{PH} = \phi_H \pm \sin^{-1}[(F_{PH}{}^+ - F_{PH}{}^-)/2F_{PH}''] \quad (7.7)$$

Equation (7.7), like Eq. (7.6), gives two possible solutions for ϕ_{PH} but this can be used in place of a second heavy atom derivative, to give a unique solution for the phase angle (Figure 7.30). Once again, the Gaussian plane representation of the phase relationships show that the structure factor of the protein vector, $\mathbf{F}_P$, lies on a circle of radius F_P centred at O, the structure factors of the heavy metal derivatives, $\mathbf{F}_{PH}{}^+$ and $\mathbf{F}_{PH}{}^-$, lie on circles of radii $F_{PH}{}^+$ and $F_{PH}{}^-$ with centres at the tips of the vectors $-\mathbf{F}_H{}^{-''}$ and $-\mathbf{F}_H{}^{+''}$, and the intersection of the three circles represents a unique solution to position of $\mathbf{F}_P$, corresponding to a single phase angle. Naturally, data from two or more different anomalous scatterers, or data compiled from one anomalous scatterer at different wavelengths (MAD), can also be used in a similar way.

7.16.2 The Crystallinity Problem: SFX

For a satisfactory crystallographic study of the sort described above, it is vital to start with a high-quality crystal. This is not easy for proteins. Most proteins will yield good nanometre-sized crystallites in solution, but as the crystals grow

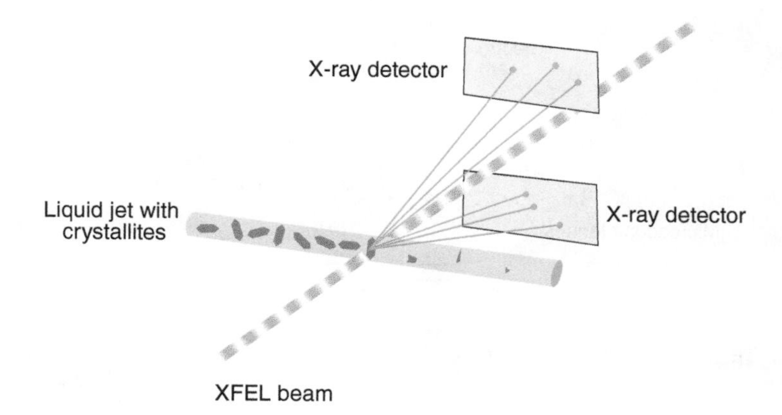

Figure 7.31 Schematic of serial femtosecond crystallography (SFX).

the large molecular complexes that make up the protein structure frequently trap solvent, such that even a reasonable protein crystal may contain between 30–80% solvent. Because the solvent is disordered, it is not registered as such by diffraction data, but large quantities of solvent do degrade the quality of the data and limit the resolution of the final structure.

Until recently the major protein structures have been obtained by using synchrotron generated X-ray beams, which are far more intense than the beams available from X-ray tubes, allowing slightly smaller crystals to be used with much shorter exposure times for data recording. However, this technique still requires crystals of good quality. Commencing in 2009, a new technique for deriving X-ray diffraction patterns has been developed that is able to utilise the nanometre-sized crystallites whilst still in solution. This uses an X-ray free-electron laser (XFEL) to generate a beam intensity of the order of 10^7 times that of synchrotron radiation, which is delivered as a sequence of short femtosecond pulses. These pulses are fired at a jet of a solution containing these nanocrystals. An X-ray pulse that hits a crystallite produces an instantaneous diffraction pattern before the crystallite is destroyed by the X-ray beam itself. This instantaneous diffraction pattern 'snapshot' is recorded (Figure 7.31). Thus, the X-ray beam generates tens of thousands of single crystal diffraction pattern snapshots, from randomly oriented crystals as they are carried into the X-ray beam before being destroyed, over a short period of time. This technique is called **serial femtosecond crystallography** (SFX). The fragmentary diffraction patterns from these crystals are combined by powerful computational techniques to give a composite single crystal diffraction pattern of the crystalline protein. These data allow the structure of the protein to be solved.

7.16.3 The Crystallinity Problem: Single Particle Cryo-EM

One way of avoiding the need for a crystal of any size is to obtain an electron microscope image of the structure itself, at an atomic resolution, as has been done for many inorganic materials (Section 7.15). However, proteins are delicate structures, often hydrated or containing integral water molecules. When these are introduced into the vacuum of the electron microscope, degradation of the protein structure will invariably take place.

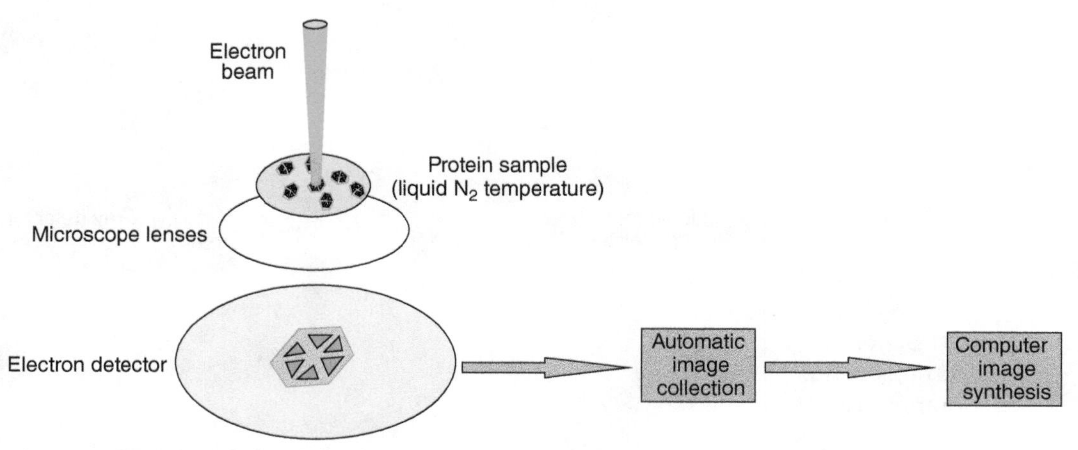

Figure 7.32 Schematic of single particle cryo-EM.

Furthermore, the powerful electron beam is sufficient to damage the structure irreparably. Lowering the beam intensity is possible, but this leads to longer exposure and severely reduces the image quality, preventing anything like atomic resolution to be achieved.

These problems have gradually been solved over the past 40 years or so, so that now (2019) electron microscopy in a form called **single particle cryo-EM**, has become a major tool in the determination of the structure of proteins at an atomic resolution, even when these are embedded in functioning cells or cell walls.

There are a number of advances that have contributed to this achievement. The preservation of the integrity of the protein in the vacuum of the electron microscope is achieved by the technique of plunge freezing. A drop of solution of a protein is placed on an electron microscope grid, which is covered by a perforated thin carbon film. The surplus liquid is blotted away and the grid and remaining solution is plunged into liquid ethane or a similar cold solvent. The remaining solution freezes such that each of the holes in the carbon film is filled with a thin layer of amorphous ice containing the protein molecules embedded in random orientations. This grid is placed into the electron microscope and maintained at a low temperature, usually that of liquid nitrogen (Figure 7.32).

A second innovation concerns the image detection system. In early studies, images were recorded on photographic film. This was a very slow process. Since then image detectors have improved enormously, so that now image recording is wholly electronic and fast enough to virtually record protein particles as a movie. This allows a single particle to be tilted during observation to obtain various projections, and large numbers of individual protein particles to be imaged, all at differing orientations. This data collection, although rapid in terms of each image recorded, is still a lengthy task. This has now been ameliorated by automatic data recording systems.

The final improvement that has allowed high-resolution images of proteins to be routinely obtained is a vast increase in computing power and construction of the algorithms that are able to compensate for sample movement and focus errors, and transform the huge number of electron microscope images into a recognisable three-dimensional model of the material with atomic resolution. This not only allows for the three-dimensional reconstruction of the protein structure, but can be used to compensate for sample movement (called drift) and focusing errors (Figure 7.33).

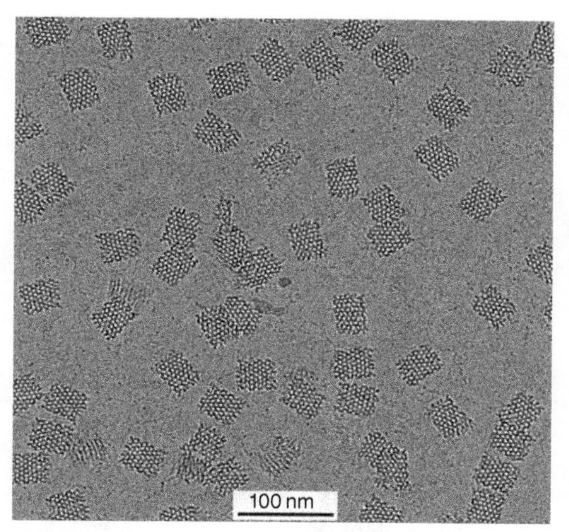

Figure 7.33 Cryo-EM image of DNA origami constructed by inducing hundreds of strands of DNA to self-assemble in precise and predictable arrangements from an agarose gel. The image was recorded at a magnification of ×120 000 on a Falcon 3EC direct detector camera using a Talos Arctica transmission electron microscope operated at 200 kV. The author would like to thank Dr Juanfang Ruan and the Electron Microscope Unit, Mark Wainwright Analytical Centre, University of New South Wales, Australia, for this micrograph.

ANSWERS TO INTRODUCTORY QUESTIONS

What crystallographic information does Bragg's law give?

A beam of radiation of a suitable wavelength will be diffracted when it impinges upon a crystal. Bragg's law defines the conditions under which diffraction occurs, and gives the position of a diffracted beam. Diffraction will occur from a set of (*hkl*) planes when:

$$n\lambda = 2d_{hkl} \sin\theta$$

where n is an integer, λ is the wavelength of the radiation, d_{hkl} is the interplanar spacing (the perpendicular separation) of the (*hkl*) planes, and θ is the diffraction angle or Bragg angle. Note that the angle between the direction of the incident and diffracted beam is equal to 2θ. The geometry of Bragg diffraction is identical to that of reflection, and diffracted beams are frequently called 'reflections' in X-ray literature.

What is an atomic scattering factor?

The atomic scattering factor is a measure of the scattering power of an atom for radiation such as X-rays, electrons or neutrons. Because each of these types of radiation interact differently with an atom, the scattering factor of an atom is different for X-rays, electrons and neutrons.

X-rays are mainly diffracted by the electrons on each atom. The scattering of the X-ray beam increases as the number of electrons, or equally, the atomic number (proton number), Z, of the atom increases. Thus, heavy metals scatter X-rays far more strongly than light atoms, whilst the scattering from neighbouring atoms is similar in magnitude.

The X-ray scattering factor is strongly angle-dependent, this being expressed as a function of $(\sin\theta)/\lambda$. The scattering factor curves for all atoms have a similar form, and at $(\sin\theta)/\lambda = 0$, the X-ray scattering factor is equal to the atomic number, Z, of the element in question.

What are the advantages of neutron diffraction over X-ray diffraction?

One of the greatest advantages of neutron diffraction over X-ray diffraction is that it can be used to locate atoms in a structure that are difficult to place using X-rays. This is because X-ray scattering factors are a function of the atomic number of the elements, but this is not true for neutrons. Of particular importance is the fact that the neutron scattering factors of light atoms, such as hydrogen, H, carbon, C, and nitrogen, N, are similar to those of heavy atoms, making it easier to determine their positions compared with X-ray diffraction. Neutron diffraction is the method of choice for the precision determination of the positions of light atoms in a structure, especially hydrogen atoms. Studies of hydrogen bonding often rely upon neutron diffraction data for accurate interatomic distances when one atom is hydrogen.

A second advantage is that in some instances neighbouring atoms have quite different neutron scattering capabilities, making them easily distinguished. This is not true for X-ray diffraction, in which neighbouring atoms always have very similar atomic scattering factors.

A third advantage is that neutrons have a spin and so interact with the magnetic structure of the solid, which arises from the alignment of unpaired electron spins in the structure. The magnetic ordering in a crystal is quite invisible to X-rays, but is revealed in neutron diffraction as extra diffracted beams and hence a change in unit cell.

PROBLEMS AND EXERCISES

Quick Quiz

1. In crystallography, Bragg's law is written:
   a. $2\lambda = d_{hkl} \sin \theta$
   b. $\lambda = 2d_{hkl} \sin \theta$
   c. $\lambda = nd_{hkl} \sin \theta$
2. The X-ray reflections from a mostly amorphous sample are:
   a. Very sharp
   b. Moderately well defined
   c. Poorly defined or absent
3. The intensity of a beam of radiation diffracted by a set of (*hkl*) planes does NOT depend upon:
   a. The time of irradiation
   b. The nature of the radiation
   c. The chemical composition of the crystal
4. The atomic scattering factor for X-rays is greatest for:
   a. Metals
   b. Light atoms, whether metals or not
   c. Heavy atoms, whether metal or not
5. When the diffracted beams scattered from two different atoms are exactly in step, the phase difference is:
   a. 2π
   b. π
   c. $\pi/2$
6. The X-ray intensity scattered by a unit cell is given by the modulus of:
   a. The structure factor
   b. The square of the structure factor
   c. The square root of the structure factor
7. X-ray reflections with zero intensity because of the symmetry of the unit cell are called:
   a. Systematic absences
   b. Structural absences
   c. Symmetry absences
8. The compound SrF_2 crystallises with the same (*fluorite*) structure as CaF_2. The X-ray powder patterns will be:
   a. Identical
   b. Almost identical
   c. Very similar but easily distinguished
9. In order to locate very light atoms accurately in a crystal it is preferable to use:
   a. Electron diffraction
   b. Neutron diffraction
   c. X-ray diffraction
10. In protein crystallography, the technique of IR is used to:
    a. Determine the space group of the crystal
    b. Determine the structure factors of the reflections
    c. Determine the phases of the reflections

Calculations and Questions

7.1. The oxide $NiAl_2O_4$ adopts the *spinel* structure, with a cubic lattice parameter of $a = 0.8048$ nm. The structure is derived from a face-centred cubic lattice. Making use of Table 7.4, calculate the angles of diffraction of the first six lines expected on a powder diffraction pattern.

7.2. A small particle of CrN, which has the *halite* structure (see Chapter 1) with $a = 0.4140$ nm, gives a reflection from the (200) planes at an angle of 33.60°. The angular range over which the reflection occurs is approximately 1.55°. (a) Calculate the wavelength of the radiation used. (b) Estimate the size of the particle in a direction normal to the (200) planes.

7.3. Electron diffraction patterns recorded by tilting a crystal of niobium, Nb, which has the cubic A2 structure (see Chapter 1) with $a = 0.3300$ nm, are reproduced in the figure below. Using the partly indexed pattern, (a) complete the indexing; (b) determine the camera constant λl for the microscope; (c) index the remaining patterns. (Note that not every lattice point will be present on these patterns, see Section 7.9. It may be helpful to determine first of all which reflections will be present using Table 7.4.)

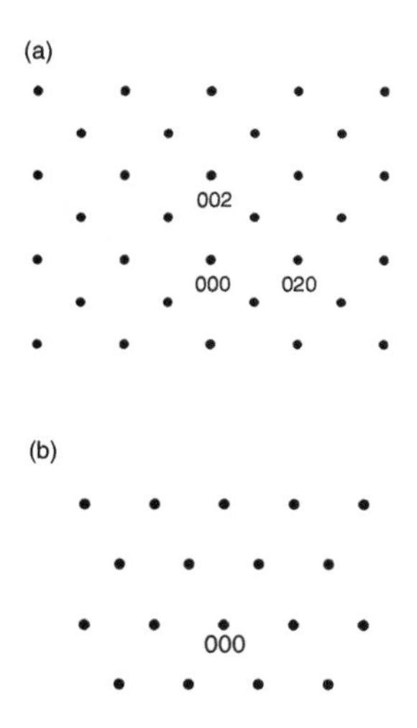

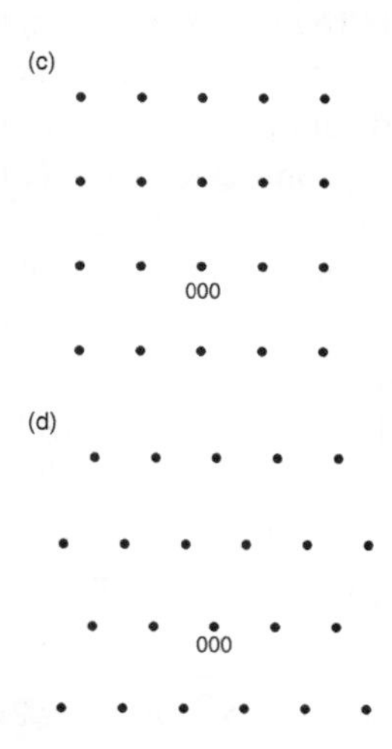

7.4. The Cromer-Mann coefficients for niobium (λ in nm) are given in the table. Plot the scattering factor for niobium as a function of $\sin \theta/\lambda$.

Coefficient	**Index**			
	1	**2**	**3**	**4**
a	17.614	12.014	4.042	3.533
b	0.01189	0.11766	0.00205	0.69796
c	3.756			

These values are adapted from http://www-structure.llnl.gov.

7.5. Repeat the calculation in Question 7.4 if the value of the atomic temperature factor B for niobium is (a) 0.1 nm^2 and (b) 0.2 nm^2. What are the mean displacements of the niobium atoms corresponding to these values of B?

7.6. Calculate the value of the scattering factor (magnitude and phase) for the 110 reflection for rutile, TiO_2, using the data in Section 7.7. The scattering factors are f(Ti): 17.506; f(O): 6.407. Use the data to compute the same values using vector diagrams.

7.7. Calculate the intensities of (a) the 200 and (b) the 110 reflections of rutile, TiO_2, on an X-ray powder pattern, using the data in Section 7.8 and Question 7.6. Take the temperature factor, B, to be 0.1 nm^2 for each reflection and the wavelength of the X-rays as 0.15406 nm (Cu radiation).

7.8. Many inorganic crystals contain one-dimensional chains of atoms that endow the material with interesting electronic properties. Using a computer, plot the electron density function:

$$\rho(x) = F(000) + \sum_{h=-\infty}^{+\infty} F(h) \cos 2\pi(xh - \phi_h)$$

for a chain of alternating Pt and Cl atoms, using first just $F(000)$ 100 and $\bar{1}00$, then, using the first four reflections, $F(000)$ up to 200 and $\bar{2}00$, and so on, to see how the electron density pattern changes as the amount of information supplied increases. The repeat distance along the chain, a, is 1 nm, and the atomic positions are Cl, 0; Pt, ½. (The 'lattice' fringe patterns observed in an electron microscope using these reflections will evolve in a similar manner.)

Structure factors and phase angles for a chain of Pt and Cl atoms

h	$\lvert F(h)\rvert = F(h)$	ϕ_h
8	60	0
7	46	π
6	68	0
5	52	π
4	78	0
3	57	π
2	89	0
1	60	π
0	95	0
−1	60	$-\pi$
−2	89	0
−3	57	$-\pi$
−4	78	0
−5	52	$-\pi$
−6	68	0
−7	46	$-\pi$
−8	60	0

Chapter 8

The Depiction of Crystal Structures

What is the size of an atom?
How does the idea of bond valence help in structure determination?
What is the secondary structure of a protein?

Lists of atomic positions are not very helpful when a variety of structures have to be compared. This chapter describes attempts to explain and systematise the vast amount of structural data now available. The original aim of ways of comparing similar structures was to provide a set of empirical guidelines for use in the determination of a new crystal structure. With sophisticated computer methods now available, this approach is rarely employed, but one empirical rule, the bond-valence method (see Section 8.8 below), is widely used in structural studies.

Most data are available for inorganic solids, as these have been studied longest and in greatest detail. The commonest ways of describing these structures is either as built up by packing together spheres, or else in terms of polyhedra linked by corners and edges. The description of structures by nets or by tilings is also of value for some categories of structure. The study of large organic molecules, especially proteins, is of increasing importance. These are often described by stylised 'cartoons' in which sections of structure are drawn as coiled or folded ribbons.

8.1 THE SIZE OF ATOMS

Quantum mechanics makes it clear that no free atom has a fixed size. Electron orbitals extend from the nucleus to a greater or lesser extent, depending upon the chemical and physical environment in the locality of the atomic nucleus. It may be thought that the scattering of X-rays and electrons can give absolute values for the sizes of atoms in crystals, but this is incorrect. The scattering of both these types of radiation is due to interaction with the electrons surrounding each atomic nucleus. A diffraction experiment thus gives details of a varying electron density throughout the unit cell

Crystals and Crystal Structures, Second Edition. Richard J. D. Tilley.

volume, as described in the previous chapter. Because the electron density is highest nearer to atomic cores, a diffraction experiment yields reasonable positions for the atomic cores rather than the atomic sizes, and *interatomic distances* are obtained, *not* atomic sizes.

Nevertheless, although it may be more correct to think of the contents of a unit cell in terms of a variation in electron density, the concept of an atom possessing a definite and fixed size is attractive, and is a useful starting point for the discussion of many chemical and physical properties. The simplest model is to suppose that each atom is spherical with a characteristic atomic radius. This information is used to derive structural detail, such as bond lengths, bond angles, coordination number and molecular geometry, in advance of any structural determination. However, it is important to stress again that structure determination gives interatomic distances, and the division of these into parts 'belonging' to the individual atoms is to some extent an arbitrary procedure. For example, the distances between similar atoms that are nearest neighbours, linked by strong chemical bonds, are different from the distances between the same atoms when they are linked by weaker non-bonded interactions. Such considerations lead to several different size scales.

The structures of most metallic elements are simple (Section 1.5) and the size of metallic atoms in these crystals can be obtained by assuming that the metal atoms are spherical and in contact, leading to a set of *metallic radii* (Section 8.3). Inorganic chemists have, by and large, used *ionic radii*, which are derived from the notion that anions are relatively large, spherical, and in contact in an inorganic crystal (Section 8.4). The smaller cations, also spherical, are then situated in the spaces between the large anion array. Organic chemists attempting to model complex molecules use *covalent radii* (Section 8.5) for neighbouring atoms, again presumed to be spherical, and *van der Waals* radii (Section 8.6) for atoms at the periphery of molecules. Details of the derivation of these and other atomic radii, and critical discussions as to the rationale and self-consistency of the resulting values, are given in the Bibliography.

8.2 SPHERE PACKING

The structure of many crystals can conveniently be described in terms of an ordered packing of spheres, and the earliest sets of atomic radii were derived via this type of model. Although there are an infinite number of ways to pack spheres, only two main arrangements, called closest (or close) packing, are sufficient to describe many crystal structures. These structures can be thought of as built from layers of close-packed spheres. Each close-packed layer consists of a hexagonal arrangement of spheres just touching each other to fill the space as much as possible (Figure 8.1).

These layers of spheres can be stacked in two principal ways to generate the structures. In the first of these, a second layer fits into the dimples in the first layer, and the third layer is stacked in dimples on top of the second layer to lie over the first layer (Figure 8.2). This sequence

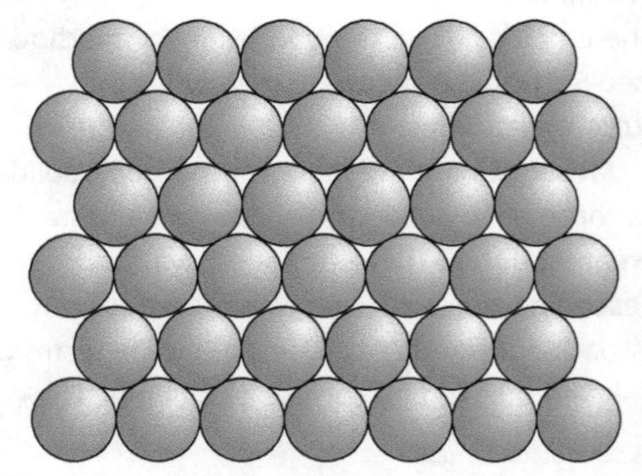

Figure 8.1 A single close-packed layer of spheres.

is repeated indefinitely. If the position of the spheres in the first layer is labelled A, and the positions of the spheres in the second, B, the complete stacking is described by the sequence ... ABABAB .....

The structure so formed has a **hexagonal** symmetry and unit cell. Each sphere has 12 nearest neighbours, six in the same plane and three in the layers above and below. The **a**- and **b**-axes lie in the close-packed A sheet, and the hexagonal **c**-axis is perpendicular to the stacking and runs from one A sheet to the next above it (Figure 8.3). There are two spheres (two atoms) in a unit cell, at positions 0, 0, 0, and 1/3, 2/3, ½. If the spheres just touch, the relationship between the sphere radius, r, and the lattice parameter a, is:

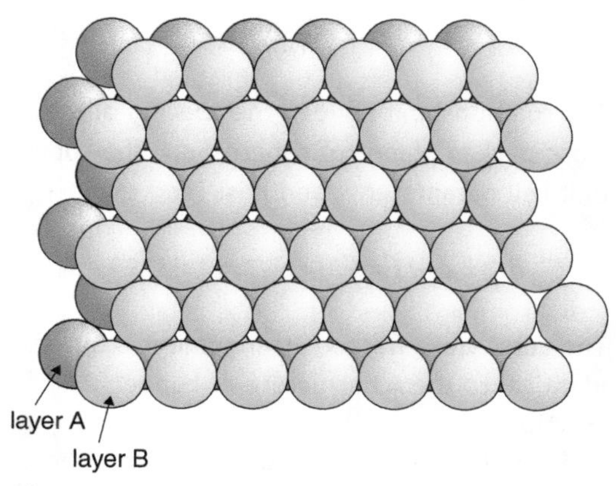

Figure 8.2 Hexagonal (...ABAB...) close-packing of spheres.

$$2r = a = b$$

The structural repeat normal to the stacking is two layers of spheres. The spacing of the close-packed layers, d, is:

$$d = \sqrt{8}r/\sqrt{3} = 1.63299\,r$$

so that the vertical repeat, the c-lattice parameter, is given by:

$$c = 2 \times 1.63299\,r = 2 \times 1.63299 \times a/2 \approx 1.633\,a$$

The ratio of the hexagonal lattice parameters, c/a, in this ideal sphere packing is thus 1.633.

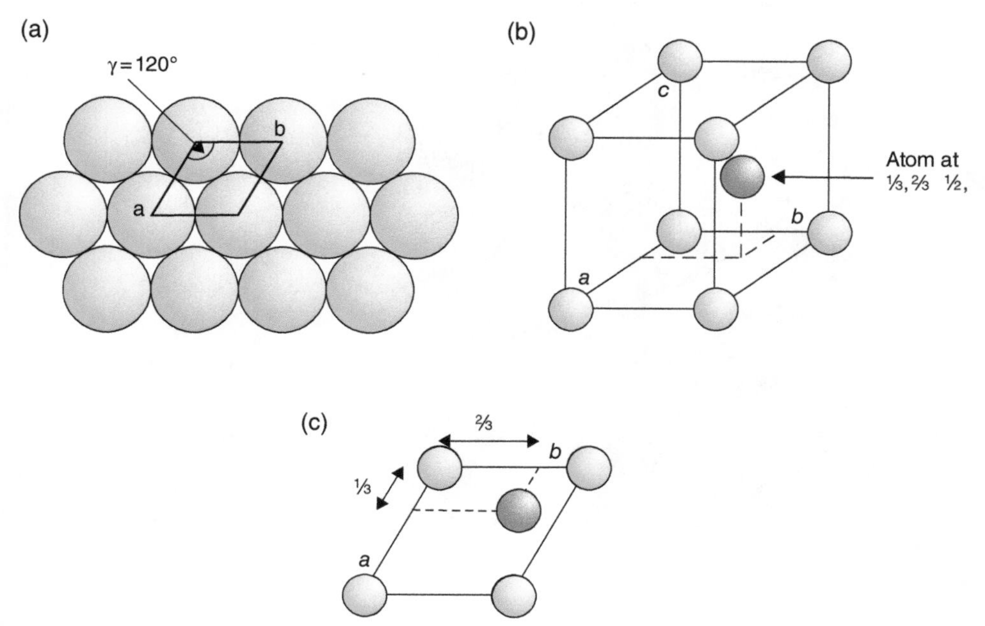

Figure 8.3 Hexagonal unit cell of hexagonal close-packing of spheres.

The separation of the close-packed atom planes is $c/2$, and the ratio of c/a in an ideal close-packed structure is $\sqrt{8}/\sqrt{3} = 1.633$. The c/a ratio departs from the ideal value of 1.633 in most real structures. This model forms an idealised representation of the A3 structure of magnesium (Section 1.5).

The second structure of importance also starts with two layers of spheres, A and B, as before. The difference lies in the position of the third layer. This fits onto the preceding B layer, but occupies the set of dimples that is not above the lower A level. This set is given the position label C (Figure 8.4), and the three-layer stacking is repeated indefinitely to give the sequence … ABCABC …

Although this structure can be described in terms of a hexagonal unit cell, the structure is **cubic**, and this description is always chosen. In terms of the cubic unit cell, there are spheres at the corners of the cell and in the centre of each of the faces. The close-packed layers lie along the [111] direction (Figure 8.5). As in hexagonal close-packing, each sphere has 12 nearest neighbours. The spacing of the close-packed planes for an ideal packing, d, is 1/3 of the body diagonal of the cubic unit cell, which is equal to the spacing of the (111) planes, i.e. $a/\sqrt{3}$. If the spheres just touch, the relationship between the sphere radius, r, and the cubic lattice parameter a, is:

$$r = a/\sqrt{8}$$

The relationship between the spacing of the close-packed planes of spheres, d_{cp}, the cubic unit cell parameter a, and r, is:

$$d_{cp} = a/\sqrt{3} = (\sqrt{8}r)/\sqrt{3} = 1.63299\, r$$

The cubic close-packed arrangement is almost identical to the A1 structure of copper (Section 1.5).

Both the hexagonal close-packing of spheres and the cubic close-packing of spheres result in

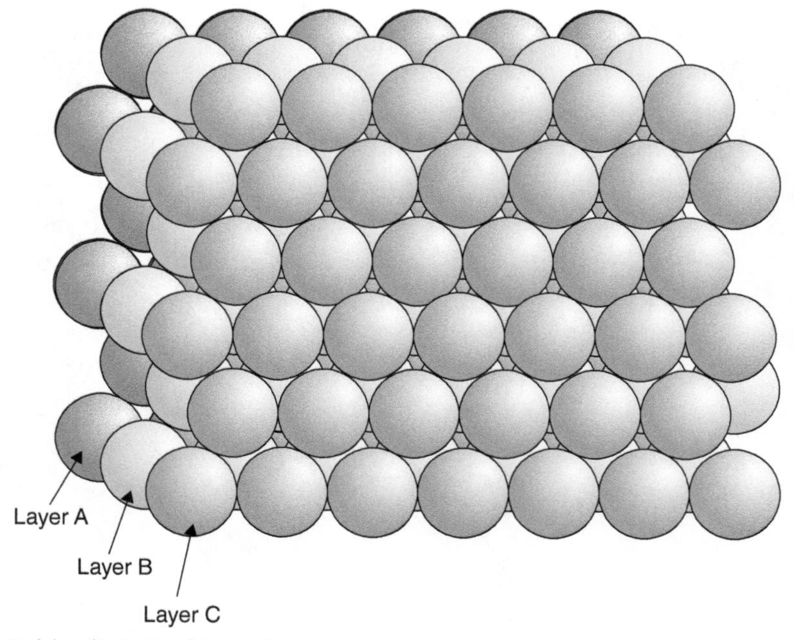

Figure 8.4 Cubic close-packing of spheres.

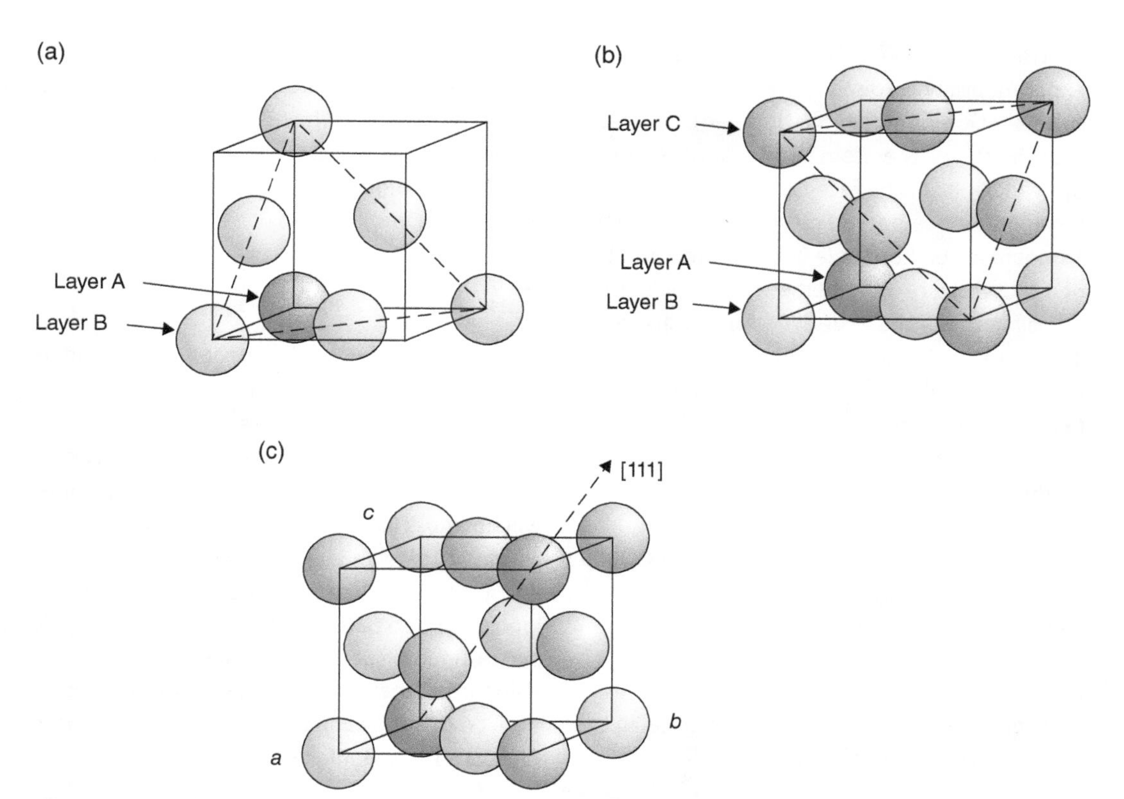

Figure 8.5 Cubic unit cell of cubic close-packing of spheres.

equally dense packing of the spheres. The fraction of the total volume occupied by the spheres, when they touch, is 0.7405.

8.3 METALLIC RADII

Most pure metals adopt one of three crystal structures: A1, copper structure (cubic close-packed); A2, tungsten structure (body-centred cubic); or A3, magnesium structure (hexagonal close-packed) (Chapter 1). If it is assumed that the structures of metals are made up of touching spherical atoms, it is quite easy, knowing the structure type and the size of the unit cell, to work out their radii, which are called **metallic radii**.

For the face-centred cubic (A1, copper structure), the atoms are in ...ABCABC... packing, and in contact along a cube-face diagonal, so that the metallic radius, r is given by:

$$4r = \sqrt{2}a; \quad r = a/\sqrt{8}$$

For the body-centred cubic (A2, tungsten structure) the atoms are in contact along a cube body diagonal. The body diagonal of the unit cell, $3a^2$ is thus equal to $4r$, hence:

$$r = \sqrt{3}a/4$$

For the hexagonal close-packed (A3, magnesium structure) the atoms are in ...ABAB... packing, and in contact along the **a**-axis direction, hence

$$r = a/2$$

Determination of these atomic radii, even in such an apparently clear-cut situation, leads to contradictions. The radius determined experimentally is found to depend upon the **coordination number** (the number of nearest neighbours) of the atom in question. The radii derived for both the face-centred cubic and the hexagonal close-packed structures, which have 12 nearest neighbours (a coordination number of 12, CN12), are in agreement with each other, but differ from those derived for metals that adopt the body-centred cubic structure, in which each atom has eight nearest neighbours (a coordination number of 8, CN8). The conversion between the two sets of radii can be made using the empirical formula:

$$\text{radius [CN12]} = 1.032 \times (\text{radius [CN8]}) - 0.0006$$

where the radii are measured in nm. Metallic radii appropriate to CN12 are given in Figure 8.6.

There are a number of trends to note. In the well-behaved alkali metals and alkaline earth metals, the radius of an atom increases smoothly as the atomic number increases. The transition metals all have rather similar radii, which increase slightly with atomic number going down a group. The same is true for the lanthanoids and actinoids.

8.4 IONIC RADII

X-ray structures only give a precise knowledge of the distances between the atoms. Whilst this is not so important for pure metals, as the interatomic distance is simply divided by two to obtain the metallic radius, this simple method will not work for ionic compounds. To begin, it is assumed that the individual ions are spherical and in contact. The strategy then used to derive ionic radii is to take the radius of one commonly occurring ion, such as the oxide ion, O^{2-}, as a standard. Other consistent radii can then be derived by subtracting the standard radius from measured inter-ionic distances.

Li 0.1562	Be 0.1128											B	C
Na 0.1911	Mg 0.1602											Al 0.1432	Si
K 0.2376	Ca 0.1974	Sc 0.1641	Ti 0.1462	V 0.1346	Cr 0.1282	Mn 0.1264	Fe 0.1274	Co 0.1252	Ni 0.1246	Cu 0.1278	Zn 0.1349	Ga 0.1411	Ge
Rb 0.2546	Sr 0.2151	Y 0.1801	Zr 0.1602	Nb 0.1468	Mo 0.1400	Tc 0.1360	Ru 0.1339	Rh 0.1345	Pd 0.1376	Ag 0.1445	Cd 0.1568	In 0.1663	Sn 0.1545
Cs 0.2731	Ba 0.2243	La 0.1877	Hf 0.1580	Ta 0.1467	W 0.1408	Re 0.1375	Os 0.1353	Ir 0.1357	Pt 0.1387	Au 0.1442	Hg	Tl 0.1716	Pb 0.1750

radii in mn.

Figure 8.6 Metallic radii for coordination number 12 [CN12]. *Source*: Data from Teatum, Gschneidner and Waber, cited by Pearson, W. (1971). *The Crystal Chemistry and Physics of Metals and Alloys*, 151. New York: Wiley-Interscience.

The ionic radius quoted for any species depends upon the standard ion by which the radii were determined. This has resulted in a number of different tables of ionic radii. Although these are all internally self-consistent, they have to be used with thought. Moreover, as with metallic radii, ionic radii are sensitive to the surrounding coordination geometry. The radius of a cation surrounded by six oxygen ions in octahedral coordination is different than the same cation surrounded by four oxygen ions in tetrahedral coordination. Similarly, the radius of a cation surrounded by six oxygen ions in octahedral coordination is different than the same cation surrounded by six sulfur ions in octahedral coordination. Ideally, tables of cationic radii should apply to a specific anion and coordination geometry. Representative ionic radii are given in Figure 8.7.

On this basis, several trends in ionic radius are apparent.

(i) Cations are usually regarded as smaller than anions, the main exceptions being the largest alkali metal and alkaline earth metal cations. The reason for this is that removal of electrons to form cations leads to a contraction of the electron orbital clouds due to the relative increase in nuclear charge. Similarly, addition of electrons to form anions leads to an expansion of the charge clouds due to a relative decrease in the nuclear charge.

(ii) The radius of an ion increases with atomic number. This is simply a reflection that the electron cloud surrounding a heavier ion is of a greater extent than that of the lighter ion.

(iii) The radius decreases rapidly with increase of positive charge for a series of isoelectronic ions such as Na^+, Mg^{2+}, Al^{3+}, all of which have the electronic configuration (Ne), as the increase in the effective nuclear charge acts so as to draw the surrounding electron clouds closer to the nucleus.

(iv) Successive valence increases decrease the radius. For example, Fe^{2+} is larger than Fe^{3+}, for the same reason as in (iii).

(v) Increase in negative charge has a smaller effect than increase in positive charge. For example, F^- is similar in size to O^{2-}, and Cl^- is similar in size to S^{2-}.

When making use of these statements, it is important to be aware that the potential experienced by an ion in a crystal is quite different from that of a free ion. This implies that all of the above statements need to be treated with caution. Cationic and anionic sizes in solids may be significantly different to those that free ion calculations suggest.

Whilst the majority of the ions of elements can be considered to be spherical, a group of ions, the **lone pair ions**, found in the lower part of groups 13, 14 and 15 of the Periodic Table, are definitely not so. These ions all take two ionic states. The high charge state, M^{n+} can be considered as spherical, but the lower valence state, $M^{(n-2)+}$, is definitely not so. For example, tin, Sn, has an outer electron configuration [Kr] $4d^{10}$ $5s^2$ $5p^2$. Loss of the two *p* electrons will leave the ion with a series of closed shells that is moderately stable. This is the Sn^{2+} state, with a configuration of [Kr] $5s^2$ $4d^{10}$. The pair of *s* electrons – the *lone pair* – imposes important stereochemical constraints on the ion. However, loss of the lone pair *s* electrons will produce the stable configuration [Kr] $4d^{10}$ of Sn^{4+}, which can be considered spherical. The atoms that behave in this way are characterised by two valence states, separated by a charge difference of 2+. The atoms involved are indium, In (1+, 3+), thallium, Tl (1+, 3+), tin, Sn (2+, 4+), lead, Pb (2+, 4+), antimony, Sb (3+, 5+) and bismuth, Bi (3+, 5+). These lone pair ions tend to be

+1	+2	+3	+4 (+3)	+5 (+4) [3+]	+6 (+4) [3+]	+6 (+4) [3+] {2+}	+4 (+3) [2+]	+4 (+3) [2+]	+4 (+2)	+2 (+1)	+2	+3	+4 (2+)	+5 (3+)	-2 (6+)	-1
Li 0.088	Be 0.041(t)											B 0.026 (t)	C** 0.006	N*** 0.002	O 0.126	F 0.119
Na 0.116	Mg 0.086											Al 0.067	Si 0.040 (t)	P	S 0.170	Cl 0.167
K 0.1521	Ca 0.114	Sc 0.0885	Ti 0.0745 (0.081)	V 0.068 (0.073) [0.078]	Cr [0.0755]	Mn (0.068) [0.079*] {0.097*}	Fe (0.079*) [0.092*]	Co (0.075*) [0.089*]	Ni (0.083)	Cu 0.087 (0.108)	Zn 0.089	Ga 0.076	Ge 0.068	As 0.064	Se (0.043 t)	Br 0.182
Rb 0.163	Sr 0.1217	Y 0.104	Zr 0.086	Nb 0.078	Mo 0.074	Tc (0.078)	Ru 0.076	Rh 0.0755	Pd (0.100)	Ag (0.129)	Cd 0.109	In 0.094	Sn 0.083 (0.105)	Sb 0.075	Te (0.068)	I 0.206
Cs 0.184	Ba 0.150	La 0.1185	Hf 0.085	Ta 0.078	W 0.074 (0.079)	Re 0.066	Os 0.077	Ir 0.077	Pt 0.077 (0.092)	Au	Hg 0.116	Tl 0.1025	Pb 0.0915 (0.132)	Bi 0.086 (0.116)	Po	At

* high spin configuration; ** C in carbonate, CO_3^{2-}; *** N in nitrate, NO_3^-; radii in nm.

Figure 8.7 Ionic radii. Values marked * are for high-spin states, which have maximum numbers of unpaired electrons. *Source*: Data mainly from Shannon, R.D. and Prewitt, C.T. (1969). *Acta Crystallographica* **B25**: 925–946; ibid, 1046–1048 (1970), with additional data from Muller, O. and Roy, R. (1974). *The Major Ternary Structural Families*, 5. Berlin: Springer-Verlag.

surrounded by an irregular coordination polyhedron of anions. This is often a distorted trigonal bipyramid, and it is hard to assign a unique radius to such ions.

8.5 COVALENT RADII

Covalent radii are mostly used in organic chemistry. The simplest way to obtain a set of covalent radii is to half the distance between atoms linked by a covalent bond in a homonuclear molecule, such as H_2. Covalent radii defined in this way frequently do not reproduce the interatomic distances in organic molecules very well, because these are influenced by double bonding and electronegativity differences between neighbouring atoms. In large molecules such as proteins, this has important structural consequences.

The estimation of covalent radii using data from large molecules poses another problem. The atomic positions given by different techniques are not identical. For example, neutron diffraction gives information on the positions of the atomic nuclei, whereas X-ray diffraction gives information on the electron density in the material. This is not serious for many atoms, but in the case of light atoms, especially hydrogen, the difference can be quite large, of the order of 0.1 nm. As hydrogen atoms play an important role in protein chemistry, such differences can be important. For this reason, it is rather more difficult to arrive at a set of self-consistent radii than it is for either ionic or metallic systems. Nevertheless, the large numbers of organic compounds studied have produced a value for the covalent radius of single bonded carbon as 0.767 nm. Taking this as standard, other covalent radii can be calculated by subtracting the carbon radius from the bond

H 0.0299						
	Be 0.106	B 0.083	C 0.0767 0.0661* 0.0591**	N 0.0702 0.0618* 0.0545**	O 0.0659 0.049*	F 0.0619
		Al 0.118	Si 0.109	P (III) 0.1088	S 0.1052	Cl 0.1023
		Ga 0.1411	Ge 0.122	As (III) 0.1196	Se (II) 0.1196	Br 0.1199
		In 0.141	Sn 0.139	Sb (III) 0.137	Te (II) 0.1391	I 0.1395

* double bond radius ** triple bond radius
radii in nm.

Figure 8.8 Covalent radii. *Source*: Data mainly taken from Alcock, N.W. (1990). *Structure and Bonding*. Ellis Horwood.

length in organic molecules. Those best suited for this are the simple tetrahedral molecules, with carbon as the central atom, such as carbon tetrachloride, CCl_4. In this way a consistent set of carbon-based single bond **covalent radii** has been derived (Figure 8.8). A comparison with molecules containing double or triple bonds then allows values for multiple bond radii to be obtained.

These radii are applicable really to isolated molecules. To a large extent, bond lengths between a pair of atoms A–B in such molecules are fairly constant, because the energy of bond stretching and compression is usually high. Bond angles are also fairly constant, but less so than bond lengths, as the energy to distort the angle between three atoms is much less. Skeletal and 'ball and stick' models make use of this constancy to built molecular structures with reliable shapes. However, bond distances and angles can become modified in crystals, where other interactions may dominate.

8.6 VAN DER WAALS RADII

The van der Waals bond is a weak bond caused by induced dipoles on otherwise neutral atoms. Atoms linked by this interaction have much larger interatomic distances, and hence radii, compared with those linked by strong chemical bonds. A measure of the van der Waals radius that is widely used, especially in organic chemistry, is the idea of a **non-bonded** radius. In this concept, the bond distance between an atom, X, and its *next nearest* neighbour, Z, in the configuration X–Y–Z, is measured. The van der Waals or non-bonded radius is then determined by assuming that the non-bonded atoms are hard spheres that just touch. Some values are given in Figure 8.9.

Organic chemists make considerable use of **space-filling models**, also called **Corey-Pauling-Koltun** or **CPK** models, to represent organic molecules. The atoms in such models are given van der Waals radii, and are designed to give an idea of the crowding that may take place in a molecule. For example, a space-filling model of an α-helix in a protein shows the central core of the helix to be fairly full, whilst a description of the helix as a coiled ribbon suggests that it is hollow (also see Section 8.15).

H 0.120							
			C 0.170	N 0.155	O 0.152	F 0.147	
			Si 0.210	P 0.180	S 0.180	Cl 0.175	
			Ge 0.195	As 0.185	Se 0.190	Br 0.185	
			Sn 0.210	Sb 0.206	Te 0.206	I 0.198	Xe 0.200
				Bi 0.215			

radii in nm.

Figure 8.9 Van der Waals (non-bonded) radii. *Source*: Data from Alcock, N.W. (1990). *Structure and Bonding*. Ellis Horwood.

8.7 IONIC STRUCTURES AND STRUCTURE BUILDING RULES

Ionic bonding is non-directional. The main structural implication of this is that ions simply pack together to minimise the total lattice energy. (Note that the real charges on cations in solids are generally smaller than the formal ionic charges on isolated ions.) There have been many attempts to use this simple idea, coupled with chemical intuition, to derive the structures of inorganic solids. These resulted in a number of structure building 'rules', most famously Pauling's rules, which were used to derive model structures that were used as the starting point for the refinement of X-ray diffraction results in the early years of X-ray crystallography. These rules are no longer widely used for this purpose, as sophisticated computing techniques are able to derive crystal structures from raw diffraction data with minimal input of model structures. Nevertheless, these ideas are still useful in some circumstances. For example, although cation coordination is not determined quantitatively by the relative radii of anions and cations, it is generally found that large cations tend to be surrounded by a cubic arrangement of anions, medium-sized cations by an octahedral arrangement of anions, and small cations by a tetrahedron of anions. The smallest cations are surrounded by a triangle of anions. In addition, the bond-valence model, described in the following section, which is an extension of

Pauling's second rule, is widely used to elucidate cation location in inorganic structures.

8.8 THE BOND VALENCE MODEL

The intensities of the X-rays scattered by atoms that are neighbours in the Periodic Table are almost identical, so that it is often difficult to discriminate between such components of a structure. The same is true for different ionic states of a single element. Thus, problems such as the distribution of Fe^{2+} and Fe^{3+} over the available sites in a crystal structure may be unresolved by conventional structure determination methods. The bond valence model,[1] an empirical concept, is of help in these situations. The model correlates the notion of the relative strength of a chemical bond between two ions with the length of the bond. A short bond is *stronger* than a long bond. Because crystal structure determinations yield accurate interatomic distances, precise values of these relative bond strengths, called experimental **bond valence** values, can be derived.

Imagine a cation, i, surrounded by anions. The formal valence of the central ion, V_i, is equal to the formal charge on the cation. Thus, the value of V_i for an Fe^{3+} ion is +3, for an Nb^{5+} ion is +5, and so on. This value, V_i, is also taken to be equal to the sum of all of the bond valences, v_{ij}, of the ions j of opposite charge within the first coordination shell of ion i thus:

$$\sum_j v_{ij} = V_i = \text{formal charge on the cation} \tag{8.1}$$

[1] The bond valence model should not be confused with the quantum mechanical model of chemical bonding called the valence bond model, which describes covalent bonds in compounds.

To make use of this simple idea, it is necessary to relate the experimentally observable bond length, r_{ij}, between cation i and anion j with the bond valence, v_{ij}. Two empirical equations have been suggested, either of which can be employed.

$$v_{ij} = \left(\frac{r_{ij}}{r_0}\right)^{-N}$$

$$v_{ij} = e^{(r_0 - r_{ij})/B} \tag{8.2}$$

where r_0, N and B are empirical parameters.[2] These are derived from crystal structures that are considered to be particularly reliable. In general, Eq. (8.2) is most frequently used, and the value of B is often taken as 0.037 nm (0.37 Å) for all bonds, so that this equation takes the practical form:

$$v_{ij} = e^{(r_0 - r_{ij})/0.037} \tag{8.3}$$

Figure 8.10 gives graphs of this function for Zn^{2+}, Ti^{4+}, and H^+ when bonded to O^{2-} ions. It is

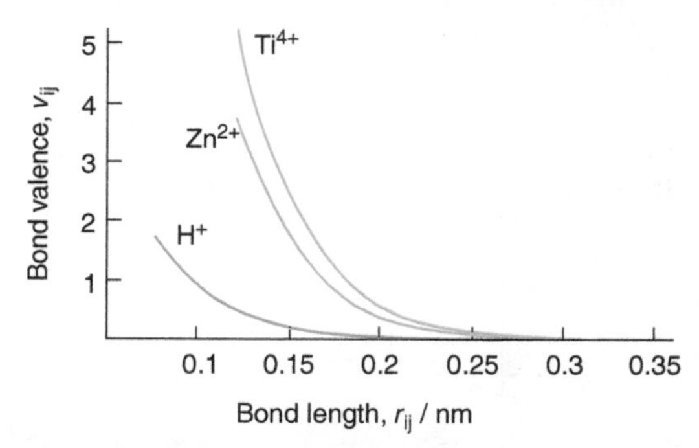

Figure 8.10 The relationship between bond valence, v, and bond length (nm) for Ti^{4+}, Zn^{2+} and H^+, when bonded to oxide, O^{2-}.

[2] The units used here for r_{ij}, r_0, N and B are nm. Crystallographers often prefer the non-SI unit Å, where 1 nm = 10 Å.

seen that the bond valence varies steeply with bond distance.

To illustrate the relationship between the valence of an ion and the bond length, imagine a Ti^{4+} ion surrounded by a regular octahedron of oxide ions (Figure 8.11a). Ideally each bond will have a bond valence, *v*, of 4/6, or 0.6667. Using a value of r_0 for Ti^{4+} of 0.1815 nm, each Ti–O bond is computed by Eq. (8.3) to have a length, *r*, of 0.1965 nm. If some of the bond lengths, *r*, are shorter than this (Figure 8.11b), it can be seen from Figure 8.10, that the values of the bond valences, *v*, of these short bonds will be greater than 0.6667 each. When the six bond valences are added, they will come to more than 4.0. In this case the Ti^{4+} ion is said to be **overbonded**. In the reverse case, when some of the bond lengths, *r*, are longer than 0.1965 nm (Figure 8.11c), the bond valences, *v*, will be less than 0.6667. When the six bond valences are added, the result will be less than 4.0, and the ion is said to be **underbonded**.

In order to apply the method to a crystal structure, take the following steps:

(i) Values of r_0 (and *B*, if taken as different from 0.037) can be found from tables (see Bibliography).

(ii) A list of bond valence (v_{ij}) values for the bonds around each cation are calculated, using the crystallographic bond lengths, r_{ij}.

(iii) The bond valence (v_{ij}) values are added for each cation to give a set of ionic valences.

(iv) If the structure is correct, the resulting ionic valence values should be close to the normal chemical valence.

The technique can be illustrated by describing the site allocation of cations in a crystal. The oxide $TiZn_2O_4$ crystallises with the AB_2O_4 **spinel** structure. In this structure, the two different cations, *A* and *B*, are distributed between two sites, one with octahedral coordination by oxygen, and one with tetrahedral coordination by oxygen. If all of the cations *A* are in the tetrahedral sites, (*A*), and all of the *B* cations are in octahedral sites, [*B*], the formula is written (*A*)[B_2]O_4 and the compound is said to be a **normal** spinel. If all of the *A* cations are in the octahedral sites, [*A*], and the *B* cations are distributed over the remaining octahedral and tetrahedral sites, (*B*) and [*B*], the formula is written (*B*)[*AB*]O_4, and the compound

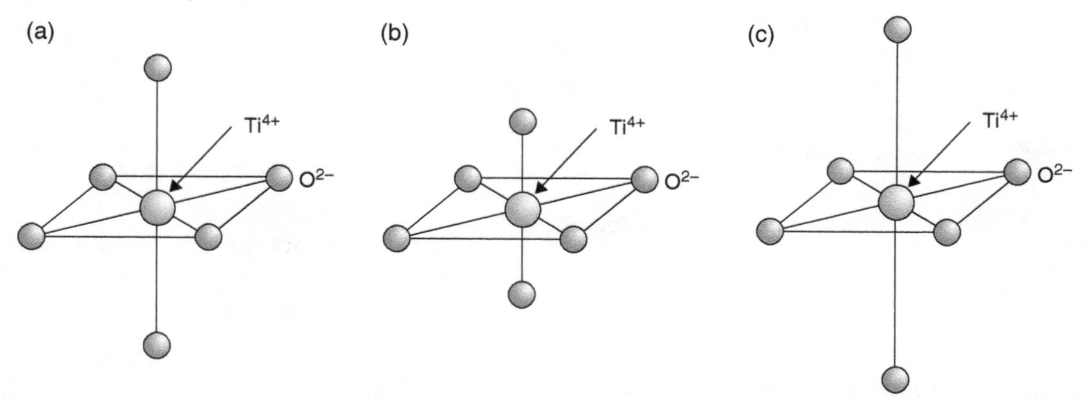

Figure 8.11 TiO_6 octahedra: (a) an ideal TiO_6 octahedron, with equal Ti–O bond lengths of 0.1965 nm; (b) an overbonded TiO_6 octahedron, with two shorter bonds; (c) an underbonded TiO_6 octahedron with two longer bonds.

is said to be an **inverse** spinel. The bond valence method can be used to differentiate between these two possibilities.

The inverse spinel structure (Zn)[TiZn]O_4 has Zn1 placed in octahedral sites, Zn2 placed in tetrahedral sites, and Ti placed in octahedral sites. The normal spinel structure, (Ti)[Zn_2]O_4, exchanges the Ti in octahedral sites with Zn2 in tetrahedral sites. The bond valence sums for the two alternative structures are given in Table 8.1. The cation valences for the inverse structure, Zn1 = 2.13, Zn2 = 1.88, Ti = 3.85, are reasonably close to those expected (Zn1 = 2.0, Zn2 = 2.0, Ti = 4.0). In the case of the normal structure, the fit is nothing like so good, with Zn1 = 2.13, Zn2 = 2.85, Ti = 2.53. It is clear that the bond distances in the structure correspond more closely to the inverse cation distribution, (Zn)[TiZn]O_4. (Note that it may be possible to improve the fit further by assuming that the structure is not entirely inverse, but that a distribution holds so that most of the Ti ions occupy octahedral sites, but a small percentage occupy tetrahedral sites.)

Table 8.1 Bond valence data for $TiZn_2O_4$

Atom pair	Bond length, r_{ij}, nm	Number of bonds	Bond valence, v_{ij}	Sum (cation valence)
Inverse spinel, (Zn)[TiZn]O_4				
[Zn1] octahedral				
Zn1–O1	0.2091	2	0.3514	2.13
Zn1–O1	0.2112	2	0.3320	
Zn1–O2	0.2059	2	0.3831	
(Zn2) tetrahedral				
Zn2–O1	0.2010	2	0.4373	1.88
Zn2–O2	0.1960	2	0.5006	
[Ti] octahedral				
Ti–O1	0.1873	2	0.8558	3.85
Ti–O2	0.1996	2	0.6131	
Ti–O2	0.2106	2	0.4559	
Normal spinel, (Ti)[Zn_2]O_4				
[Zn1] octahedral				
Zn1–O1	0.2091	2	0.3514	2.13
Zn1–O1	0.2112	2	0.3320	
Zn1–O2	0.2059	2	0.3831	
(Ti) tetrahedral				
Ti–O1	0.2010	2	0.5904	2.53
Ti–O2	0.1960	2	0.6758	
[Zn2] octahedral				
Zn2–O1	0.1873	2	0.6340	2.85
Zn2–O2	0.1996	2	0.4542	
Zn2–O2	0.2106	2	0.3378	

Source: Data taken from Marin, S.J., O'Keeffe, M. and Partin, D.E. (1994). *Journal of Solid State Chemistry* **113**: 413–419. The values of r_0 are Zn: 0.1704 nm; Ti, 0.1815 nm, from Breese, N.E. and O'Keeffe, M. (1991). *Acta Crystallographica* **B47**: 192–197.

8.9 STRUCTURES IN TERMS OF NON-METAL (ANION) PACKING

The geometric problem of packing spheres is described in Section 8.2 above. In such structures, the spheres do not fill all the available volume as there are small holes between the spheres that occur in layers between the sheets of spheres. These holes, which are called **interstices**, **interstitial sites** or **interstitial positions**, are of two types (Figure 8.12). In one type of position, three spheres in the lower layer are surmounted by one sphere in the layer above, or vice versa. The geometry of this site is that of a **tetrahedron**. The other position is made up of a lower layer of three spheres and an upper layer of three spheres. The shape of the enclosed space is not so easy to see, but is found to have an **octahedral** geometry.

In the two closest packed sequences, ...ABCABC... and ...ABAB..., there are $2N$ tetrahedral interstices and N octahedral interstices for every N spheres. Ionic structures can be modelled on this arrangement by assuming that the anions pack together like spheres. The structure is made electrically neutral by placing cations into some of the interstices, making sure that the total positive charge on the cations is equal to the total negative charge on the anions. The formula of the structure can be found by counting up the numbers of each sort of ion present.

Consider the structures that arise from a cubic close-packed array of X anions. If every

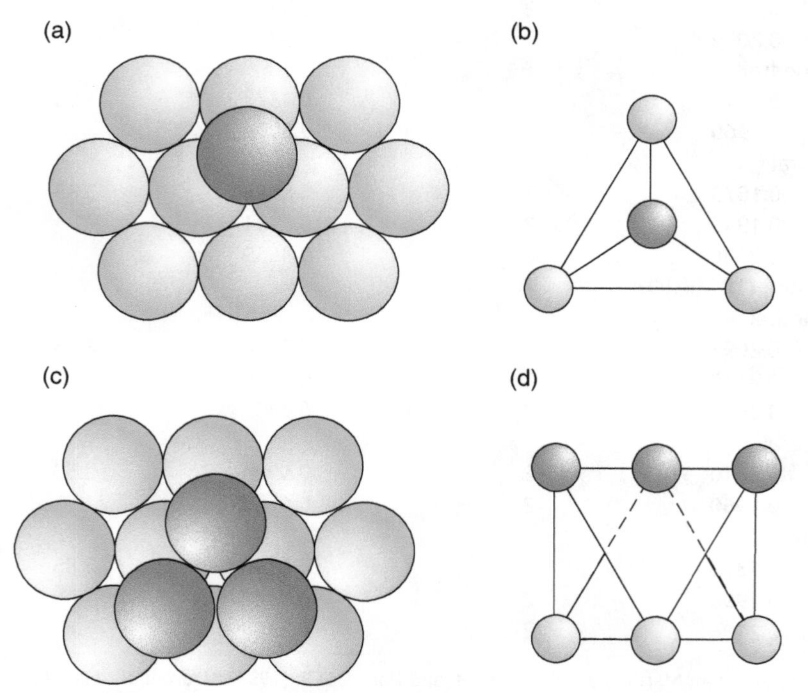

Figure 8.12 Tetrahedral and octahedral sites in close-packed arrays of spheres: (a, b) a tetrahedral site; (c, d) an octahedral site.

octahedral position contains an M cation, there are equal numbers of cations and anions in the structure. The formula of compounds with this structure is MX, and the structure corresponds to the **halite** (NaCl, B1) structure. If halide anions, X^-, form the anion array, to maintain charge balance, each cation must have a charge of +1. Compounds that adopt this arrangement include AgBr, AgCl, AgF, KBr, KCl, KF, NaBr, NaCl and NaF. Should oxygen anions, O^{2-}, form the anion array, the cations must necessarily have a charge of 2+, to ensure that the charges balance. This structure is adopted by a number of oxides, including MgO, CaO, SrO, BaO, MnO, FeO, CoO, NiO and CoO.

Should the anions adopt hexagonal close-packing and all of the octahedral sites contain a cation, the hexagonal analogue of the halite structure is produced. In this case, the formula of the crystal is again MX. The structure is the **nicolite** (NiAs) structure, and is adopted by a number of alloys and metallic sulfides, including NiAs, CoS, VS, FeS and TiS.

If only a fraction of the octahedral positions in the hexagonal packed array of anions is filled, a variety of structures results. The **corundum** structure is adopted by the oxides α-Al_2O_3, V_2O_3, Ti_2O_3 and Fe_2O_3. In this structure, two-thirds of the octahedral sites are filled in an ordered way. Of the structures that form when only half of the octahedral sites are occupied, those of rutile (TiO_2) and α-PbO_2 are best known. The difference between the two structures lies in the way in which the cations are ordered. In the rutile form of TiO_2 the cations occupy straight rows of sites, whilst in α-PbO_2 the rows are staggered.

A large number of structures can be generated by the various patterns of filling either the octahedral or tetrahedral interstices. The number can be extended if both types of position are occupied. One important structure of this type is the spinel ($MgAl_2O_4$) structure, discussed above. The oxide lattice can be equated to a cubic close-packed array of oxygen ions, and the cubic unit cell contains 32 oxygen atoms. There are, therefore, 32 octahedral sites and 64 tetrahedral sites for cations. The unit cell contains 16 octahedrally coordinated cations and 8 tetrahedrally coordinated cations, which are distributed in an ordered way over the available positions to give the formula $A_8B_{16}O_{32}$ or, as usually written, AB_2O_4. Provided that the charges on the cations A and B add to 8, any combination can be used. The commonest is $A = M^{2+}$, $B = M^{3+}$, as in spinel itself, $MgAl_2O_4$, but a large number of other possibilities exist, including $A = M^{6+}$, $B = M^+$, as in Na_2WO_4, or $A = M^{4+}$, $B = M^{2+}$, as in Zn_2SnO_4. In practice, the distribution of the cations over the octahedral and tetrahedral sites is found to depend upon the conditions of formation. The two extreme cation distributions, the normal and inverse spinel structures, are described above.

Structures containing cations in tetrahedral sites can be described in exactly the same way. In this case, there are twice as many tetrahedral sites as anions, and so if all sites are filled the formula of the solid will be M_2X. When half are filled this becomes MX, and so on.

A small number of the structures that can be described in this way is given in Table 8.2.

8.10 STRUCTURES IN TERMS OF METAL (CATION) PACKING

A consideration of metallic radii will suggest that an alternative way of modelling structures is to consider that metal atoms pack together like spheres, and small non-metal atoms fit into interstices so formed. In structural terms, this is the opposite viewpoint to that described in the previous section. In fact, metallurgists have

Table 8.2 Structures in terms of non-metal (anion) packing

Fraction of tetrahedral sites occupied	Fraction of octahedral sites occupied	Sequence of anion layers	
		...ABAB...	...ABCABC...
0	1	NiAs (nicolite)	NaCl (halite)
½		ZnO, ZnS (wurtzite)	ZnS (sphalerite or zincblende)
0	⅔	Al_2O_3 (corundum)	—
0	½	TiO_2 (rutile), α-PbO_2	TiO_2 (anatase)
⅛	½	Mg_2SiO_4 (olivine)	$MgAl_2O_4$ (spinel)

used this concept to describe the structures of a large group of materials, the **interstitial alloys**, for many years. The best known of these are the various steels, in which small carbon atoms occupy interstitial sites between large iron atoms in the A1 (cubic close-packed) structure adopted by iron between the temperatures of 912 and 1394°C.

The simplest structures to visualise in these terms are those that can be derived directly from the structures of the pure metals, the A1 (cubic close-packed) or the A3 (hexagonal close-packed) structures. The halite structure of many solids of composition MX, described above in terms of anion packing, can be validly regarded in terms of metal packing. For example, Ni, Ca and Sr adopt the A1 copper structure. The corresponding oxides, NiO, CaO and SrO, can be built, at least conceptually, by the insertion of oxygen into every octahedral site in the metal atom array. Similarly, Zn adopts the A3 magnesium structure, and ZnO can be built by filling half of the tetrahedral sites, in an ordered way, with oxygen atoms. The mechanism of oxidation of these metals often follows a route in which oxygen at first diffuses into the surface layers of the metal via octahedral sites, supporting a view of the structure in terms of large metal atoms and small non-metal atoms.

Many other inorganic compounds can be derived from metal packing in similar ways. For example, the structure of the mineral fluorite, CaF_2, can be envisaged as the A1 structure of the metal Ca, with every tetrahedral site filled by fluorine. Each Ca atom is surrounded by a cube of fluorine atoms.

Metal packing models can account for more structures if alloys are chosen as the starting metal atom array. For example, the AB_2 metal atom positions in the normal spinel structure are identical to the metal atom positions in the cubic alloy $MgCu_2$. In spinel itself, $MgAl_2O_4$, the metal atom arrangement is identical to that of a cubic alloy of formula $MgAl_2$. The spinel structure results if oxygen atoms are allocated to every tetrahedral site present in $MgAl_2$.

8.11 CATION-CENTRED POLYHEDRAL REPRESENTATIONS OF CRYSTALS

A considerable simplification in describing structural relationships can be obtained by drawing a structure as a set of linked polyhedra. The price paid for this simplification is that important structural details are often ignored, especially when the polyhedra are idealised. The polyhedra selected for such representations are generally metal–non-metal coordination polyhedra, composed of a central metal surrounded by a number of neighbouring non-metal atoms. To construct a metal-centred polyhedron, the non-metal atoms are reduced to points, which form the vertices of the anion polyhedron. These polyhedra are then linked together to build up the complete structure. As an example, the fluorite structure (Section 1.6) can be depicted as a three-dimensional packing of calcium-centred CaF_8 cubes (Figure 8.13). Each cube shares all its edges with neighbours, to create a 'three-dimensional chess board' array. A model can easily be constructed from sugar cubes.

Octahedral coordination is frequently adopted by the important 3*d* transition-metal ions. In this coordination polyhedron, each cation is surrounded by six anions to form an octahedral $[MO_6]$ group (Figure 8.14). The cubic structure of rhenium trioxide, ReO_3, $a = 0.3750$ nm, is readily visualised in terms of corner-linked $[ReO_6]$ octahedra (Figure 8.15a). This structure is similar to that of tungsten trioxide, WO_3, but in the latter compound, the octahedra are distorted slightly, so that the symmetry is reduced from cubic in ReO_3 to monoclinic in WO_3.

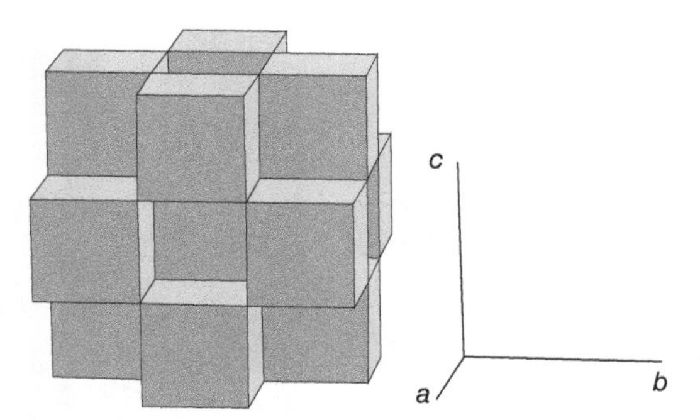

Figure 8.13 The fluorite (CaF_2) structure as an array of edge-shared CsF_8 cubes.

Just as the mineral spinel, $MgAl_2O_4$, has given its name to a family of solids AB_2X_4 with topologically identical structures, the mineral perovskite, $CaTiO_3$, has given its name to a family of closely related phases of ABX_3 structure, where A is a large cation, typically Sr^{2+}, B is a medium-sized cation, typically Ti^{4+} and X is most often O^{2-}. The structures of these phases are often idealised to a cubic framework with a unit cell edge of approximately 0.375 nm. It is seen that the skeleton of the idealised structure, composed of corner-linked BO_6 octahedra, is identical to the ReO_3 structure. The large A cations sit in the cages that lie between the octahedral framework (Figure 8.15b). Most real *perovskites*, including perovskite ($CaTiO_3$) itself, and barium titanate, $BaTiO_3$, are built of slightly distorted $[TiO_6]$ octahedra, and as a consequence the crystal symmetry is reduced from cubic to tetragonal, orthorhombic or monoclinic.

The complex families of silicates are best compared if the structures are described in terms of linked tetrahedra. The tetrahedral shape used is the idealised coordination polyhedron of the $[SiO_4]$ unit (Figure 8.16). Each silicon atom is linked to four oxygen atoms by tetrahedrally directed sp^3-hybrid bonds. As an example, Figure 8.17 shows the structure of diopside, $MgCaSi_2O_6$, drawn in this way. The $[SiO_4]$ tetrahedra form chains, each linked by two corners, running in (100) layers parallel to the **c**-axis. Between the chains, the medium-sized Mg^{2+} ions are in octahedral (sixfold) coordination and the larger Ca^{2+} ions in cubic (eightfold) coordination. For many purposes this depiction is clearer than that given in Figure 6.9. In particular, the $[SiO_4]$ units are very strong and persist during physical and chemical reactions. Cation replacement can be imagined to occur without significant rearrangement of the $[SiO_4]$ chains reactions, whilst structural transformations of diopside involving

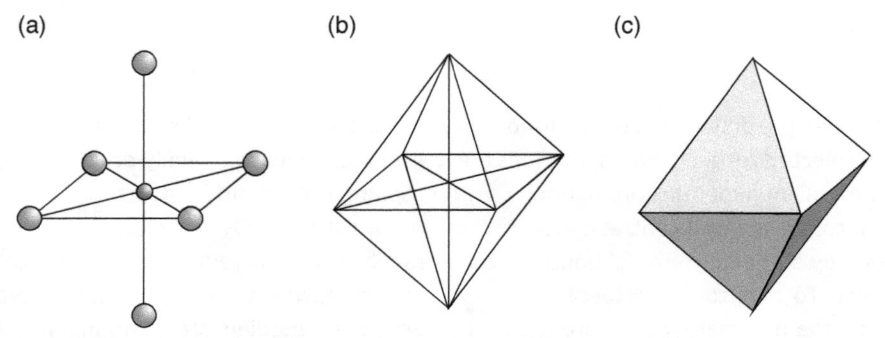

Figure 8.14 Representations of octahedra: (a) 'ball and stick', in which a small cation is surrounded by six anions; (b) a polyhedral framework; (c) a 'solid' polyhedron.

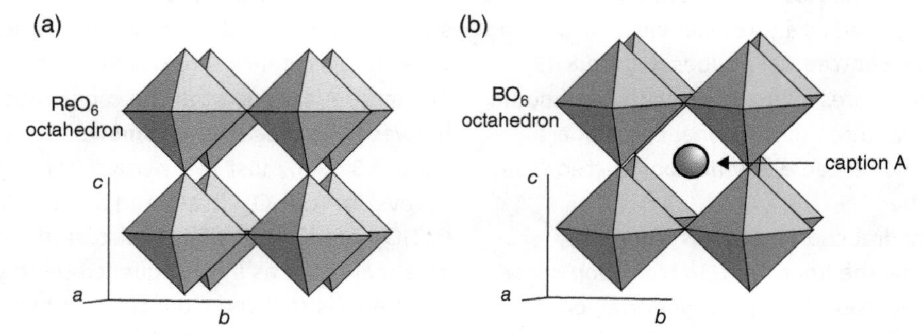

Figure 8.15 (a) Perspective view of the cubic structure of ReO_3, drawn as corner-linked ReO_6 octahedra; (b) the idealised cubic ABO_3 *perovskite* structure.

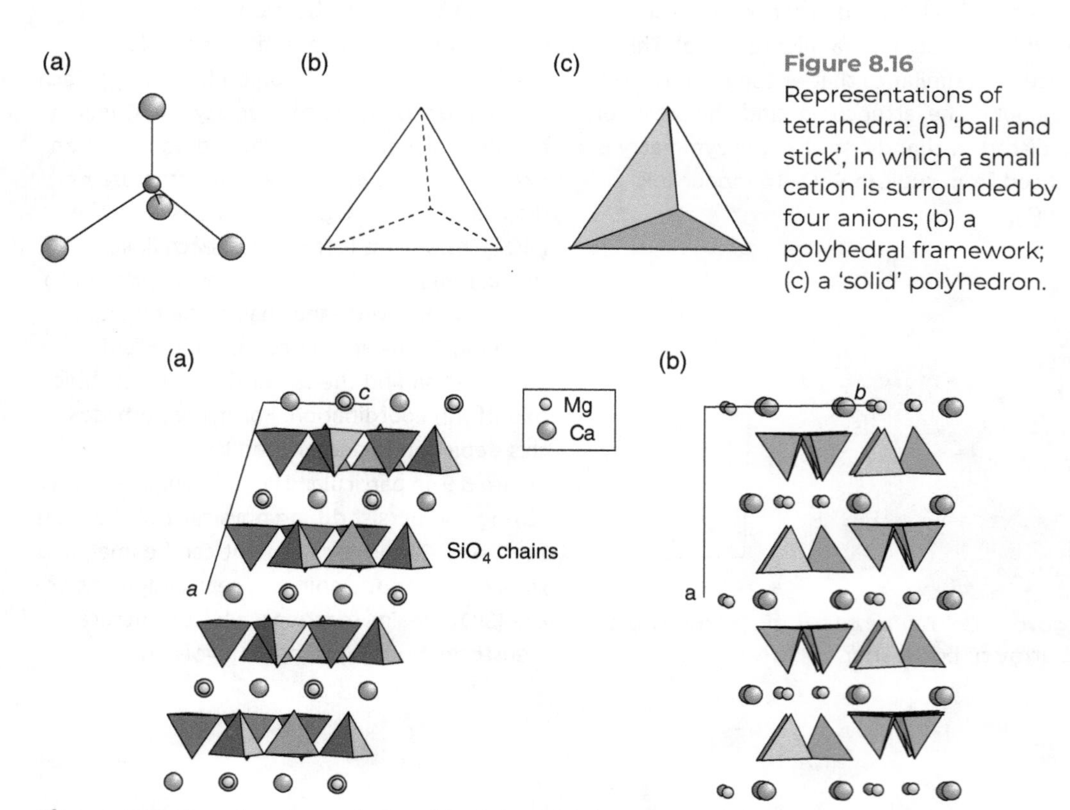

Figure 8.16 Representations of tetrahedra: (a) 'ball and stick', in which a small cation is surrounded by four anions; (b) a polyhedral framework; (c) a 'solid' polyhedron.

Figure 8.17 The structure of diopside: (a) projection down the monoclinic **b**-axis; (b) projection down the monoclinic **c**-axis.

rearrangement of the $[SiO_4]$ chains that may result in sheets of linked tetrahedra are also understandable.

8.12 POLYHEDRAL REPRESENTATIONS OF CRYSTALS AND DIFFUSION PATHS

Although the polyhedral representations described are generally used to depict structural relationships, they can also be used to depict diffusion paths in a compact way. Diffusion paths in solids are of importance in the design and function of fast ionic conductors and battery materials.

The edges of a cation-centred polyhedron represent the paths that a diffusing anion can take in a structure, provided that anion diffusion takes place from one normal anion site to another. Thus, the shortest anion diffusion path in crystals with the fluorite structure will lie localised along the cube edges of Figure 8.13. Diffusion jumps across a cube face are also allowed, but diffusion across a cube body centre is only possible for the 'empty' anion cubes that do not contain a cation.

In order to visualise cation diffusion more widely, viewing structures as made up of linked anion-centred polyhedra rather than cation-centred polyhedra can be useful. The anion-centred polyhedron formed by taking all of the possible cation sites in a face-centred cubic packing of anions as vertices is a rhombic dodecahedron (Figure 8.18). The polyhedron can be related to the cubic axes by comparing Figure 8.18a with Figure 2.10d. However, it is often convenient to relate this representation to that of close-packed anion layers, described in Section 8.9, and to do this it is best to make the direction normal to the layers, the [111] direction, vertical (Figure 8.18b) by rotating the polyhedron anticlockwise.

The resulting anion-centred rhombic dodecahedron can be used to describe the possible structures derived from a face-centred cubic packing of anions. The octahedral sites in the structure are represented by vertices at which four edges meet, whilst tetrahedral sites are represented by vertices at which three edges meet. The anion-centred polyhedral representation of the *halite* structure has all of the octahedral sites filled (Figure 8.18c), whilst that of the cubic zincblende (sphalerite) structure of ZnS has half of the tetrahedral sites filled (Figure 8.18d). The spinel structure, a cubic structure in which an ordered arrangement of octahedral and tetrahedral sites is filled, needs two identical blocks, one rotated with respect to the other (Figure 8.18e, f), for construction. One block in the orientation shown in Figure 8.18e is united with three blocks in the orientation shown in Figure 8.18f, formed by rotations of 0°, 120° and 240° about the [111] direction through the tetrahedrally coordinated cation. The faces that are shaded are joined in the composite polyhedron.

The analogous anion-centred polyhedron for a hexagonal close-packed anion array (Figure 8.19a) is similar to a rhombic dodecahedron, but has a mirror plane normal to the vertical axis, which is perpendicular to the close-packed planes of anions and so forms the hexagonal **c**-axis. The octahedral sites in the structure are again found at the vertices where four edges meet, and the tetrahedral sites at the vertices where three edges meet. The relationship between structures formed by an ordered filling these sites, such as corundum, Al_2O_3, (Figure 8.19b), in which two-thirds of the octahedral sites are filled in an ordered manner, and (idealised) rutile, TiO_2 (Figure 8.19c), in which half of the octahedral sites are filled in an

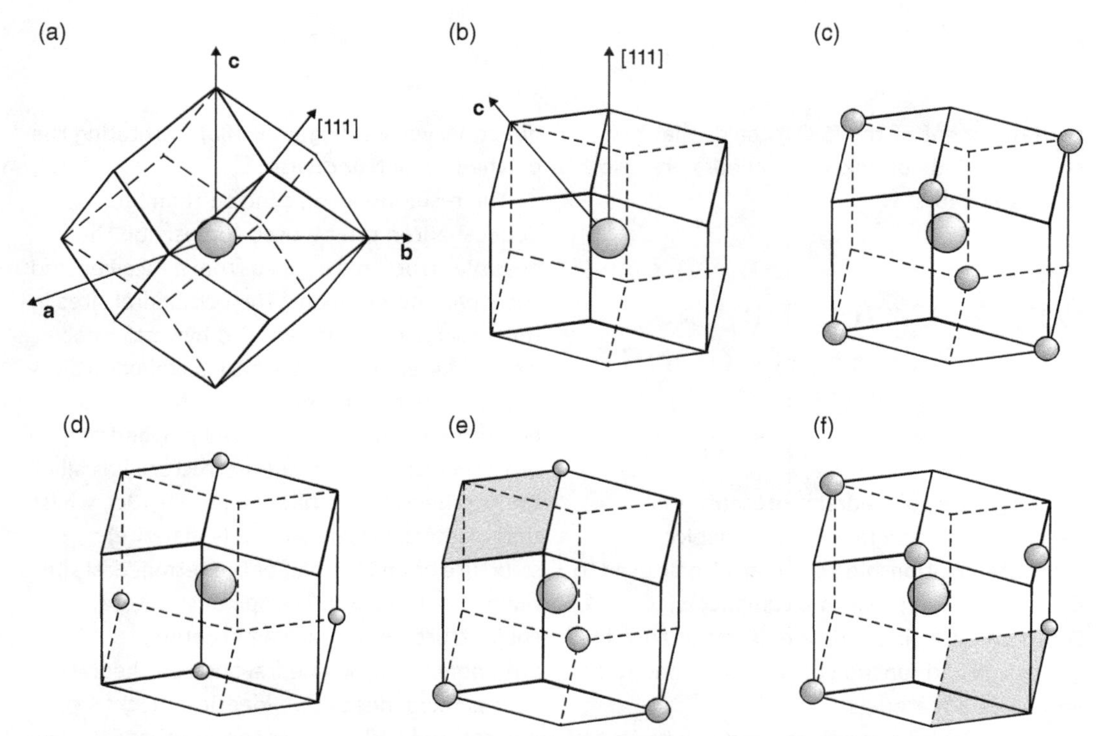

Figure 8.18 The anion-centred polyhedron (rhombic dodecahedron) found in the cubic close-packed structure: (a) oriented with respect to cubic axes; (b) oriented with [111] vertical; (c) octahedral cation positions occupied in the halite, NaCl, structure; (d) tetrahedral cation positions occupied in the zincblende (sphalerite), ZnS, structure; (e, f) the two anion-centred polyhedra needed to create the spinel, $MgAl_2O_4$, structure. *Source*: Adapted from Gorter, E.W. (1960). Proceedings of the International Conference on Pure and Applied Chemistry, Munich, 1959, Butterworths, London, p. 303.

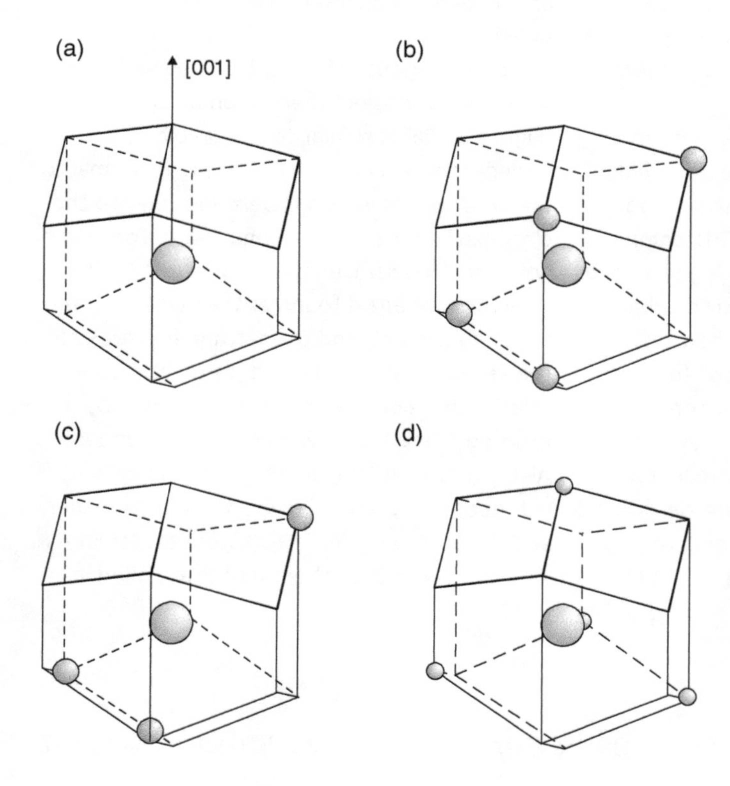

Figure 8.19 The anion-centred polyhedron found in the hexagonal closest-packed structure: (a) oriented with the hexagonal **c**-axis vertical; (b) octahedral cation positions occupied in the ideal corundum, Al_2O_3, structure; (c) octahedral cation positions occupied in the idealised rutile, TiO_2, structure; (d) tetrahedral cation positions occupied in the wurtzite, ZnS, structure. *Source*: Adapted from Gorter, E. W. (1960). Proceedings of the International Conference on Pure and Applied Chemistry, Munich, 1959, Butterworths, London, p. 303.

ordered manner, is readily seen. Filling of half the tetrahedral sites leads to the hexagonal wurtzite form of ZnS (Figure 8.19d).

Just as the edges of cation-centred polyhedra represent possible anion diffusion paths in a crystal, the edges of anion-centred polyhedra represent possible cation diffusion paths. The polyhedra shown in Figures 8.18 reveal that cation diffusion in cubic close-packed structures will take place via alternative octahedral and tetrahedral sites. (Direct pathways, across the faces of the polyhedron, are unlikely, as these mean that a cation would have to squeeze directly between two anions.) There is no preferred direction of diffusion. For ions that avoid either octahedral or tetrahedral sites, for bonding or size reasons, diffusion will be slow compared with ions that are able to occupy either site. In solids in which only a fraction of the available metal atom sites are filled, such as the spinel structure, clear and obstructed diffusion pathways can be delineated.

The situation is different in the hexagonal close-packed structures. The octahedral and tetrahedral sites are arranged in chains parallel to the **c**-axis (Figure 8.19). A diffusing ion in one of these sites, either octahedral or tetrahedral, can jump to another without changing the nearest-neighbour coordination geometry. However, diffusion perpendicular to the **c**-axis must be by way of alternating octahedral and tetrahedral sites. Thus a cation that has a preference for one site geometry, (for example, Zn^{2+} prefers tetrahedral coordination and Cr^{3+} prefers octahedral coordination), will find diffusion parallel to the **c**-axis easier than diffusion parallel to the **a**- or **b**-axes. This effect is enhanced in structures in which only a fraction of the cation sites are occupied. For example, cation diffusion parallel to the **c**-axis in the rutile form of TiO_2 (Figure 8.19c) is of the order of 100 times faster than perpendicular to the **c**-axis.

The use of anion-centred polyhedra can be particularly useful in describing diffusion in fast ion conductors. These materials, which are solids that have an ionic conductivity approaching that of liquids, find use in batteries and sensors. An example is the high-temperature form of silver iodide, α-AgI. In this material, the iodide anions form a body-centred cubic array (Figure 8.20a). The anion-centred polyhedron of this structure is a truncated octahedron (Figure 8.20b). Each vertex of the polyhedron represents a tetrahedral cation site in the structure. Octahedral sites are found at the centres of each square face, and trigonal pyramidally coordinated cation sites occur at the midpoints of each edge (Figure 8.20b). In α-AgI, two silver atoms occupy the 12 available tetrahedral sites on a statistical basis. These cations are continually diffusing from one tetrahedral site to another, along paths represented by the edges of the polyhedron in a sequence: ...tetrahedral – trigonal bipyramid – tetrahedral – trigonal bipyramid.... Movement from one tetrahedral site to another via octahedral sites, by jumping across the centres of each square face of the polyhedron is not forbidden. However, at an octahedral site it is seen that the large silver ions must be sandwiched between large iodide ions, which would be energetically costly. Diffusion along the polyhedron edges avoids this problem. Due to the high diffusion rate of the cations, the silver ion array is described as a 'molten sublattice' (better termed a molten substructure) of Ag^+ ions. It is this feature that leads to the anomalously high values for the ionic conductivity. Similar conclusions are obtained for other fast-ion conductors and many solid-state battery materials.

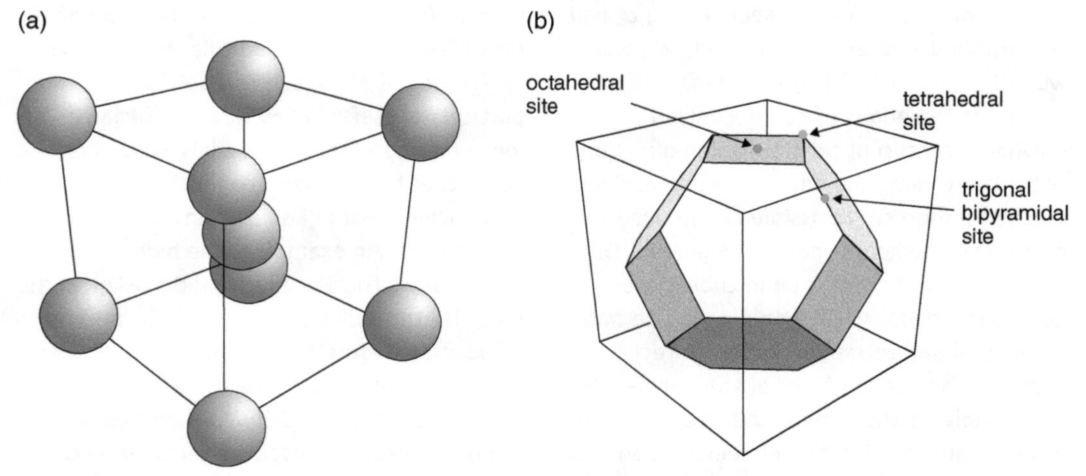

Figure 8.20 Perspective view of the structure of α-AgI: (a) the body-centred cubic arrangement of iodide ions in the unit cell; (b) the anion-centred polyhedron, (a truncated octahedron) around the iodide ion at the unit cell centre.

8.13 STRUCTURES AS NETS

The chemical bonds between the atoms of a crystal structure can be thought of as defining a net. The atoms of the structure lie at the nodes of the net. For many layer structures, nets can also be regarded as tessellations (Section 3.8). For example, both graphite and boron nitride can be described as a stacking of hexagonal nets. In the graphite structure (Figure 8.21a and b), the nets are staggered in passing from one layer to the other, so as to minimise electron–electron repulsion between the layers. In the case of boron nitride (Figure 8.21c and d), the layers are stacked vertically over each other, with boron atoms lying over and under nitrogen atoms in each layer, reflecting the different chemical bonding in this material. Naturally, these atomic differences may be lost when structures are represented as nets.

It is more difficult to represent truly three-dimensional structures by nets in plane figures, but the tetrahedral bond arrangements around carbon in diamond or around zinc in zinc sulfide and zinc oxide, are well suited to this representation. In the cubic diamond structure (Figure 8.22a), each carbon atom is bonded to four others, at the vertices of a surrounding tetrahedron. The structure can also be described as a net with tetrahedral connections (Figure 8.22b). If the carbon atoms are replaced by alternating sheets of zinc (Zn) or sulfur (S), the structure is that of cubic zincblende (sphalerite) (Figure 8.22c). The net that represents this structure (Figure 8.22d) is exactly the same as the diamond net. Zinc sulfide also adopts a hexagonal symmetry in the wurtzite structure, identical to that of zincite, ZnO (Figure 8.22e). The net that represents this structure is the structure of 'hexagonal diamond' (Figure 8.22f).

If the puckered nets formed by the bases of the tetrahedra, indicated by brackets in Figure 8.22d and f, are flattened, they form hexagonal sheets identical to those in the boron nitride and graphite nets.

The representation of crystal structures as nets is often helpful in demonstrating structural

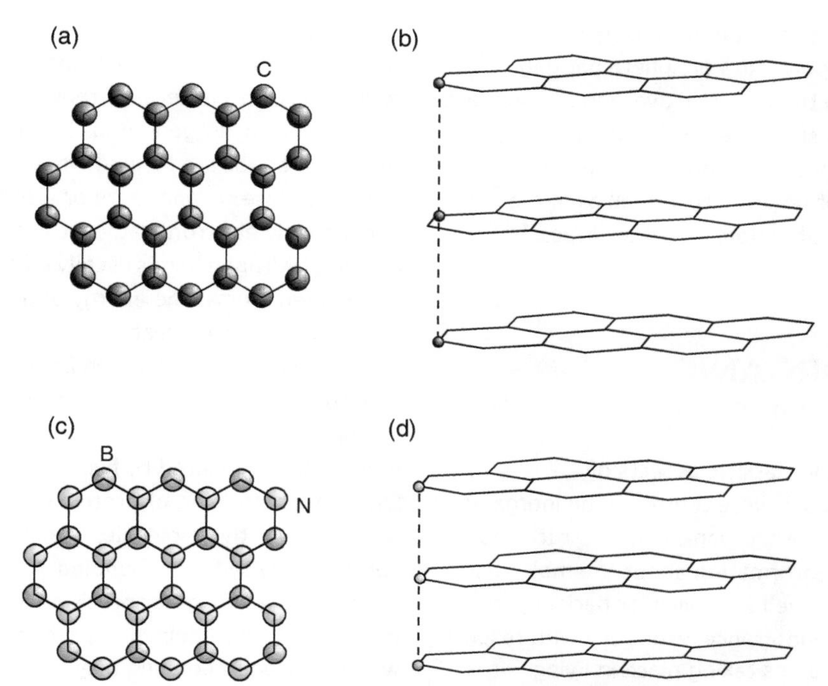

Figure 8.21 The structures of graphite and boron nitride: (a) a single layer of the graphite structure; (b) the stacking of layers, represented as nets, in graphite; (c) a single layer of the boron nitride structure; (d) the stacking of layers, represented as nets, in boron nitride.

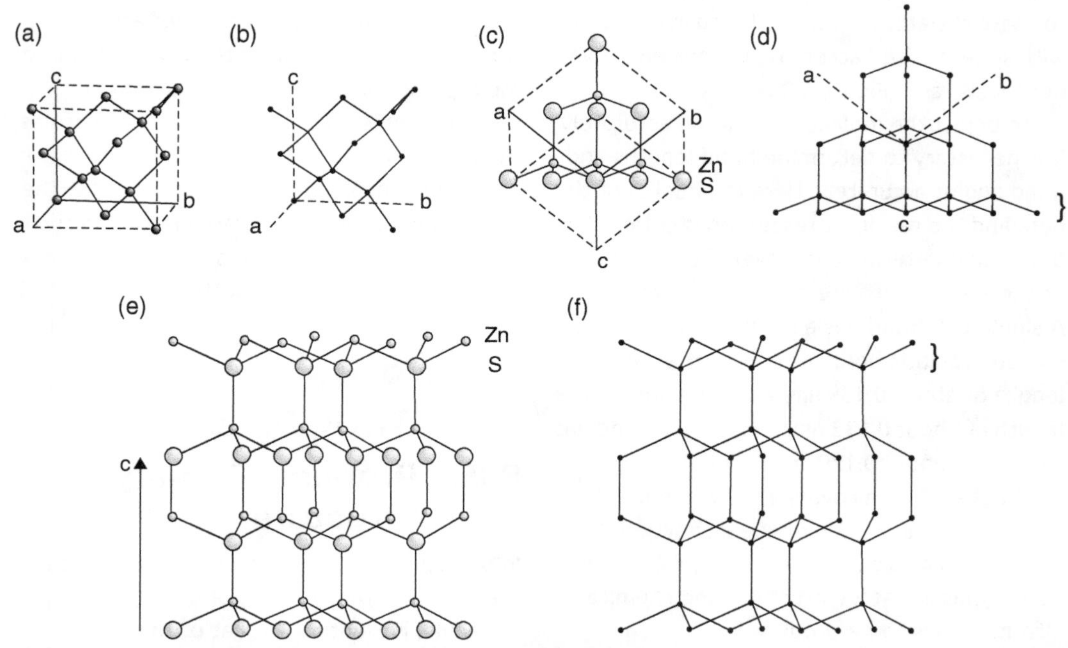

Figure 8.22 Three-dimensional nets: (a) the cubic diamond structure; (b) the net equivalent to (a); (c) the cubic zincblende (sphalerite) structure; (d) the net equivalent to (c), which is identical to that in (b); (e) the hexagonal wurtzite structure; (f) the net equivalent to (e).

relationships that are not lucidly portrayed by the use of polyhedra or sphere packing. Research in this area is active, and the portrayal of complex structures as (frequently very beautiful) nets is important in the design and synthesis of new solids, as well as in the correlation of various structure types.

8.14 ORGANIC STRUCTURES

Although the classical aspects of crystallography were centred upon inorganic or metallic crystals, organic crystal structures, which give information about the molecular structure as well as molecular packing, are of increasing importance, especially with respect to the processes taking place in living organisms. Organic molecules are often represented by net-like fragments, in which the links between carbon atoms are represented by 'sticks' and the atoms by 'balls'. A more realistic impression of the molecular geometry is given to these skeletal outlines by filling in the atoms with spheres sized according to covalent or van der Waals radii (Figure 8.23).

To depict the molecular skeleton accurately, it is necessary to determine bond lengths and bond angles accurately. Here the quality of the data and the resultant resolution is critical. The bonds that determine the overall configuration of these molecules are between carbon atoms. A single C–C bond has a length of about 0.154 nm, an aromatic bond, as in benzene, has a length of about 0.139 nm, a double bond has a length of about 0.133 nm, and a triple bond has a length of about 0.120 nm. That is, a total range of about 0.034 nm covers all of these bond lengths. The resolution of the diffraction data must be such as to reliably distinguish these bond types, if useful structural and chemical information is to be produced.

The disposition of hydrogen atoms is another vital aspect of the crystallography of organic compounds. This is because they control the formation of hydrogen bonds in the crystal. Hydrogen bonds are formed when a hydrogen atom lies close to, and more or less between, two strongly electronegative atoms. In organic crystals, hydrogen bonds usually involve oxygen or nitrogen atoms. The energy of a hydrogen bond is not all that great, of the order of 20 kJ mol^{-1}. Despite this, hydrogen bonding has a vital role to play in biological activity. For example, the base pairing in DNA and related molecules is mediated by hydrogen bonding. Similarly, the folding of protein chains into biologically active molecules is almost totally controlled by hydrogen bonding. As hydrogen is the lightest atom, great precision in diffraction data is needed to delineate probable regions where hydrogen bonding may occur. Indeed, neutron diffraction is often needed for this purpose, as neutrons are more sensitive to hydrogen than X-rays (see Section 7.12).

The ball and stick or space-filling approach to the depiction of organic structures is satisfactory when small molecules are concerned, but it becomes too cumbersome for the representation of large molecules. In the case of complex naturally occurring molecules, this level of detail may even obscure the way in which the molecule functions biologically. The problem of structural characterisation in such large molecules is described with respect to proteins in the following section.

8.15 PROTEIN STRUCTURES

8.15.1 Proteins: Primary Structure

In principle, protein crystallography, or indeed, the crystallography of any large organic molecule, is identical to that of the

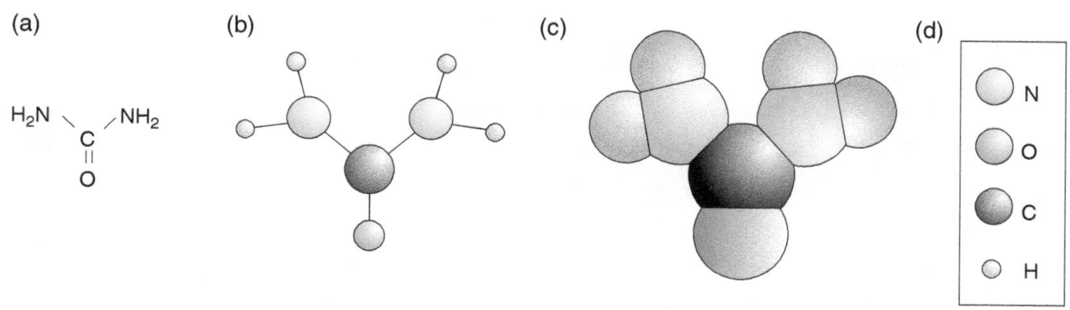

Figure 8.23 Depiction of urea: (a) chemical structural formula; (b) using covalent radii from Figure 8.8; (c) using van der Waals radii from Figure 8.9.

crystallography of inorganic solids. The most significant difference between proteins and the structures considered earlier is the enormous increase in the complexity between the two. For example, one of the smaller proteins, insulin, has a molar mass of approximately 5700 g mol^{-1}, and a run of the mill (in size terms) protein, haemoglobin, has a molar mass of approximately 64 500 g mol^{-1}. When it is remembered that proteins are mostly made up of light atoms, H, C, N and O, the atomic complexity can be appreciated. This difficulty is reflected in the difference between the rate of development of inorganic crystallography compared with protein crystallography. The first inorganic crystal structures were solved by W.H. and W.L. Bragg in 1913, but it was not until the middle of the same century that the structures of macromolecules and proteins began to be resolved.

Even in very simple materials, the overall pattern of structure in the crystal is not always easy to appreciate just by plotting the contents of a single unit cell. This problem of visualisation is far greater when complex structures such as proteins are considered. One of the more important tasks for the scientists that determined the first protein structures was how to depict the data in such a way that biological activity could be inferred without a loss of chemical or structural information. The way in which this is achieved is to describe protein structures by way of a hierarchy of levels, starting with the **primary structure**.

Proteins are polymers of amino acids. Amino acids, typified by alanine (Section 6.7), have an acidic –COOH group and a basic $-NH_2$ group on the same molecule. The naturally occurring amino acids are often called α-amino acids, meaning that the $-NH_2$ group is bonded to the carbon atom (the α carbon) next to the –COOH group. Naturally occurring proteins are built from 20 natural α-amino acids, and in all but one, glycine (NH_2CH_2COOH), the α carbon atom is a chiral centre, so that the molecules exist as enantiomers. However, only one enantiomer is used to construct proteins in cells, a configuration often called the L-form. Thus, the naturally occurring form of the α-amino acid alanine is often called L-alanine.

Amino acids are linked together to form polymers via the reaction of the acidic –COOH group on one molecule with the basic $-NH_2$ group on another. The resulting link is called a **peptide bond** (Scheme 8.1). If there are fewer than about 50 amino acids in the chain, the molecules are referred to as **polypeptides**, whilst if there are more than this number, they are usually called **proteins**. As H_2O is lost during the formation of a peptide bond, the remaining sections of amino acids are called amino acid **residues**. The sequence of amino acid residues,

$$H_2N - R - COOH + H_2N - R' - COOH \longrightarrow H_2N - \underline{R - CO} - NH - R' - COOH + H_2O$$

amino acid 1 amino acid 2 peptide bond

R and R′ represent the rest of the amino acid molecule.

Scheme 8.1 The formation of a peptide bond.

(a) E - K - F - D - K - F - L - T - A - M - K

(b) Glu - Lys - Phe - Asp - Lys - Phe - Leu - Thr - Ala - Met - Lys -

(c) Glutamic acid - Lysine - Phenylalanine - Aspartic acid - Lysine - Phenylalanine - Leucine - Threonine - Alanine - Methionine - Lysine -

Scheme 8.2 A primary sequence of amino acid residues in a protein; (a) single-letter abbreviations for the amino acids; (b) three-letter abbreviations; (c) full names.

which can be determined by chemical methods, is then called the primary structure of the protein (Scheme 8.2). The primary structure is written from left to right, starting with the amino (–NH_2) end of the molecular chain and ending with the carboxyl (–COOH) end.

8.15.2 Proteins: Secondary, Tertiary and Quaternary Structure

Although the primary structure of a protein is of importance, the way in which the polypeptide chain folds controls the chemical activity of the molecule. The folding is generally brought about by hydrogen bonds between oxygen and nitrogen atoms in the chains, and mistakes in the folding lead to the malfunction of the protein and often to illness. The overall complexity of a protein molecule can be dissected by making use of the fact that, in all proteins, relatively short stretches of the polypeptide chains are organised into a small number of particular conformations. These local forms make up the **secondary structure** of the protein. They are often represented in a 'cartoon' form by coiled or twined ribbons. The most common secondary structures are helices and β-sheets.

The polypeptide chain can coil into a number of different helical arrangements, but the commonest is the **α-helix**, a right-handed coil. The backbone of the helix is composed of the repeating sequence: (i) carbon atom double-bonded to oxygen, C=O; (ii) carbon atom bonded to a general organic group, C–R; (iii) nitrogen, N (Figure 8.24a). The helical conformation is a result of the geometry of the bonds around the carbon atoms and hydrogen bonding between hydrogen atoms on N and oxygen atoms on C=O groups. The repeat distance between any two corresponding atoms, the **pitch** of the helix, is 0.54 nm, and there are 3.6 amino acid residues in each turn (Figure 8.24b). The –NH_2 end of the helix has a positive charge relative to the –COOH end, which is negative. The 'cartoon' depiction of the α-helix as a coiled ribbon is drawn in Figure 8.24c. The structure of the protein myoglobin, the first protein structure to be solved using X-ray diffraction, is built of a compact packing of eight α-helices (Figure 8.25).

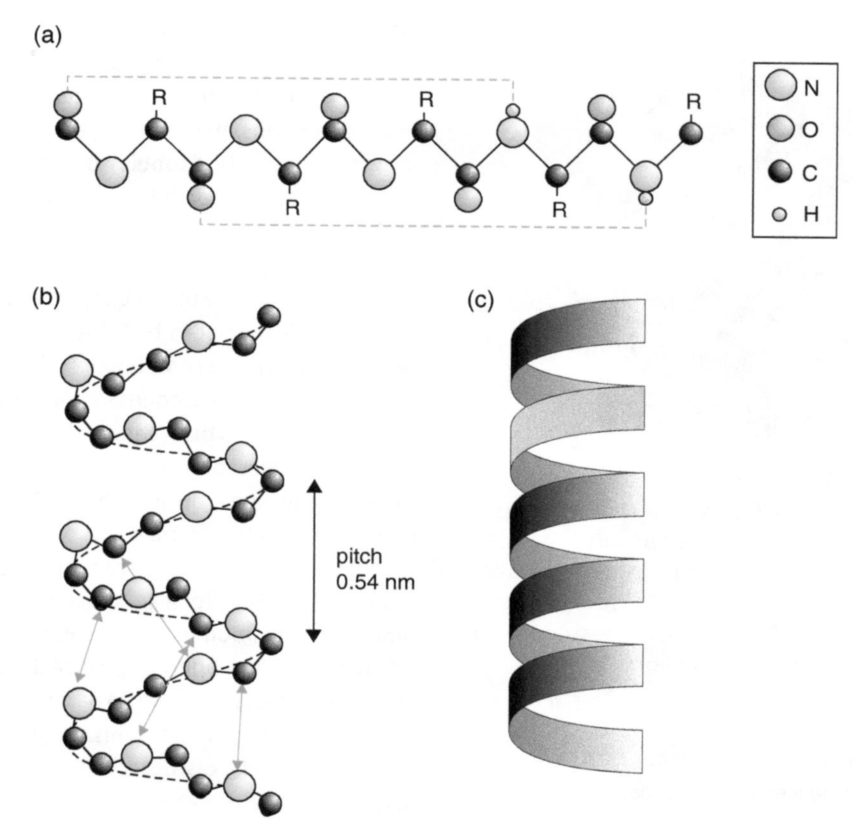

Figure 8.24 The α-helix: (a) the succession of carbon and nitrogen atoms along the backbone of an α-helix, hydrogen bonds form between O atoms on carbon and H atoms on nitrogen linked by dashed lines, R represents a general organic side-group, and for clarity not all hydrogen atoms are shown; (b) schematic depiction of the coiled α-helix, with four hydrogen bonds indicated by double-headed arrows; (c) 'cartoon' depiction of an α-helix as a coiled ribbon.

Sheets are formed by association of lengths of polypeptide chain called **β-strands**. The backbone of a β-strand is exactly the same as that of an α-helix (Figure 8.26a). A single β-stand is not stable alone, but links to a parallel strand to form a pair, via hydrogen bonding. The direction of the strands runs from the $-NH_2$ termination to the –COOH termination. Pairs of strands can run in the same direction, to give **parallel chains**, or in opposite directions to give **antiparallel chains** (Figure 8.26b and c). The antiparallel arrangement is more stable than the parallel arrangement, but both occur in protein structures. A β-strand is represented in 'cartoon' form by a broad ribbon arrow, with the head at the –COOH end (Figure 8.26d). Several parallel or antiparallel β-strands in proximity form a **β-sheet** or **β-pleated sheet** (Figure 8.26e). Stable sheets can consist of just two or more antiparallel β-strands, or, due to the lower stability of the parallel arrangement, four or more parallel β-strands. A typical

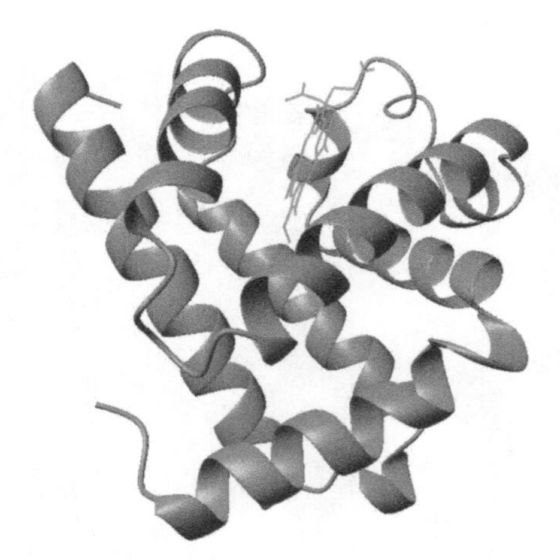

Figure 8.25 The structure of the protein myoglobin. This is composed of a compact packing of eight α-helices around the planar heme group. *Source*: This figure provided courtesy of Whitford, D. (2005). *Proteins, Structure and Function*. Chichester: Wiley, reproduced with permission.

secondary structure fragment of a protein, containing helices and strands, is drawn in Figure 8.27a.

The **tertiary structure** of proteins, also called the **super-secondary structure**, denotes the way in which the secondary structures are assembled in the biologically active molecule (Figure 8.27b). These are described in terms of **motifs**[3] that consist of a small number of secondary structure elements linked by **turns**, such as (α-helix – turn – α-helix). Motifs are further arranged into larger arrangements called **folds**, which can be, for example, a collection of β-sheets arranged in a cylindrical fashion to form a β-barrel, or **domains**, which may be made up of helices, β-sheets or both.

Fibrous proteins are those in which the polymer chains are arranged so as to form long fibres. These are common constituents of muscle, tendons, hair and silk, and are both strong and flexible. **Globular proteins** have the polymer chains coiled into a compact, roughly spherical shape.

This tertiary organisation governs molecular activity and is of the utmost importance in living systems. It is mediated by fairly weak chemical bonds between the components of the polymer chains, including hydrogen bonding, van der Waals bonds, links between sulfur atoms, called disulfide bridges, and so on. The bonds need to be relatively weak, because biological activity often requires the tertiary structure of the protein to change in response to biologically induced chemical signals. A mistake at a single amino acid position may prevent these weak interactions from operating correctly, leading to folding errors that can have disastrous implications for living organisms. The tertiary structure is easily disrupted, and the protein is said to become **denatured**. A typical example is given by the changes that occur when an egg is cooked. The white insoluble part of the cooked egg is denatured albumin. In denatured proteins, the primary structure of the polymer chains is maintained, but the coiling into a tertiary form is destroyed, and biological activity is lost.

Many proteins are built from more than one polypeptide chain or **subunit**. The bonding and arrangement of these is the **quaternary structure** of the protein (Figure 8.27c). For example, the globular protein haemoglobin is composed of two distinct subunits, called α- and β-chains. These link to form αβ-dimers, and two of these, related by a twofold axis, form the molecule, which thus consists of four subunits in all. When the molecule gains or loses oxygen, both the ternary and quaternary structures change significantly.

[3] Note that this use of the term motif is quite different from that used normally in crystallography, which indicates the unit of structure to be placed at each lattice point to recreate the whole structure.

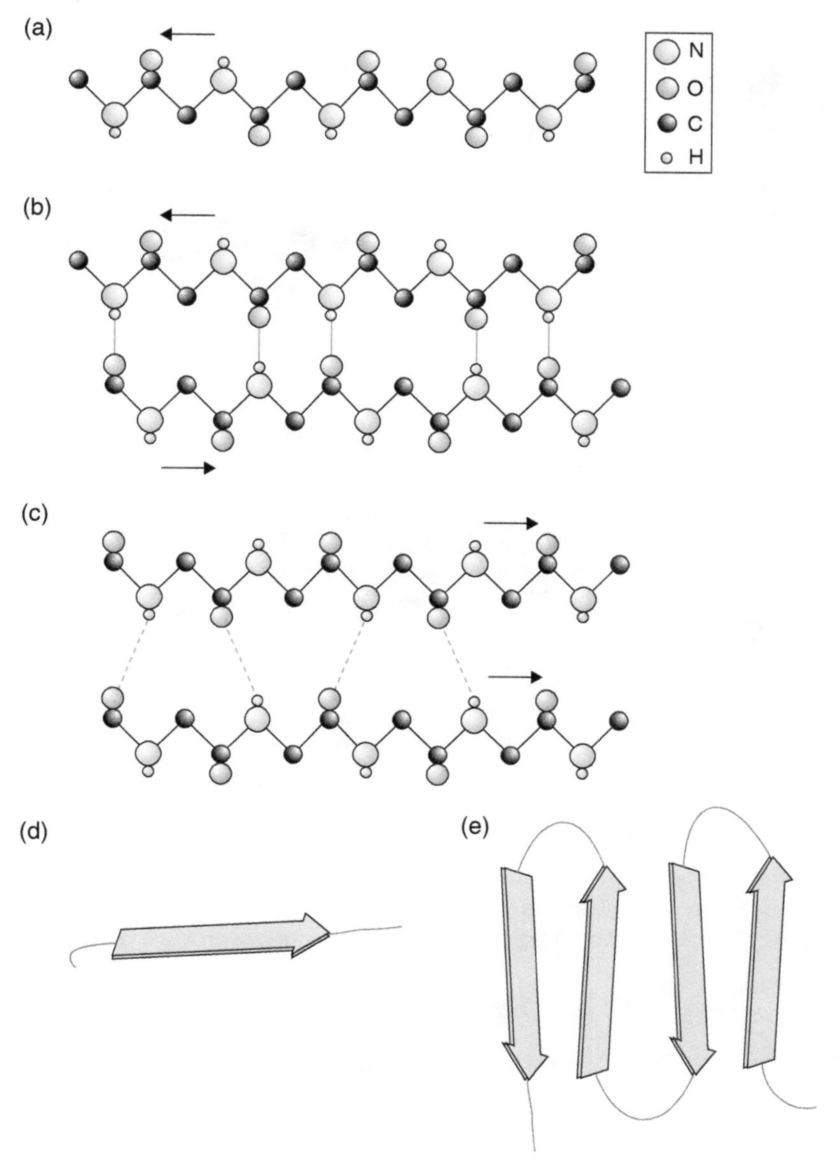

Figure 8.26 The β-sheet: (a) the succession of carbon and nitrogen atoms along the backbone of a β-strand, the arrow indicates the direction of the strand, from the $-NH_2$ termination to the –COOH termination (for clarity not all hydrogen atoms and side groups are shown); (b) schematic depiction of a pair of antiparallel β-strands, hydrogen bonds form between O atoms on carbon and H atoms on nitrogen linked by lines; (c) schematic depiction of a pair of parallel β-strands, hydrogen bonds form between O atoms on carbon and H atoms on nitrogen linked by dashed lines; (d) 'cartoon' depiction of a β-strand as a ribbon, in which the arrowhead represents the –COOH termination; (e) a β-sheet composed of four antiparallel β-strands.

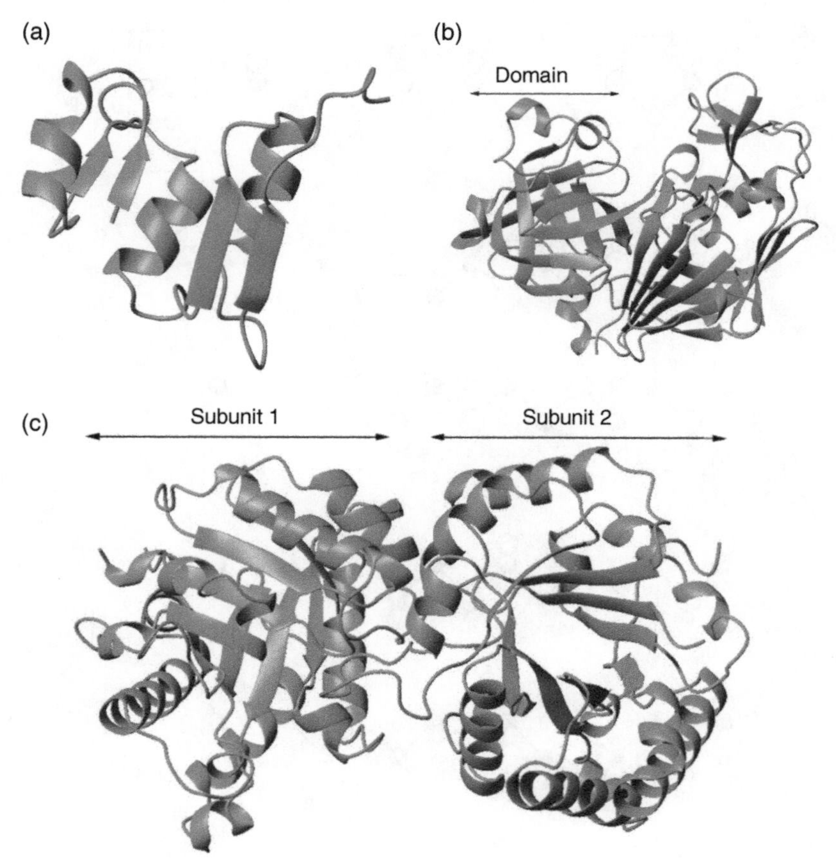

Figure 8.27 Organisational levels in protein crystallography: (a) secondary structure; (b) tertiary structure; (c) quaternary structure. *Source*: This figure provided courtesy of Whitford, D. (2005). *Proteins, Structure and Function*. Chichester: Wiley, reproduced with permission.

ANSWERS TO INTRODUCTORY QUESTIONS

What is the size of an atom?

Quantum mechanics makes it clear that atoms do not have a fixed size. Electron orbitals extend from the nucleus to a greater or lesser extent, depending upon the chemical and physical environment in the locality of the atomic nucleus. The scattering of X-rays and electrons gives details of the varying electron density throughout the unit cell volume. As the electron density is highest nearer to atomic cores, and least well away from atomic cores, these diffraction experiments yield the positions for the atomic cores, that is, interatomic distances are obtained, not atomic sizes. The tables of atomic radii found in textbooks are derived by dividing up the interatomic distances derived by diffraction methods to allocate radii to each atom involved. To some extent these radii are arbitrary, being chosen to reflect chemical or physical aspects of the crystal, which accounts for the differing values to be found for the radius of any atom.

Thus, an atom does not have a size that can be defined categorically, and applied to all chemical and physical regimes, but can be given an empirical size that is of value in restricted circumstances.

How does the idea of bond valence help in structure determination?
X-ray diffraction does not separate atoms that are Periodic Table neighbours well, since the scattering factors of these species are so similar. Thus, problems such as the distribution of Fe^{2+} and Fe^{3+} over the available sites in a crystal structure may be unresolved by conventional structure determination methods. The bond valence model is an empirical concept that correlates the strength of a chemical bond between two atoms and the length of the bond. Because crystal structure determinations yield accurate interatomic distances, precise values of the bond strength, called the experimental bond valence, can be derived.

Knowledge of the interatomic distances allows the experimental valence of an atom to be calculated using tables of bond-valence parameters. If the apparent valence of an atom is higher than that expected for an ionic bonding model, say 4.2 for a nominal Ti^{4+} ion, the ion is said to be overbonded, and if less, say 3.8, the ion is said to be underbonded. The cation and anion distribution in the structure is considered to be correct if none of the ions are significantly over- or underbonded.

What is the secondary structure of a protein?
The secondary structure of a protein is the arrangement in space of relatively short sequences of amino acids that make up the polypeptide chain. These conformations are often represented in a 'cartoon' form by coiled or extended ribbons. The most common secondary structures are α-helices and β-sheets. The helix has a repeat distance between any two corresponding atoms, the pitch of the helix, of 0.54 nm, and there are 3.6 amino acids, called residues, in each turn. The $-NH_2$ end of the helix has a positive charge relative to the –COOH end, which is negative. Sheets are formed by lengths of polypeptide chain called β-strands. The direction of the strands runs from the $-NH_2$ termination to the –COOH termination. A single β-strand is not stable alone, but links to parallel strands to form sheets, via hydrogen bonding.

PROBLEMS AND EXERCISES

Quick Quiz

1. X-ray diffraction is able to quantify:
   a. The volume of an atom
   b. The radius of an atom
   c. The distance between two atoms
2. The sphere packing that gives rise to a cubic structure is described as:
   a. ...ABABAB...
   b. ...ABCABCABC...
   c. ...ABACABAC...
3. The radius of a cation M^{4+} is:
   a. Greater than the radius of M^{2+}
   b. Less than the radius of M^{2+}
   c. The same as the radius of M^{2+}
4. Space-filling models of organic molecules use:
   a. Van der Waals radii
   b. Covalent radii
   c. Ionic radii
5. The largest cations tend to be surrounded by anions at the vertices of:
   a. A tetrahedron
   b. An octahedron
   c. A cube
6. In a cubic close-packed array of *N* spheres there are:
   a. *N* octahedral interstices
   b. 2*N* octahedral interstices
   c. 4*N* octahedral interstices

7. In a solid represented by packing of anion-centred polyhedra, the polyhedron edges represent possible:
   a. Cation diffusion paths
   b. Anion diffusion paths
   c. Both cation and anion diffusion paths
8. The resolution quoted for a protein structure corresponds to:
   a. The wavelength of the radiation used in the diffraction experiment
   b. The smallest d_{hkl} value for which reliable intensity data are available
   c. The lowest diffracted intensity measured
9. Protein structures are described in a hierarchy of:
   a. Two levels
   b. Three levels
   c. Four levels
10. In a protein structure, a β-pleated sheet is regarded as part of the:
   a. Primary structure
   b. Secondary structure
   c. Tertiary structure

Calculations and Questions

8.1. Estimate the ideal lattice parameters of the room-temperature and high-temperature forms of the following metals, ignoring thermal expansion: (a) Fe [r(CN12) = 0.1274 nm] adopts the A2 (body-centred cubic) structure at room temperature and the A1 (face-centred cubic) structure at high temperature; (b) Ti [r(CN12) = 0.1462 nm] adopts the A3 (hexagonal close-packed) structure at room temperature and the A2 (body-centred cubic) structure at high temperature; (c) Ca [r(CN12) = 0.1974 nm] adopts the A1 (face-centred cubic) structure at room temperature and the A2 (body-centred cubic) structure at high temperature. The radii given, r(CN12), are metallic radii appropriate to 12 coordination.

8.2. The following compounds all have the *halite* (NaCl) structure (see Chapter 1). LiCl, a = 0.512 954 nm; NaCl, a = 0.563 978 nm; KCl, a = 0.629 294 nm; RbCl, a = 0.658 10 nm; CsCl, a = 0.7040 nm. Make a table of the ionic radii of the ions assuming that the anions are in contact in the compound with the smallest cations.

8.3. Because the X-ray scattering factors of Mg and Al are similar, it is not easy to assign the cations in the mineral spinel, $MgAl_2O_4$, to either octahedral or tetrahedral sites (see Section 8.8 for more information). The bond lengths around the tetrahedral and octahedral positions are given in the table. Use the bond valence method to determine whether the spinel is normal or inverse. The values of r_0 are: r_0 (Mg^{2+}) = 0.1693 nm, r_0 (Al^{3+}) = 0.1651 nm, B = 0.037 nm, from Breese, N.E. and O'Keeffe, M. (1991). *Acta Crystallographica* **B47**: 192–197.

	Number of bonds	r_{ij}, nm
Tetrahedral site		
M–O	4	0.19441
Octahedral site		
M–O	6	0.19124

Source: Data from Zorina, N.G. and Kvitka, S.S. (1968). *Kristallografiya* **13**: 703–705, provided by the EPSRC's Chemical Database Service at Daresbury.

8.4. The tetragonal tungsten bronze structure is adopted by a number of compounds including $Ba_3(ZrNb)_5O_{15}$. There are two different octahedral positions in the structure, which are occupied by the ions Zr^{4+} and Nb^{5+}. The compound metal–oxygen bond lengths, r_{ij}, for the two octahedral sites are given in the table. Use the bond-valence method to determine which ion occupies which site. The values of r_0 are: r_0 (Nb^{5+}) = 0.1911 nm, r_0 (Zr^{4+}) = 0.1937 nm, B = 0.037 nm, from Breese, N.E. and O'Keeffe, M. (1991). *Acta Crystallographica* **B47**: 192–197.

	Number of bonds	r_{ij}, nm[a]
Site 1		
M–O1	1	0.1858
M–O2	1	0.2040
M–O3	4	0.2028
Site 2		
M–O1	1	0.1982
M–O2	1	0.1960
M–O3	1	0.2125
M–O4	1	0.2118
M–O5	1	0.2123
M–O6	1	0.2098

[a]These bond lengths are averages derived from a number of different crystallographic studies.

8.5. The following solids are described in terms of close-packed anion arrays. What is the formula of each?

a. Lithium oxide: anion array, cubic close-packed; Li^+ in all tetrahedral sites.

b. Iron titanium oxide, ilmenite: anion array, hexagonal close-packed; Fe^{3+} and Ti^{4+} share 2/3 octahedral sites.

c. Gallium sulfide: anion array, hexagonal close-packed; Ga^{3+} in 1/3 tetrahedral sites.

d. Cadmium chloride: anion array, cubic close-packed; Cd^{2+} in every octahedral site in alternate layers.

e. Iron silicate (fayalite): anion array, hexagonal close-packed; Fe^{2+} in half the octahedral sites, Si^{4+} in ⅛ tetrahedral sites.

f. Chromium oxide: anion array, cubic close-packed; Cr^{3+} in ⅙ octahedral sites, Cr^{6+} in ⅛ tetrahedral sites.

8.6. Draw the urea molecule CH_4N_2O (see Chapter 1), using the metallic, ionic, covalent and van der Waals radii, bond angles and bond lengths given in the table below. The crystallographic data are from: Zavodnik, V., Stash, A., Tsirelson, V. et al. (1999), *Acta Crystallographica* **B55**: 45. The ionic radii are from Muller, O. and Roy, R. (1974), *The Major Ternary Structural Families*, 5. Berlin: Springer-Verlag. The metallic radii are from Teatum, Gschneidner and Waber, cited in *The Crystal Chemistry and Physics of Metals and Alloys* (1972), Wiley-Interscience, New York, p. 151. Other data are from Figures 8.7 and 8.8. Comment on the validity of using metallic or ionic radii in such depictions.

Data for urea

Bond lengths: N–H, 0.100 nm; C–N, 0.134 nm; C–O, 0.126 nm.

Bond angles: O–C–N, 121.6°; N–C–N, 116.8°; H–N–H, 174.0°.

Atom	Metallic radius, nm	Ionic radius, nm	Covalent radius, nm	Van der Waals radius, nm
H	0.078	0.004 H^+	0.0299	0.120
C	0.092	0.006 C^{4+}	0.0767	0.170
N	0.088	0.002 N^{5+}	0.0702	0.155
O	0.089	0.121 O^{2-}	0.0659	0.152

Chapter 9

Defects, Modulated Structures and Quasicrystals

What are modular structures?
What are incommensurately modulated structures?
What are quasicrystals?

In this chapter, the concepts associated with classical crystallography are gradually weakened. Initially the effects of introducing small defects into a crystal are examined. These require almost no modification of the ideas already presented. However, structures with enormous unit cells pose more severe problems, and incommensurate structures are known in which a diffraction pattern is best quantified by recourse to higher dimensional space. Finally, classical crystallographic ideas break down when quasicrystals are examined. These structures, related to the Penrose tilings described in Chapter 2, can no longer be described in terms of the Bravais lattices described earlier.

9.1 DEFECTS AND OCCUPANCY FACTORS

In previous discussions of crystal structures, each atom was considered to occupy a crystallographic position completely. For example, in the crystal structure of Cs_3P_7 (Section 6.5), the Cs1 atoms occupied completely all of the positions with a Wyckoff symbol 4*a*. There are four equivalent Cs1 atoms in the unit cell. In such (normal) cases, the **occupancy** of the positions is said to be 1.0.

Not all materials are so well behaved. For example, many metal alloys have considerable composition ranges and a correct calculation of the intensities of diffracted beams needs inclusion of a site **occupancy factor**. For example, the disordered gold–copper alloy Au_xCu_{1-x} is able to take compositions with x varying from 0, pure gold, to 1.0, pure copper.

Crystals and Crystal Structures, Second Edition. Richard J. D. Tilley.

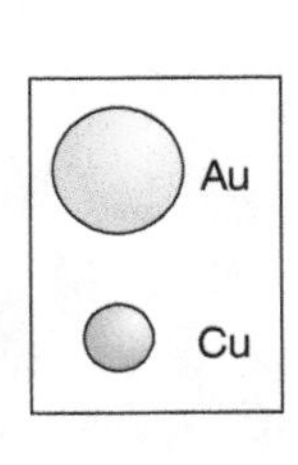

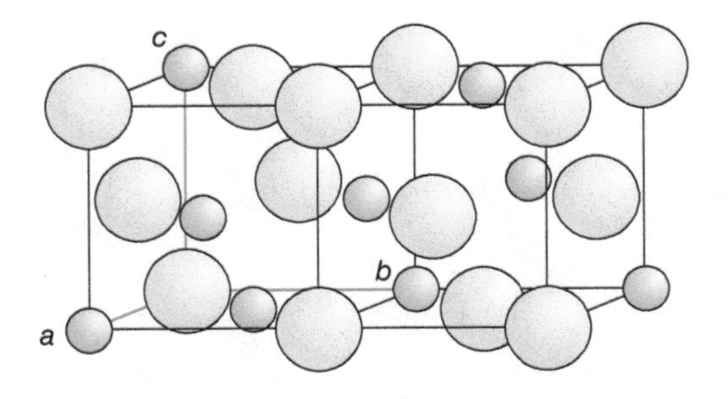

Figure 9.1 The structure of the disordered alloy CuAu. The Cu and Au atoms are distributed at random over the available metal sites.

The structure of the alloy is the copper (A1) structure (see Chapter 1), but in the alloy the sites occupied by the metal atoms contain a mixture of Cu and Au (Figure 9.1). This situation can be described by giving a site occupancy factor to each type of atom. For example, the alloy $Au_{0.5}Cu_{0.5}$ would have occupancy factors of 0.5 for each atom. The occupancy factor for an alloy of x Au and $(1 - x)$ Cu would simply be x and $(1 - x)$.

The same is true for compounds that form a **solid solution**. A solid solution, as the name implies, is a crystal in which some or all of the atoms are distributed at random over the various sites available, just as molecules are distributed throughout the bulk of a solvent. Alloys are simply metallic solid solutions. A typical example is provided by the oxides that form when mixtures of Al_2O_3 and Cr_2O_3 are heated together at high temperatures. The crystal structure of both of these oxides is the same, corundum (Al_2O_3) structure, and the final product is an oxide in which the Al and Cr atoms are distributed over the available metal atom positions (with Wyckoff symbol 12*c*), whilst the oxygen sites remain unchanged, to give a material of formula $Cr_xAl_{2-x}O_3$. The metal site occupancy factor for a material containing x Cr and $(2 - x)$ Al would be $x/2$ and $(1 - x/2)$, as the total site occupancy is 1 not 2.

In many compounds that are not obviously solid solutions or alloys, some sites will be occupied normally, with occupancy of 1, whilst other sites may accommodate a mixture of atoms. The spinel structure, described earlier (Section 8.8), can be used to illustrate this. Normal spinels are written $(A)[B_2]O_4$, and inverse spinels are written $(B)[AB]O_4$, where ( ) represents metal atoms in tetrahedral sites and [ ] represents metal atoms in octahedral sites. Now the forces that produce this ordering are not strong, and in many spinels the cation distribution is not clear-cut, so that the tetrahedral sites are not filled by one atom type alone, but by a mixture. In such cases, satisfactory structure determination will require that appropriate site occupancy factors are used. The oxygen atoms, however, are not subject to mixing, and the occupancy factor of these sites will always be 1.

The atomic scattering factor applicable to such alloys or solid solutions is an average value, referred to as the **site scattering factor**, f_{site}. In general, if two atoms, A and B, with atomic scattering factors f_A and f_B, fully occupy a single site in a structure, the site scattering factor is

$$f_{site} = x f_A + (1 - x) f_B$$

where x is the occupancy of the A sites and $(1 - x)$ is the occupancy of the B sites.

In some structures, all positions require site occupancy factors different from 1 if the diffracted intensities are to be correctly reproduced. This is the case when structures contain defects at all atomic sites in the structure, called **point defects**. Cubic calcia-stabilised zirconia, which crystallises with the fluorite (CaF_2), structure, provides an example. The parent structure is that of zirconia, ZrO_2. The stabilised phase has Ca^{2+} cations in some of the positions that are normally filled by Zr^{4+} cations, that is, cation substitution has occurred. As the Ca^{2+} ions have a lower charge than the Zr^{4+} ions, the crystal will show an overall negative charge if we write the formula as $Ca_x^{2+} Zr_{1-x}^{4+} O_2$. The crystal compensates for the extra negative charge by leaving some of the anion sites unoccupied. The number of vacancies in the anion substructure needs to be identical to the number of calcium ions present for exact neutrality, and the correct formula of the crystal is $Ca_x^{2+} Zr_{1-x}^{4+} O_{2-x}$. In order to successfully model the intensities of the diffracted beams, it is necessary to include a site occupancy factor of less than 1 for the oxygen sites, as well as fractional site occupancies for the Ca and Zr ions.

Many other examples of the use of site occupancy factors could be cited, especially in mineralogy, where most natural crystals have a mixed population of atoms occupying the various crystallographic positions.

9.2 DEFECTS AND UNIT CELL PARAMETERS

The positions of the lines or spots on the diffraction pattern of a single phase are directly related to the unit cell dimensions of the material. The unit cell of a solid with a fixed composition varies with temperature and pressure, but under normal circumstances is regarded as constant. If the solid has a composition range, as in a solid solution or an alloy, the unit cell parameters will vary. **Vegard's law,** first propounded in 1921, states that the lattice parameter of a solid solution of two phases with similar structures will be a *linear function* of the lattice parameters of the two end-members of the composition range (Figure 9.2a).

$$x = (a_{ss} - a_1)/(a_2 - a_1)$$

$$\text{i.e. } a_{ss} = a_1 + x(a_2 - a_1)$$

where a_1 and a_2 are the lattice parameters of the parent phases, a_{ss} is the lattice parameter of the solid solution, and x is the mole fraction of the parent phase with lattice parameter a_2. (The relationship holds for the *a*-, *b*- and *c*-lattice parameters, and any interaxial angles.) This 'law' is simply an expression of the idea that the cell parameters are a direct consequence of the sizes of the component atoms in the solid solution. Vegard's law, in its ideal form, is almost never obeyed exactly. A plot of cell parameters that lies below the ideal line (Figure 9.2b) is said to show a **negative** deviation from Vegard's law, and a plot that lies above the ideal line (Figure 9.2c) is said to show a **positive** deviation from Vegard's law. In these cases, atomic

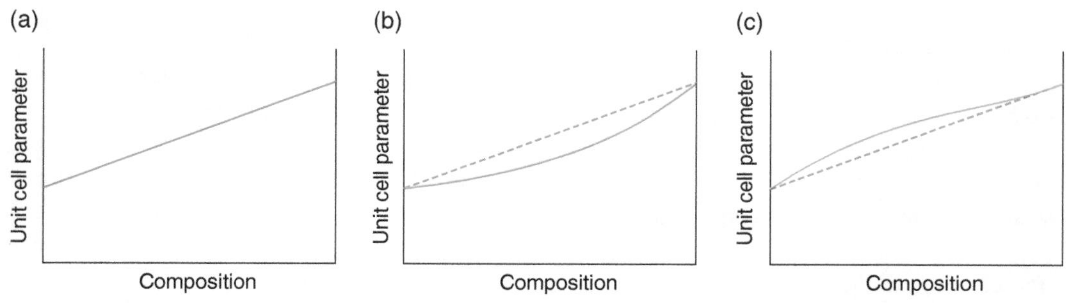

Figure 9.2 Vegard's law: (a) ideal Vegard's law behaviour; (b) negative deviation from Vegard's law; (c) positive deviation from Vegard's law.

interactions, which modify the size effects, are responsible for the deviations. In all cases, a plot of composition versus cell parameters can be used to determine the composition of intermediate compositions in a solid solution.

9.3 DEFECTS AND DENSITY

The theoretical density of a solid with a known crystal structure can be determined by dividing the mass of all the atoms in the unit cell by the unit cell volume (Section 1.9). This information, together with the measured density of the sample, can be used to determine the notional species of point defect present in a solid that has a variable composition. However, as both methods are averaging techniques, they say nothing about the real organisation of the point defects.

The general procedure is:

(i) Measure the composition of the solid.
(ii) Measure the density.
(iii) Measure the unit cell parameters.
(iv) Calculate the theoretical density for alternative point defect populations.
(v) Compare the theoretical and experimental densities to see which point defect model best fits the data.

The method can be illustrated by reference to a classic study of the defects present in iron monoxide.[1] Iron monoxide, often known by its mineral name of wüstite, has the halite (NaCl) structure. In the normal halite structure, there are four metal and four non-metal atoms in the unit cell, and compounds with this structure have an ideal composition $MX_{1.0}$ (Section 1.6). Wüstite has an oxygen-rich composition compared with the ideal formula of $FeO_{1.0}$. Data for an actual sample found an oxygen: iron ratio of 1.058, a density of 5728 kg m^{-3}, and a cubic lattice parameter, a, of 0.4301 nm. Because there is more oxygen present than iron, the real composition can be obtained by assuming either that there are extra oxygen atoms in the unit cell or that there are iron vacancies present.

Model A: Assume that the iron atoms in the crystal are in a perfect array, identical to the metal atoms in halite, and an excess of oxygen is due to interstitial oxygen atoms being present, over and above those on the normal anion positions. The ideal unit cell of the structure contains 4 Fe and 4 O, and so, in this model, the unit cell must contain 4 atoms of Fe and 4 $(1 + x)$ atoms of oxygen. The unit cell contents are Fe_4O_{4+4x} and the composition is $FeO_{1.058}$.

The mass of 1 unit cell is m_A:

$$\begin{aligned} m_A &= [(4 \times 55.85) + (4 \times 16 \times (1 + x))]/N_A \\ &= [(4 \times 55.85) + (4 \times 16 \times 1.058)]/N_A \quad \text{grams} \\ &= 4.834 \times 10^{-25}\ \text{kg} \end{aligned}$$

where N_A is Avogadro's constant. The volume, V, of the cubic unit cell is given by a^3, thus:

$$\begin{aligned} V &= (0.4301 \times 10^{-9})^3\ \text{m}^3 \\ &= 7.9562 \times 10^{-29}\ \text{m}^3 \end{aligned}$$

The density, ρ, is given by the mass m_A divided by the unit cell volume, V:

$$\begin{aligned} \rho &= [(4 \times 55.85) + (4 \times 16 \times (1 + x))]/N_A \\ &= 4.834 \times 10^{-25}/7.9562 \times 10^{-29} \\ &= 6076\ \text{kg m}^{-3} \end{aligned}$$

Model B: Assume that the oxygen array is perfect and identical to the non-metal atom array in the *halite* structure and that the unit cell contains some vacancies on the iron positions. In this case, one unit cell will contain 4 atoms of oxygen and $(4 - 4x)$ atoms of iron. The unit cell contents are $Fe_{4-4x}O_4$ and the composition is $Fe_{1/1.058}O_{1.0}$ or $Fe_{0.945}O$.

[1] The data is from the classic paper: Jette, E.R. and Foote, F. (1933). *J. Chem. Phys.* **1**: 29.

The mass of one unit cell is m_B:

$$m_B = [(4 \times (1 - x) \times 55.85) + (4 \times 16)]/N_A$$
$$= [(4 \times 0.945 \times 55.85) + (4 \times 16)]/N_A \quad \text{grams}$$
$$= 4.568 \times 10^{-25} \text{ kg}$$

The density, ρ, is given by m_B divided by the volume, V, to yield

$$\rho = 4.568 \times 10^{-25}/7.9562 \times 10^{-29}$$
$$= 5741 \text{ kg m}^{-3}$$

The difference in the two values is surprisingly large. The experimental value of the density, 5728 kg m^{-3}, is in good accordance with Model B, which assumes vacancies on the iron positions. This indicates that the formula should be written $Fe_{0.945}O$, and that the point defects present are iron vacancies.

9.4 MODULAR STRUCTURES

In this and the following three sections, structures that can be considered to be built from **slabs** of one or more parent structures are described. These materials pose particular crystallographic difficulties. For example, some possess enormous unit cells, of some hundreds of nm in length, whilst maintaining perfect crystallographic translational order. They are frequently found in mineral specimens, and the piecemeal way in which early examples were discovered has led to a number of more or less synonymic terms for their description, including 'intergrowth phases', 'composite structures', 'polysynthetic twinned phases', 'polysomatic phases' and 'tropochemical cell-twinned phases'. In general, they are all considered to be **modular structures**.

Modular structures can be built from slabs of the same or different compositions, and the slab widths can be disordered, or ordered in a variety of ways. The simplest situation corresponds to a material built from slabs of only a single parent phase and in which the slab thicknesses vary widely. In this case, the slab boundaries will not fall on a regular lattice, and they then form **planar defects**, which are two-dimensional analogues of one-dimensional point defects (Figure 9.3). The unit cell of the disordered material is the same as the unit cell of the parent phase, although the diffraction pattern will reveal evidence of this disorder in the form of diffuse or streaked diffraction spots (see also Section 9.6). Note, however, that the overall composition of the crystal may change, as detailed below.

An example of such a disordered crystal is shown in Figure 9.4a. The material is $SrTiO_3$,

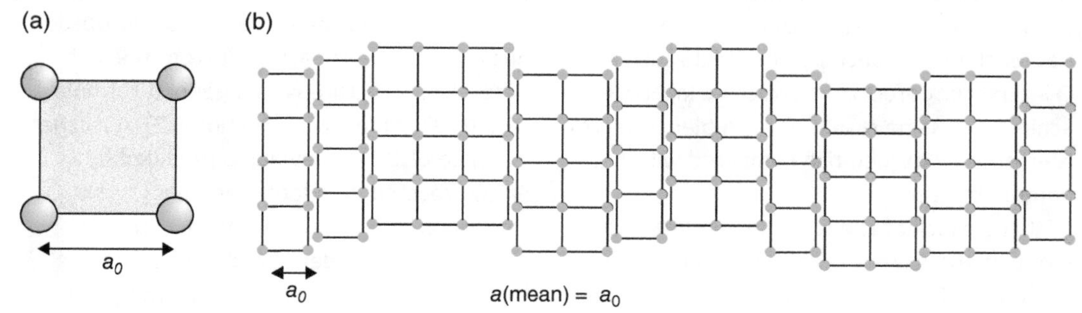

Figure 9.3 Planar faults: (a) unit cell of the parent structure; (b) crystal containing a random distribution of planar faults.

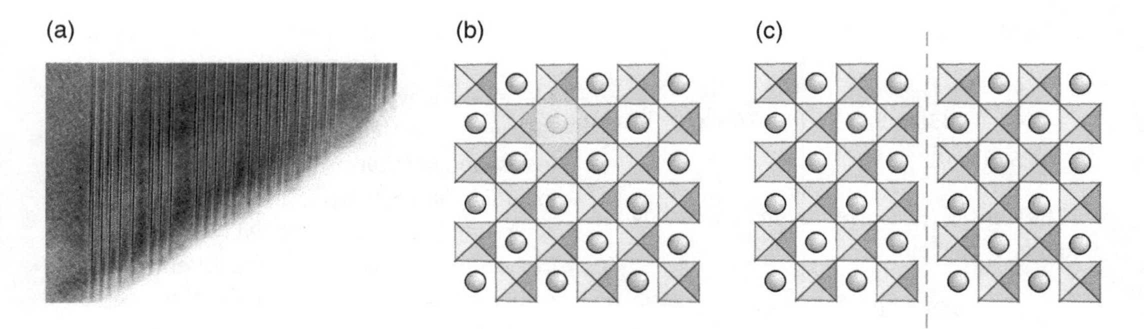

Figure 9.4 (a) Electron micrograph showing a flake of $SrTiO_3$ containing disordered planar defects lying upon (110) planes with respect to the idealised cubic unit cell. (b) The idealised structure of $SrTiO_3$ projected down the cubic **c**-axis. Each square represents a TiO_6 octahedron and the circles represent Sr atoms. The unit cell is shaded. (c) Idealised representation of a (110) fault in $SrTiO_3$.

which adopts the perovskite structure. The faulted crystal can be best described with respect to an idealised cubic perovskite structure (Section 1.7), with a lattice parameter of 0.375 nm (Figure 9.4b). The skeleton of the material is composed of corner-linked TiO_6 octahedra, with large Sr cations in the cages that lie within the octahedral framework. The planar defects, which arise where the perovskite structure is incorrectly stacked, lie upon (110) planes with respect to the idealised cubic structure (Figure 9.4c).

The faulting arises because the crystal contains some Nb_2O_5. This gives each crystal a composition $Sr(Nb_xTi_{1-x})O_{3+x/2}$. Each Nb^{5+} ion that replaces a Ti^{4+} ion requires that extra oxygen is incorporated into the crystal. Rather than introduce point defects such as O interstitial defects, the structure laminates to achieve the same result, as each planar defect is effectively oxygen-rich compared with the parent structure.

If the width of the component slabs in a modular structure is constant, no defects are present in the structure, because the unit cell includes the planar boundaries between the slabs as part of the structure. Take a case where two different slab widths of the parent structure are present (Figure 9.5). In the case where the displacement between adjacent slabs is always the same (and not equal to ½ of the cell parameter in the direction of the displacement), the unit cell will be monoclinic, and such structures are often given the prefix 'clino' in mineralogical literature (Figure 9.5b). The case in which slab displacement alternates in direction produces an orthorhombic unit cell, the 'ortho' form (Figure 9.5c).

Large numbers of these modular structures occur in mineral samples. Such structures have also been found in many systems that contain phases with the perovskite structure (Section 1.7), which include Ruddlesden-Popper phases, Aurivillius phases and Dion-Jacobson phases, all of which are built from slabs of perovskite structure with a general formula $(A_{n-1}B_nO3_{n+1})$ (also see Section 9.7). A further example of this behaviour is provided by $SrTiO_3$ reacted with larger amounts of Nb_2O_5 than described above. In this case, the boundaries are ordered and new structures form with well-defined unit cells and precise compositions. All belong to a series of perovskite-related structures related to the

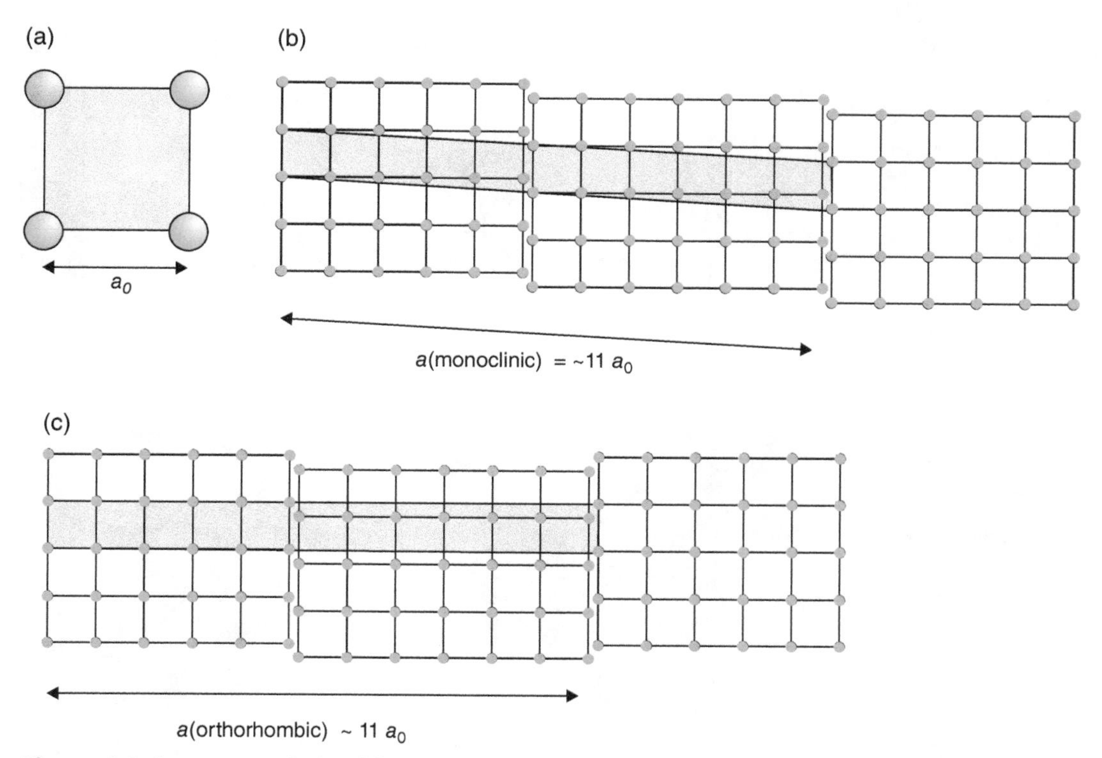

Figure 9.5 Structures derived from that of the parent phase by stacking alternating slabs of material, 5 and 6 unit cells in thickness: (a) a unit cell of the parent structure; (b) the displacement of the slabs is the same in every case, resulting in a monoclinic unit cell; (c) the displacement of the slabs alternates in direction, resulting in an orthorhombic unit cell.

phase $Ca_4Nb_4O_{14}$, with a general formula $A_nB_nO_{3n+2}$. Figure 9.6a shows the structure of the phase constructed by stacking slabs four octahedra in width, $Sr_4(Ti,Nb)_4O_{14}$, and Figure 9.6b a structure built of slabs five octahedra in width, $Sr_5(Ti,Nb)_5O_{17}$. A vast number of structures can be envisaged to form between these two. Figure 9.6c shows the simplest, consisting of alternating slabs of four and five octahedra width, written (4,5), $Sr_{4.5}(Ti,Nb)_{4.5}O_{15.5}$, but regular repeating sequences such as (4,4,4,5), (4,4,5), (4,5,5) and so on have also been characterised. The number of possible structures is increased enormously when other slab widths are also considered.

The formula of each ordered structure is given by a series formula $Sr_n(Nb,Ti)_nO_{3n+2}$, where n takes integer values from 4 upwards. The additional oxygen required to fulfil the valence requirements of the Nb^{5+} introduced into the parent structure, $SrTiO_3$, is incorporated into the crystal via the boundaries between each perovskite-structure slab.

A change in chemical composition is not a mandatory prerequisite for the formation of modular structures. A typical example of a structure built from displaced slabs but without change of composition is provided by the ordered alloy phase CuAu II (Figure 9.7a). This can be regarded as made up of slabs of another

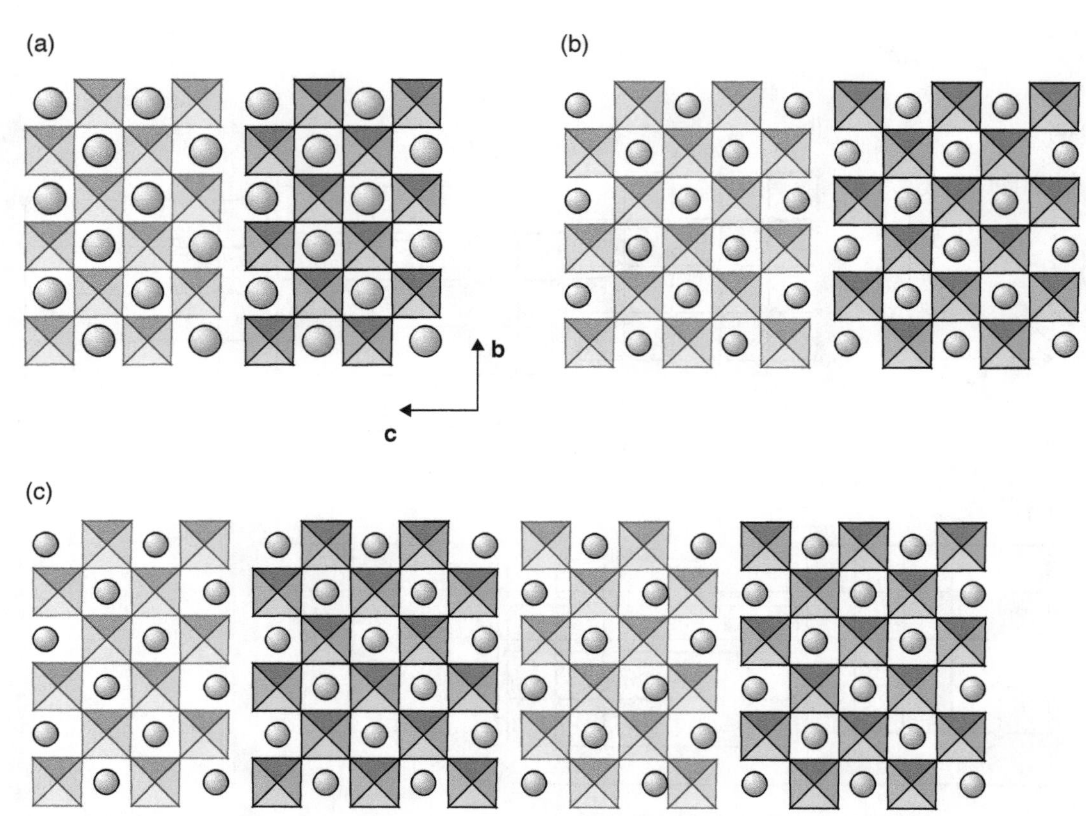

Figure 9.6 Schematic representation of the structures of the $Sr_n(Nb,Ti)_nO_{3n+2}$ phases: (a) $Sr_4Nb_4O_{14}$; (b) $Sr_5(Ti,Nb)_5O_{17}$; (c) $Sr_{4.5}(Ti,Nb)_{4.5}O_{15.5}$. The squares represent $(Nb,Ti)O_6$ octahedra and the circles represent Sr atoms.

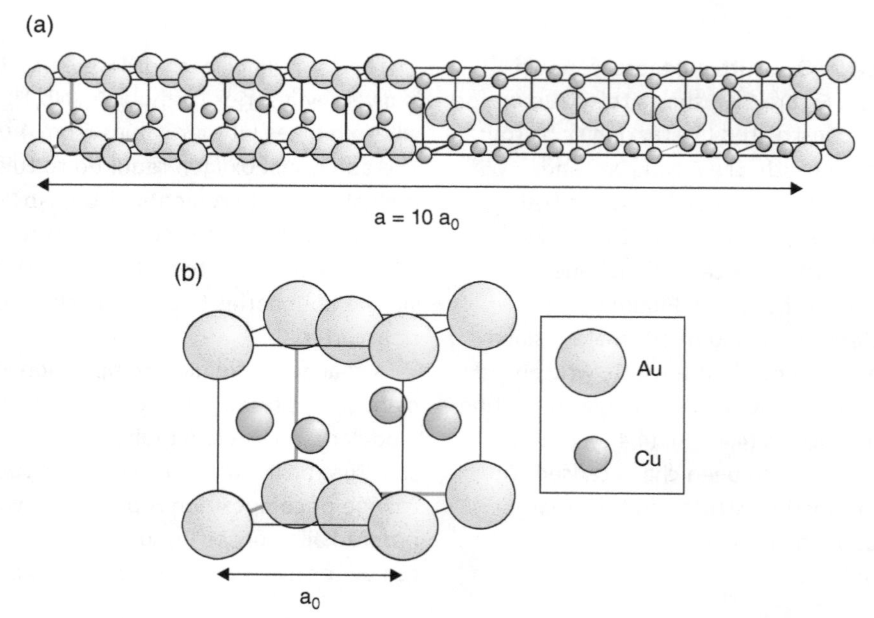

Figure 9.7 The structure of ordered copper–gold alloys: (a) CuAu II; (b) CuAu I.

ordered alloy, CuAu I, each of which is five unit cells in thickness (Figure 9.7b).

As these two examples show, besides altered unit cell dimensions, a series of modular structures may or may not show a regular variation of composition. When only one slab type is present, this will depend upon the nature of the planar boundaries involved. However, a change of composition must occur if the series of structures is composed of two or more slabs, each with a different composition. In the following three sections these aspects are described in more detail, and deal with structures built of slabs of a single parent type showing no overall composition variation, polytypes (Section 9.5); structures built of slabs of a single parent and showing regular composition variation, the crystallographic shear (CS) phases (Section 9.6); and structures built of slabs of two different structures, intergrowths or polysomes (Section 9.7).

9.5 POLYTYPES

The first long-period structures to be characterised were polymorphs of the superficially simple material silicon carbide, carborundum, SiC. These structures were called **polytypes**, to distinguish them from more normal and commonly occurring polymorphs that characterise many minerals, such as the aragonite and calcite forms of calcium carbonate, $CaCO_3$. Polytypes are now considered to be long-period structures built by stacking of layers with identical or very similar structures, and involving little or no change in composition.

Apart from silicon carbide, another apparently simple material that displays polytypism is zinc sulfide, ZnS. Silicon carbide and zinc sulfide phases exist in hundreds of structural modifications, many of which have enormous repeat distances along one direction. Despite this complexity, the composition of these compounds never strays from that of the parent phase.

The structure of both the SiC and ZnS polytypes can be illustrated with reference to the crystalline forms of ZnS. Zinc sulfide crystallises in either of two structures, one of which is cubic and given the mineral name zincblende (sphalerite) whilst the other is hexagonal and given the mineral name wurtzite. The relationship between these two structures can be understood by comparing the cubic and hexagonal close packing of spheres (Section 8.2). When the zincblende structure is viewed along the cube face-diagonal [110] direction, and oriented with the cubic [111] direction vertical, to make the zinc and sulfur layers horizontal, the layers of both zinc and sulfur are packed in the cubic closest packing arrangement ...aAbBcCaAbBcC..., where the lowercase letters stand for the Zn layers and the uppercase letters for the S layers (Figure 9.8a). The wurtzite structure, viewed with the hexagonal **c**-axis vertical and projected onto the hexagonal unit cell diagonal (i.e. the hexagonal $(11\bar{2}0)$ plane), has a layer packing ... aAbBaAbB... where the lowercase letters stand for Zn layers and the uppercase letters for S layers (Figure 9.8b). The zinc sulfide polytypes are complex arrangements of layers of both wurtzite (hexagonal) and zincblende (cubic) types. Some of the zinc sulfide polytypes are listed in Table 9.1.

Silicon carbide also crystallises in two forms, of which β-SiC has the cubic zincblende (sphalerite) structure. When viewed along the cube face-diagonal [110] direction, the layers of both silicon and carbon are packed in the cubic close-packing arrangement ... aAbBcCaAbBcC..., where the uppercase and lowercase letters stand for layers of Si and C. The other form of silicon carbide, α-SiC, is a collective name for the various silicon carbide

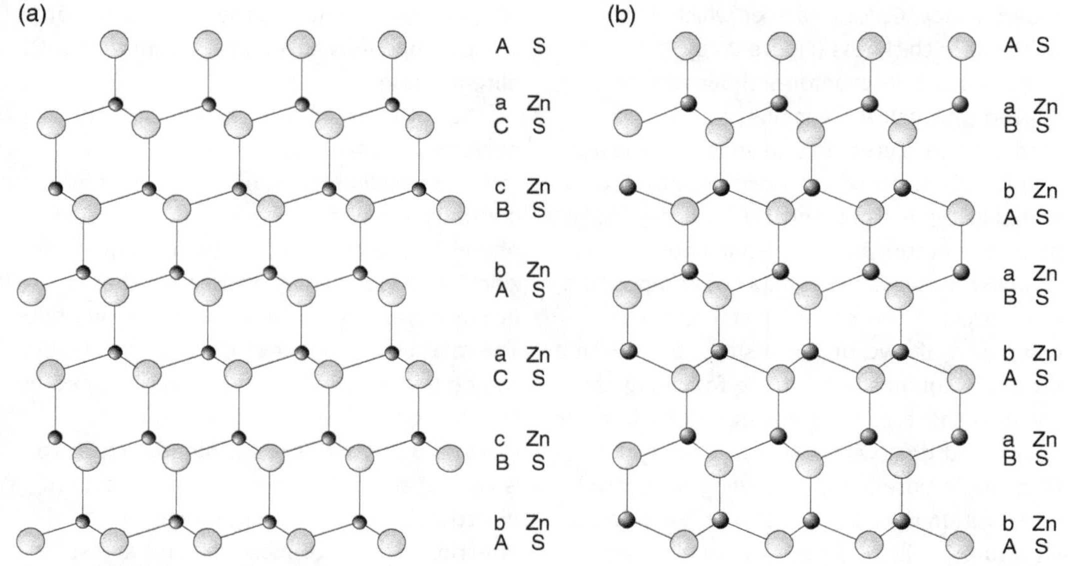

Figure 9.8 (a). The structure of the cubic form of zinc sulfide (zincblende, sphalerite). The cubic [111] direction is vertical and the structure is viewed down the cubic [110] direction. (b). The structure of the hexagonal form of zinc sulfide (wurtzite). The hexagonal **c**-axis is vertical, and the structure is viewed as if projected onto (110) = $(11\bar{2}0)$.

polytypes, which consist of complex arrangements of zincblende and wurtzite slabs. Some of these are known by names such as carborundum I, carborundum II, carborundum III, and so on. One of the simplest structures is that of carborundum I, which has a packing … aAbBaAcCaAbBaAcC…. Some silicon carbide polytypes are listed in Table 9.1.

The complexity of the structures has led to a number of proposed compact forms of nomenclature. The most widely used of these is the **Ramsdell** notation, which gives the number of layers of the stacked slabs in the crystallographic repeat, together with the symmetry of the unit cell, specified by C for cubic, H for hexagonal and R for rhombohedral. The repeating slab in the silicon carbide polytypes is taken as one sheet of Si atoms plus one sheet of C atoms, (Si + C), and in the zinc sulfide polytypes is one sheet of Zn atoms plus one sheet of S atoms, (Zn + S). The packing of the (Si + C) layers in cubic β-SiC, with the zincblende structure, is …ABC…, and the Ramsdell notation is 3C. The same symbol would apply to cubic ZnS. The Ramsdell notation for the wurtzite form of ZnS, in which the (Zn + S) layers are in hexagonal …AB… stacking, would be 2H. The same term would apply to a pure hexagonal form of SiC. The Ramsdell notation for the form of silicon carbide called carborundum I, the packing of which is described above, is 4H.

Whilst the Ramsdell notation is compact, it does not give the layer stacking sequence of the polytype, and a number of other terminologies have been invented to overcome this shortcoming. The **Zhdanov** notation, one way of specifying the stacking sequence, is derived in the following way. The translation between an A layer and a B layer, or a B layer

Table 9.1 Some polytypes of SiC and ZnS

Ramsdell notation	Zhdanov notation[a]	Stacking sequence[a]
2H	(11)	h
3C (3R)	(1)	c
4H	(22)	hc
6H	(33)	hcc
8H	(44)	hccc
10H	(82)	$(hc_7)(hc)$
	(55)	hcccc
14H	(5423)	hccccheccchchcc
	(77)	$(hc_6)_2$
16H	(88)	$(hc_7)_2$
	(14, 2)	$(hc_{13})(hc)$
	(5335)	$(hcccc)_2(hcc)_2$
	$(3223)_2$	$(hcc)_4(hc)_2$
24H	(15, 9)	$(hc_{14})(hc)_8$
24R	$(53)_2$	$[(hcccc)(hcc)]_2$
36R	$(6222)_3$	$[(hc_5)(hc)_3]_3$
48R	$(7423)_3$	$[(hc_6)(hccc)(hc)(hcc)]_3$
72R	$(6, 11, 5, 2)_3$	$[(hc_5)(hc_{10})(hc_4)(hc)]_3$

[a]Subscript numbers have the same significance as in ordinary chemical nomenclature. Thus $(53)_2$ = (5353), hc_7 = hccccccc and $(hcc)_2$ = hcchcc.

and a C layer, can be represented by a rotation of +60° or a translation of +1/3. Reverse transformations, C to B or B to A, are then represented by −60° or −1/3. The stacking sequence of any polytype can then be written as a series of + and − signs. The Zhdanov symbol of the polytype gives the number of signs of the same type in the sequence. Thus the cubic *zincblende* structure of ZnS or SiC has a stacking sequence ...+++++.... The Zhadanov symbol is written as (1), rather than (∞). The hexagonal *wurtzite* structure has a stacking sequence ...+−+−... so the Zhadanov symbol is (11). Similarly, carborundum I has a stacking sequence ...++−−..., which is represented by the Zhdanov symbol (22).

The Zhadanov notation also provides a pictorial representation of a structure. Draw structures in the same projection as used in Figure 9.8, and represent each layer pair in the stacking, say (Zn + S), by a single 'atom'. The result is a zig-zag pattern. The Zhadanov notation specifies the zig-zag sequence. Thus, the hexagonal ZnS (11) structure is a simple zig-zag (Figure 9.9a), carborundum I, (22), has a double repeat (Figure 9.9b), and the structure with a Zhdanov symbol (33) has a triple repeat (Figure 9.9c).

A third widely used terminology relies upon specifying the relative position of layers. Once again it is convenient to consider two layers, (Si + C) or (Zn + S) as a unit. A middle unit sandwiched between sheets of the same type, say *BAB* or *CAC*, is labelled *h*. Similarly, a layer sandwiched between sheets of different type, say *ABC* or *BCA*, is labelled *c*. The 4H polytype of silicon carbide is then labelled as (*hc*) (Table 9.1).

The complexity of polytypes is enormous, as can be judged from the few examples given in Table 9.1, and it must not be forgotten that polytypes exist in many other chemical systems, from chemically simple CdI_2 to the complex mica silicates. The fact that the stacking sequences are able to repeat so accurately over such long distances is still puzzling, despite the many attempts at theoretical explanations.

9.6 CRYSTALLOGRAPHIC SHEAR (CS) PHASES

Like polytypes, CS phases are built from slabs of a single parent structure, but joined in such a way as to produce a change in the overall composition of the solid. This effect can be illustrated with respect to sets of structurally simple examples – the CS phases that form when tungsten trioxide, WO_3, is reduced.

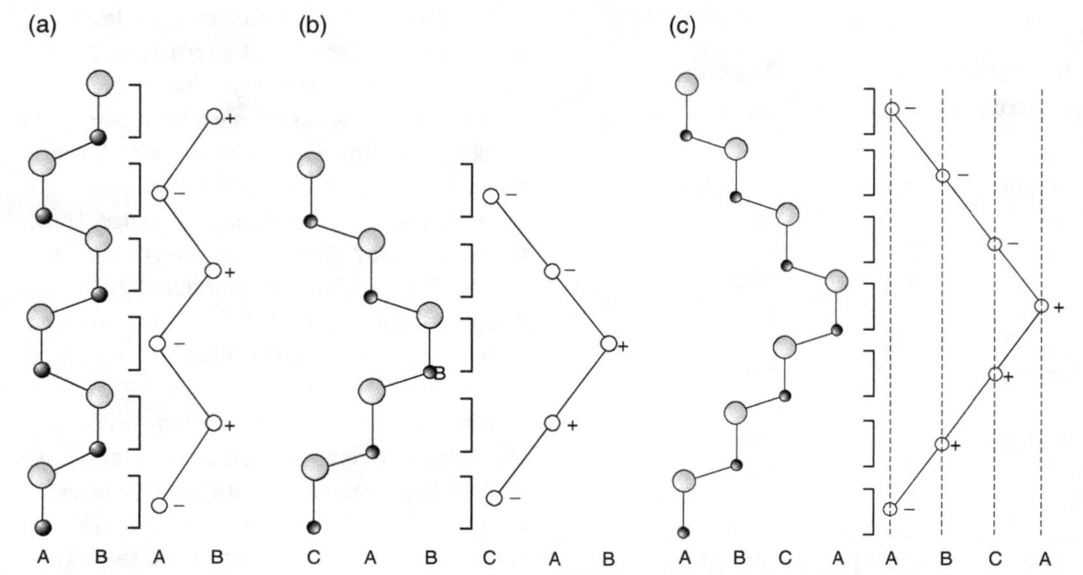

Figure 9.9 Simplified representations of polytypes: (a) wurtzite, 2H, (11); (b) carborundum I, 2H, (22); (c) 6H, (33). The zig-zag designs to the right of the structure representations show the sequence of translations in a direction, + or −, summarised in the Zhdanov symbol. In the zig-zags, each circle represents the position of a composite (Zn + S) or (Si + C) layer.

Tungsten trioxide has a unit cell that is monoclinic at room temperature, with lattice parameters $a = 0.7297$ nm, $b = 0.7539$ nm, $c = 0.7688$ nm, $\beta = 90.91°$, which is most easily visualised as a three-dimensional array of slightly distorted corner-shared WO_6 octahedra. For the present discussion, the structure can be idealised as cubic (Figure 9.10a), with a cell parameter of $a = 0.750$ nm, and viewed in projection down one of the cubic axes, as an array of corner-linked squares (Figure 9.10b). Very slight reduction, to a composition of $WO_{2.9998}$, for example, results in a crystal containing a low concentration of faults which lie on {102} planes. Structurally these faults consist of lamellae made up of blocks of four edge-shared octahedra in a normal WO_3-like matrix of corner-sharing octahedra (Figure 9.10c). A slightly reduced crystal contains CS planes lying upon {120} planes (Figure 9.11).

Continued reduction causes an increase in the density of the CS planes, and as the composition approaches $WO_{2.97}$, they tend to become ordered. In this case the structure will possess a (generally monoclinic) unit cell with a long axis, which will be more or less perpendicular to the CS planes (Figure 9.12). The length of the long axis, which can be taken as the **c**-axis, will be equal to an integral number, m, of the d_{102} spacing. The **a**-axis is roughly perpendicular to the **c**-axis, and the monoclinic **b**-axis will be perpendicular to the plane of the figure and approximately equal to the **c**-axis of WO_3.

The CS planes reduce the amount of oxygen in the structure. The composition of a crystal containing ordered CS planes is given by

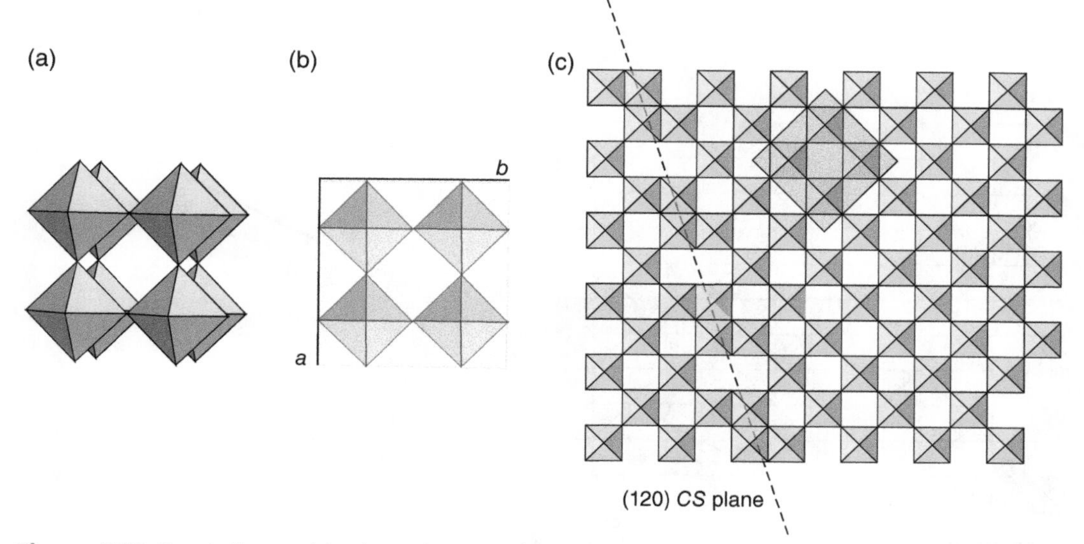

Figure 9.10 Crystallographic shear in WO_3: (a) perspective view of idealised cubic WO_3; (b) projection of the idealised structure down the cubic **c**-axis; (c) an idealised (120) CS plane. The squares represent WO_6 octahedra.

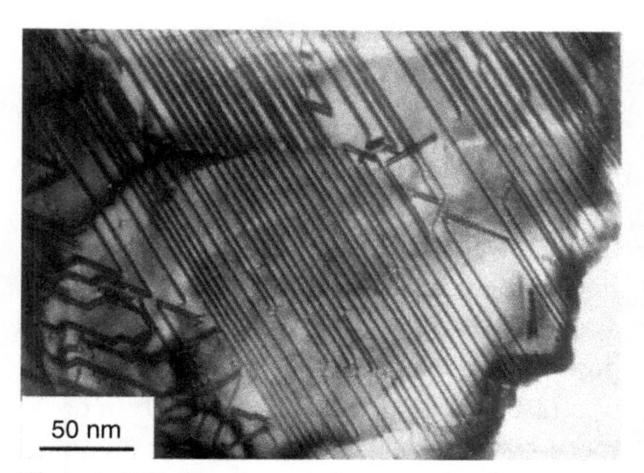

Figure 9.11 Electron micrograph of slightly reduced WO_3 showing disordered {120} CS planes, imaged as dark lines.

W_nO_{3n-1}, where n is the number of octahedra separating the CS planes (counted in the direction indicated by the arrow on Figure 9.12). The family of oxides represented by the formula W_nO_{3n-1} is known as a **homologous series**. This homologous series spans the range from approximately $W_{30}O_{89}$, with a composition of $WO_{2.9666}$, to $W_{18}O_{53}$, with a composition of $WO_{2.9444}$.

Continued reduction of the oxide causes the CS planes to adopt another configuration, oriented on cubic {130} planes (Figure 9.13). The CS plane is now made of blocks of six edge-shared octahedra, and the degree of reduction per unit length of CS plane is consequently increased. Once again, if the CS planes are ordered, the oxide will form part of a homologous series, in this instance with a series formula W_nO_{3n-2}, where n is the number of octahedra separating the CS planes counted in the direction arrowed on Figure 9.13. If the situation shown in Figure 9.13 was repeated in an ordered way throughout the crystal, the formula would be $W_{15}O_{43}$ and the composition $WO_{2.8667}$. This homologous series spans the range from approximately $W_{25}O_{73}$, with a composition of $WO_{2.9200}$, to $W_{16}O_{46}$, with a

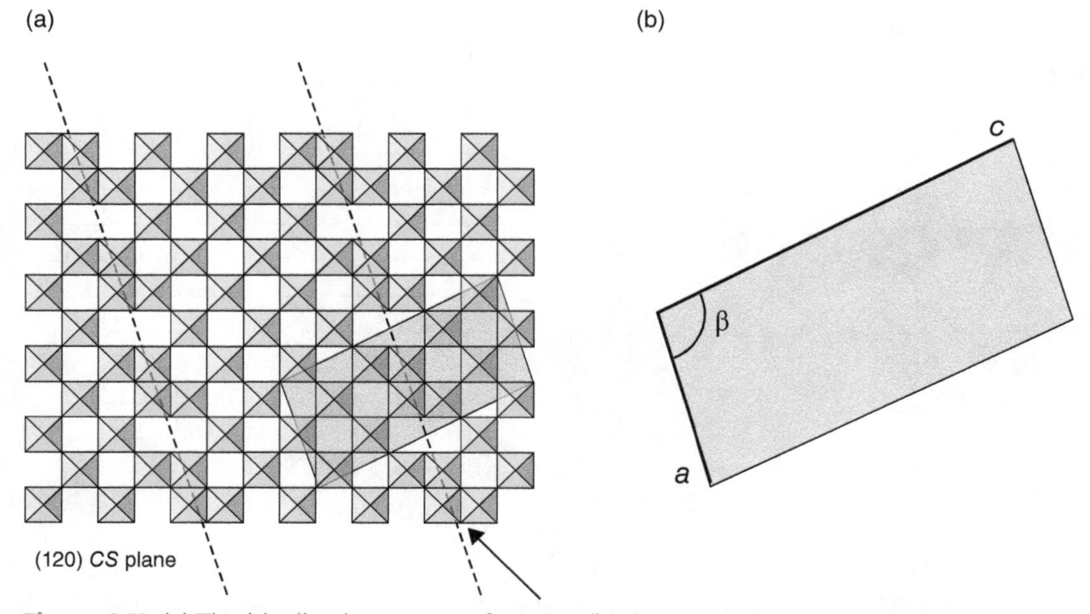

Figure 9.12 (a) The idealised structure of $W_{11}O_{32}$; (b) the monoclinic unit cell of $W_{11}O_{32}$. The squares represent WO_6 octahedra, and the CS planes are delineated by the sloping dotted lines.

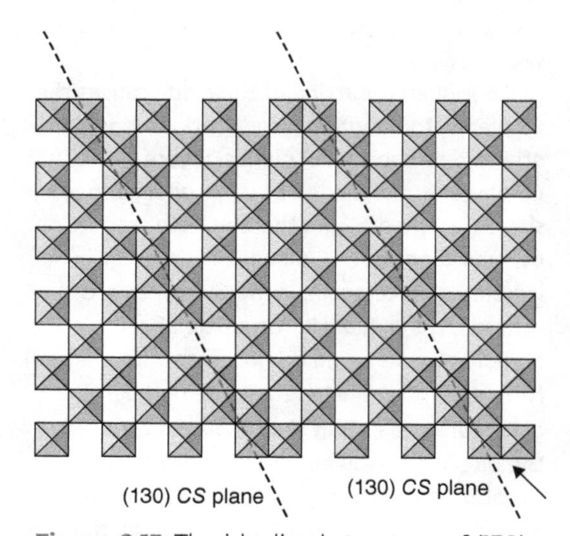

Figure 9.13 The idealised structure of (130) CS planes. The squares represent WO_6 octahedra, and the CS planes are delineated by the sloping dotted lines.

composition of $WO_{2.8750}$. The unit cells of these phases will be similar to that described in Figure 9.12.

The diffraction patterns of oxides containing CS planes evolve in a characteristic way from that of the parent structure. This can be explained using WO_3 as an example. As the parent structure forms the greatest component of a CS structure, the diffraction pattern of these materials will closely resemble that of WO_3 itself (Figure 9.14a). However, the CS planes impose a new set of diffraction conditions. Suppose that a crystal contains parallel but disordered (210) CS planes. The resulting diffraction pattern will be similar to that of the parent structure, but streaks will now appear, parallel to the line joining the origin to the 210 reflection, and passing through each reciprocal lattice point

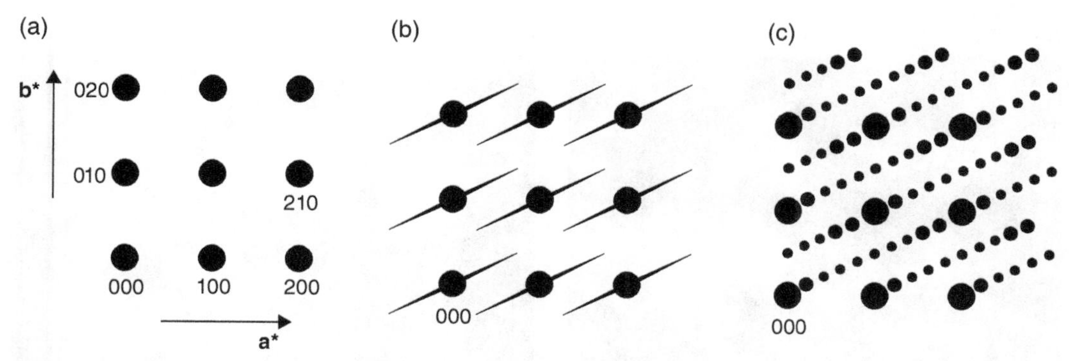

Figure 9.14 The evolution of diffraction patterns of materials containing CS planes: (a) idealised cubic WO_3; (b) idealised WO_3 containing disordered (210) CS planes, giving rise to streaking on the diffraction pattern; (c) idealised WO_3 containing ordered (210) CS planes. Note that the real structures are less symmetrical and not all reflections shown in these diagrams may be present in experimental patterns.

(Figure 8.14b). This is, in fact, an expression of the form factor for the crystal (see Sections 7.3 and 7.4). The planar faults, in effect, break the large crystal into a number of small slabs and the streaks run normal to disordered slab boundaries. As the CS planes become ordered, the streaks begin to show maxima and minima (Figure 9.14c), and fully ordered CS planes give sharp spots. The number of extra reflections will be equal to *n* in the homologous series formula, or else a multiple of it, depending upon the true symmetry of the unit cell. These extra reflections are called **superlattice** or **superstructure reflections**, and because the spacing between them will be $1/md_{210}$, where *m* is an integer, they fit exactly into the WO_3 reciprocal lattice, and are said to be **commensurate** with it. Examples of these types of diffraction pattern are given in Figure 9.15.

The discussion of these diffraction patterns is quite general and not just restricted to CS planes. This means that the polytypes and other phases described above, as well as the long period structures described in the following section, will yield similar diffraction patterns. In them, the diffraction pattern of the parent structure will be well emphasised, together with streaks or commensurate rows of superlattice reflections on the diffraction pattern, running perpendicular to the fault planes that break up the structure. To obtain the intensities of these reflections, the same structure factor calculations described earlier can be carried out.

9.7 PLANAR INTERGROWTHS AND POLYSOMES

A large number of solids encompass a range of composition variation by way of **intergrowth**. Unlike the two previous examples, in these compounds, slabs of crystalline materials with **different** compositions, say A and B, interleave to give a compound of formula A_aB_b. This implicitly entails that they each have at least one structurally compatible crystallographic plane that is shared between the two slabs and

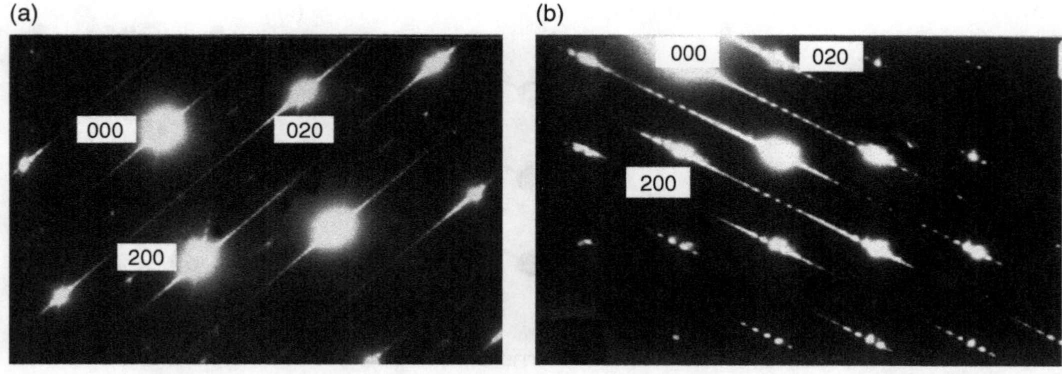

Figure 9.15 Electron diffraction patterns from: (a) a crystal containing disordered {102} CS planes, showing continuous streaks; (b) a crystal containing more ordered {120} CS planes, showing the streaks breaking up into rows of superlattice reflections.

so forms the interface between them. The composition of any phase will be given by the numbers of slabs of each kind present. When $a = \infty$ and $b = 0$, the composition of the crystal is simply A (Figure 9.16a). Similarly, when $a = 0$ and $b = \infty$, the composition is B (Figure 9.16b). An enormous number of ordered stacking sequences can be imagined, including ABAB, composition AB (Figure 9.16c), ABBABB, composition AB_2 (Figure 9.16d), and ABAAB, composition A_3B_2 (Figure 9.16e). The overall composition can vary virtually continuously from that of one of the parents to the other, depending upon the proportions of each phase present. These materials are known as **polysomes** rather than polytypes.

Examples abound, especially in minerals. A classic polysomatic series is given by the minerals that are formed by an ordered intergrowth of slabs of the ***mica***[2] and **pyroxene** structures. Both are built from silicate layers that are only weakly bonded, so that they cleave easily to form thin plates. The micas have formulae typified by phlogopite, $KMg_3(OH)_2Si_3AlO_{10}$, where the K^+ ions lie between the silicate layers, the Mg^{2+} ions are in octahedral sites, and the Si^{4+} and Al^{3+} ions occupy tetrahedral sites within the silicate layers. The pyroxenes can be represented by enstatite, $MgSiO_3$, where the Mg^{2+} ions occupy octahedral sites and the Si^{4+} ions occupy tetrahedral sites within the silicate layers. Both these materials have layer structures that fit together parallel to the layer planes. If an idealised mica slab is represented by M, and an idealised pyroxene slab by P, the polysomatic series can span the range from ...MMM..., representing pure mica, to ...PPP..., representing pure pyroxene. Many intermediates are known, for example, the sequence ...MMPMMP... is found in the mineral jimthompsonite, $(Mg,Fe)_{10}(OH)_4Si_{12}O_{32}$, and the sequence ...MPMMPMPMMP... corresponds to the mineral chesterite, $(Mg,Fe)_{17}[(OH)_6Si_{20}O_{54}]$.

There are many intergrowth phases that are composed of slabs of the ABO_3 perovskite structure. One family, $Sr_n(Ti, Nb)_nO_{3n+2}$, related

[2] Mica also forms polytypes. These occur when the component layers stack in alternative ways. In these polytypes the composition does not change from one member of the series to another, and remains that of the parent phase.

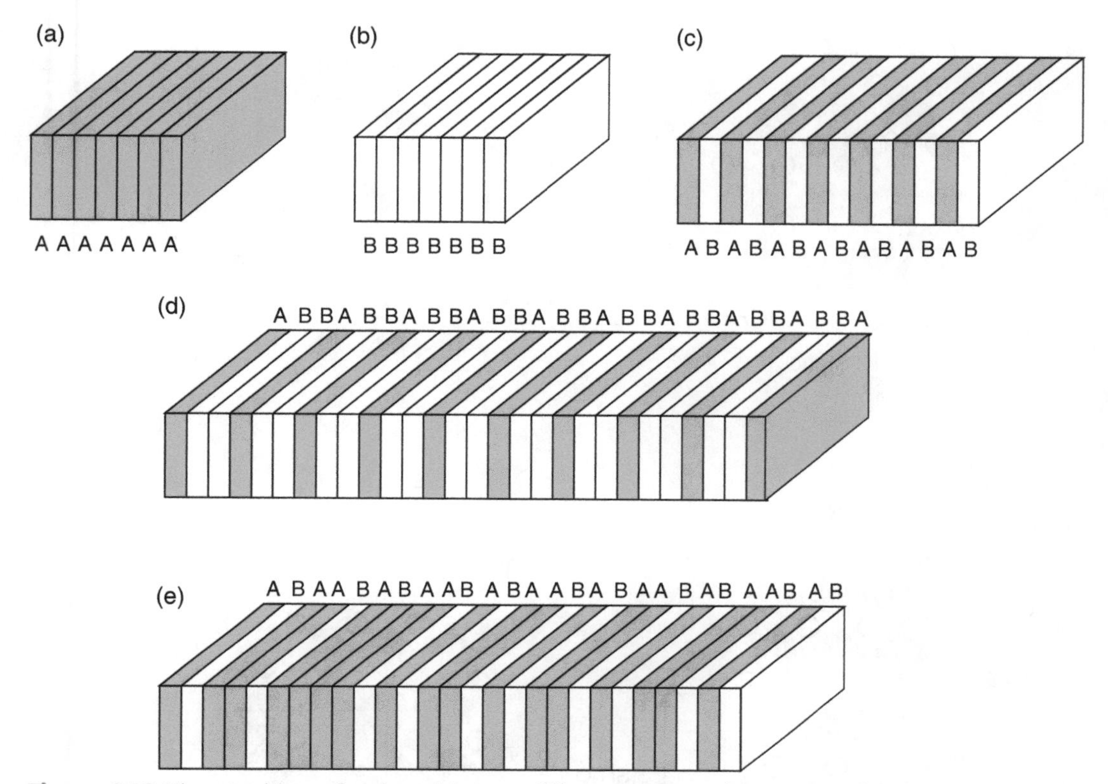

Figure 9.16 The stacking of slabs with two different compositions, A and B: (a) ...AAA..., composition A; (b) ...BBB..., composition B; (c) ...ABAB..., composition AB; (d) ...ABBABB..., composition AB_2; (e) ...ABAAB..., composition A_3B_2.

to $SrTiO_3$, was described in Section 9.4. Another structurally simple series built from slabs of $SrTiO_3$ is represented by the oxides Sr_2TiO_4, $Sr_3Ti_2O_7$ and $Sr_4Ti_3O_{10}$, with general formula $Sr_{n+1}Ti_nO_{3n+1}$, known as **Ruddlesden–Popper** phases. In these materials, slabs of $SrTiO_3$ are cut parallel to the idealised cubic *perovskite* {100} planes and stacked together, each slab being slightly displaced in the process (Figure 9.17a–d). The structure of the junction regions is identical to that of lamellae of the halite (NaCl) structure, and so the series can be thought of as being composed of intergrowths of varying thicknesses of perovskite $SrTiO_3$, linked by identical slabs of halite SrO.

Many other examples of Ruddlesden–Popper phases have been synthesised. The generalised homologous series formula for these compounds is $A_{n+1}B_nO_{3n+1}$ [= $(AO)(ABO_3)_n$ or $(H)(P)_n$], where A is a large cation, typically an alkali metal, alkaline earth or rare earth, and B is a medium-sized cation, typically a 3*d* transition metal cation, and H and P stand for halite and perovskite. The $Sr_{n+1}TiO_{3n+1}$ series is then composed of $(SrO)_a(SrTiO_3)_b$, or $(H)_a(P)_b$ where *a* is 1 and *b* runs from 1 to 3. The structure of first member of the series, corresponding to $n = 1$, is adopted by many compounds, and is often referred to as the K_2NiF_4 structure. One of these, the phase La_2CuO_4, when doped with Ba^{2+}

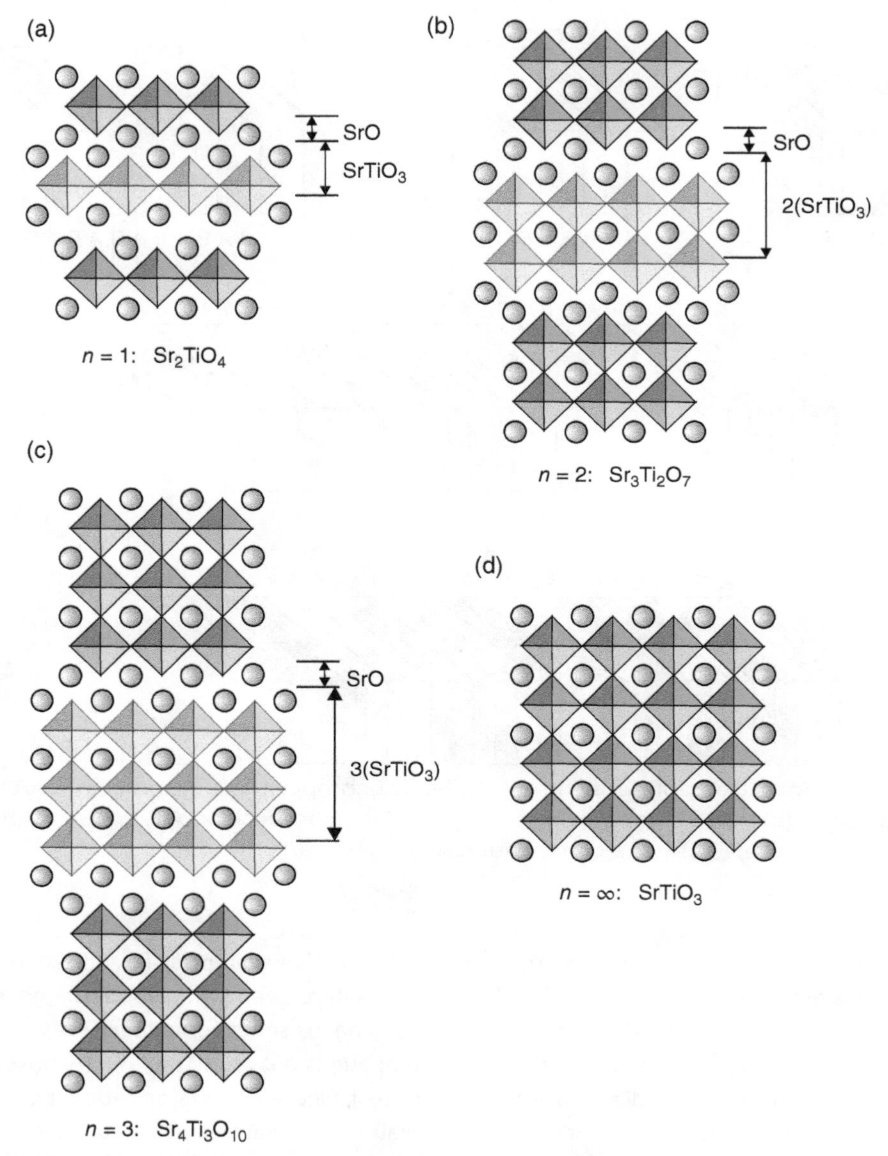

Figure 9.17 Idealised structures of the Ruddlesden–Popper phases $Sr_{n+1}Ti_nO_{3n+1}$: (a) Sr_2TiO_4, $n = 1$; (b) $Sr_3Ti_2O_7$, $n = 2$; (c) $Sr_4Ti_3O_{10}$, $n = 3$; (d) $SrTiO_3$, $n = \infty$.

to form the oxide $(La_xBa_{1-x})_2CuO_4$, achieved prominence as the first high-temperature superconducting ceramic to be characterised as such. (Note that the real structures of all of these materials have lower symmetry than the idealised structures, mainly due to temperature-sensitive distortions of the BO_6 octahedra.)

If the halite layers in the above series are replaced by a layer of composition Bi_2O_2, a

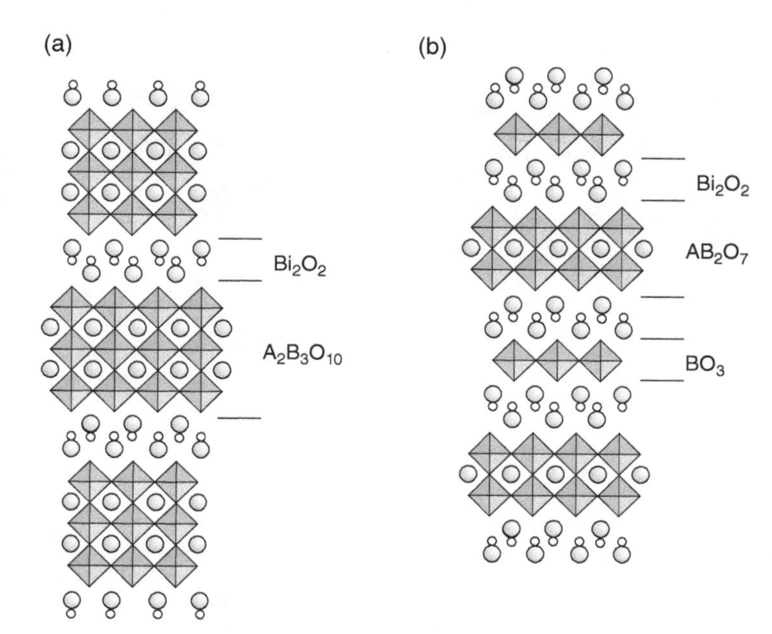

Figure 9.18 The idealised structure of the Aurivillius phases viewed along [110]: (a) $n = 3$, typified by $Bi_4Ti_3O_{12}$; (b) ordered $n = 1$, $n = 2$ intergrowth, typified by $Bi_5TiNbWO_{15}$.

series of phases is formed called **Aurivillius phases**, with a general formula $(Bi_2O_2)(A_{n-1}B_nO_{3n+1})$, where A is a large cation, and B a medium-sized cation, and the index n runs from 1 to ∞. The structure of the Bi_2O_2 layer is similar to that of fluorite (see Section 1.6), so that the series is represented by an intergrowth of fluorite and perovskite structure elements, $(F)_a(P)_b$, where a is 1 and $b = n$ runs from 1 to ∞. The $n = 1$ member of the series is represented by Bi_2WO_6 or [FP], the $n = 2$ member by $Bi_2SrTa_2O_9$, [FPP], but the best-known member of this series of phases is the ferroelectric $Bi_4Ti_3O_{12}$, [FPPP], in which $n = 3$ and A is Bi (Figure 9.18a). Ordered intergrowth of slabs of more than one thickness, especially n and $(n + 1)$, are often found in this series of phases, for example, $n = 1$ and $n = 2$, [FPFPP], typified by $Bi_5TiNbWO_{15}$ (Figure 9.18b). (As with the other structures described, the symmetry of these structures is lower than that implied by the idealised structures, mainly due to distortions of the metal–oxygen octahedra.)

The high-temperature superconductors are similar to both of these latter examples in many ways, as they are composed of slabs of perovskite structure intergrown with slabs that can be thought of as structurally related to halite or fluorite. To illustrate these materials, the idealised structures of the phases $Bi_2Sr_2CuO_6$ (= $Tl_2Ba_2CuO_6$), $Bi_2CaSr_2Cu_2O_8$ (= $Tl_2CaBa_2Cu_2O_8$) and $Bi_2Ca_2Sr_2Cu_3O_{10}$ (= $Tl_2Ca_2Ba_2Cu_3O_{10}$) are shown in Figure 9.19. The single perovskite sheets in the idealised structure of $Bi_2Sr_2CuO_6$ are complete (Figure 9.19a). These are separated by Bi_2O_2 (or Tl_2O_2) layers similar, but not identical, to those in the Aurivillius phases. In the other compounds, the oxygen structure needed to form the perovskite framework is incomplete. The nominal double layer of CuO_6 octahedra needed to form a sheet of the idealised perovskite structure is replaced by square pyramids in $Bi_2CaSr_2Cu_2O_8$ (Figure 9.19b). To make the relationship clearer, the octahedra

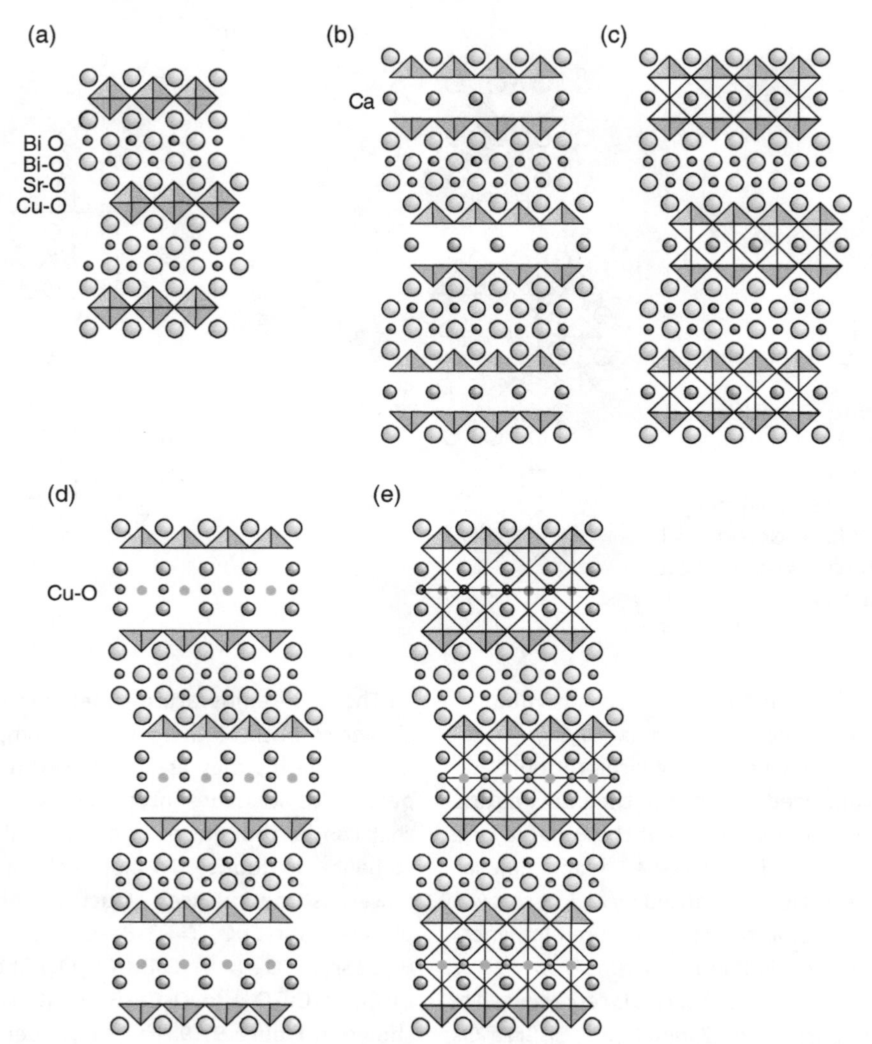

Figure 9.19 The idealised structures of some high-temperature superconductors: (a) $Bi_2Sr_2O_6$ (= $Tl_2Ba_2O_6$); (b) $Bi_2CaSr_2Cu_2O_8$ (= $Tl_2CaBa_2Cu_2O_8$); (c) as (b), but with the nominal CuO_6 octahedra completed in faint outline; (d) $Bi_2Ca_2Sr_2Cu_3O_{10}$ (= $Tl_2Ca_2Ba_2Cu_3O_{10}$); (e) as (d), but with the nominal CuO_6 octahedra completed in faint outline.

are completed in faint outline in Figure 9.19c. In $Ba_2Ca_2Sr_2Cu_3O_{10}$ the three CuO_6 octahedral perovskite layers have been replaced by two sheets of square pyramids and the middle layer by a sheet of CuO_4 squares (Figure 9.19d). The octahedra are completed in faint outline in Figure 9.19e.

9.8 INCOMMENSURATELY MODULATED STRUCTURES

In the examples just presented, the reciprocal lattices and the resulting diffraction patterns resemble that of a parent structure, say

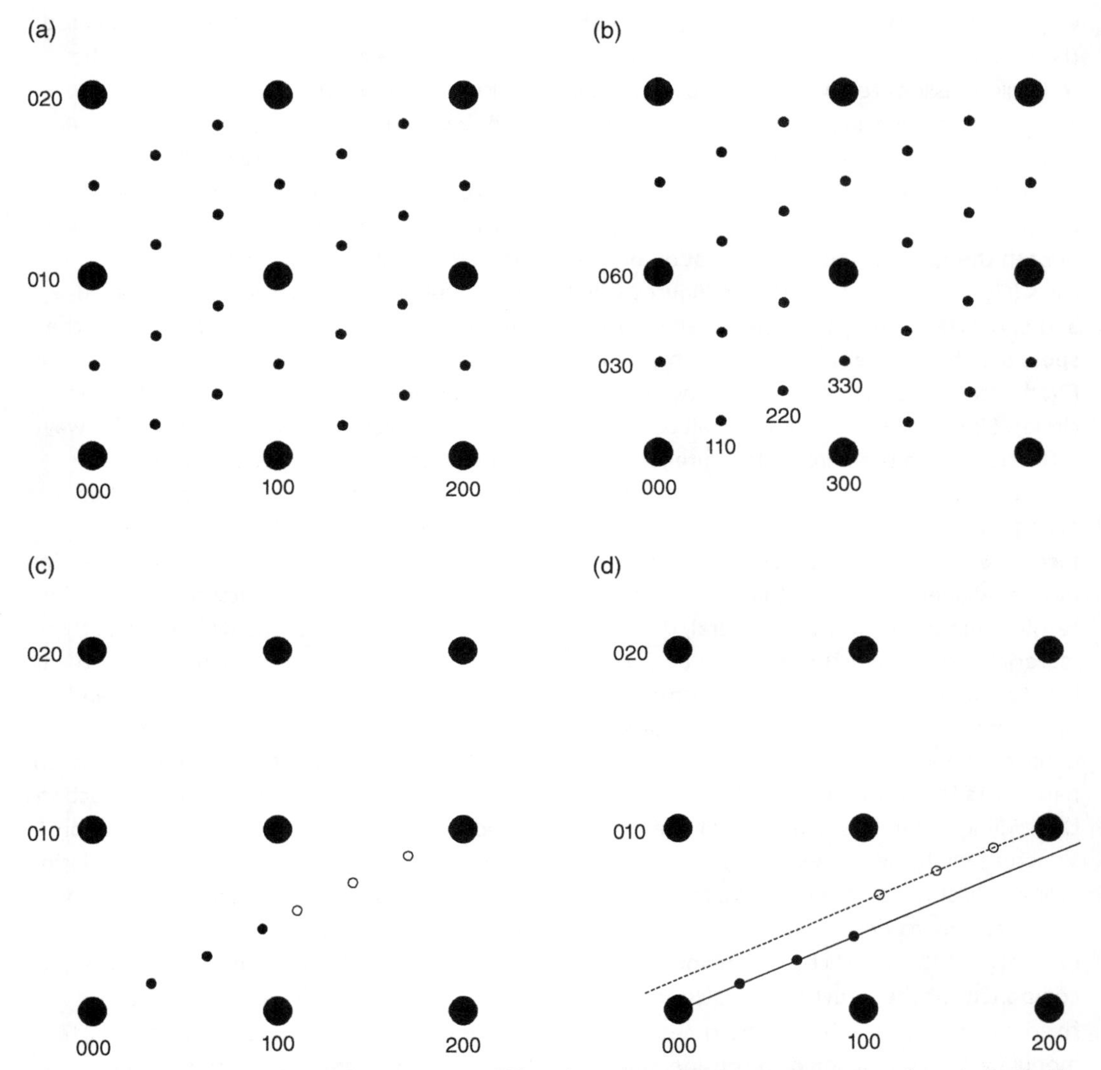

Figure 9.20 Superlattice reflections on diffraction patterns: (a) strong reflections indexed in terms of the parent structure; (b) normal superlattice reflections indexed in terms of a larger 'supercell'; (c) colinear incommensurate superlattice reflections (associated with the 000, filled circles, and 210 reflections, open circles); (d) non-colinear incommensurate superlattice reflections (associated with the 000, filled circles, and 210 reflections, open circles).

perovskite, together with arrays of 'extra' superstructure reflections, running between the main reflections (Figure 9.20a). These additional reflections can be indexed via a unit cell with one or two long cell axes, so that all reflections have conventional *hkl* values (Figure 9.20b).

An increasing number of crystalline solids have been characterised in which the spots on the diffraction pattern cannot be indexed in this way. As before, the main reflections, those of the parent structure, have rows of superlattice reflections associated with them. However, it is found that the spacing between the spots is sometimes 'anomalous', and the superlattice reflections do not quite fit in with the parent reflections, either in spacing, (Figure 9.20c), or in both the spacing and the orientation of the rows (Figure 9.20d). (Note that in Figures 9.20c and d, for clarity, only two rows of superlattice spots are shown: that associated with 000 as filled circles, and that through 210 as open circles. Similar rows pass through all of the reflections from the parent structure.)

The diffraction patterns, and the structures giving rise to this feature, are both said to be **incommensurate**. Crystallographic techniques have now been developed that are able to resolve such ambiguities. In general, these materials have a structure that can be divided into two parts. To a reasonable approximation, one component is a conventional structure that behaves like a normal crystal, but an additional part exists that is **modulated**[3] in one, two or three dimensions. The fixed part of the structure might be, for example, the metal atoms, whilst the anions might be modulated in some fashion, as described below. In some examples of these structures, the modulated component of the structure can also force the fixed part of the structure to become modulated in turn, leading to considerable crystallographic complexity.

The diffraction pattern from a normal crystal is characterised by an array of spots separated by a distance $1/a = a^*$ that arise from the parent structure, together with a set of commensurate superlattice reflections that arise as a consequence of the additional ordering. In this case the spot spacing is $1/na = a^*/n$, where n is an integer (Figure 9.21a and b). In modulated structures, the modulation might be in the position of the atoms, called a **displacive modulation** (Figure 9.21c). Displacive modulations sometimes occur when a crystal structure is transforming from one stable structure to another as a result of a change in temperature. Alternatively, the modulation might be in the occupancy of a site, called **compositional modulation**, for example, the gradual replacement of O by F in a compound M$(O,F)_2$ (Figure 9.21d). In such a case the site occupancy factor would vary in a regular way throughout the crystal. Compositional modulation is often associated with solids that have a composition range.

As with normal superlattices, the diffraction patterns of modulated structures can also be divided into two parts. The diffraction pattern of the unmodulated component stays virtually unchanged in the modulated structure and gives rise to a set of strong 'parent structure' reflections of spacing $1/a = a^*$. The modulated part gives rise to a set of superlattice reflections 'attached' to each parent structure reflection. In cases where the dimensions of the modulation are incommensurate (that is, do not fit) with the underlying structure, the phase is an **incommensurately modulated phase**. The spot spacing is equal to $1/\lambda$, where λ is the wavelength of the modulation (Figure 9.21e). The positions of these reflections change smoothly as the modulation varies. In such cases, the diffraction pattern will show rows of superlattice spots that cannot be indexed in terms of the diffraction pattern of the parent phase. In cases where the modulation wave runs at an angle to the fixed part of the structure, the extra spots will fall on lines at an angle to the main reflections, to give orientation anomalies (Figure 9.21f and g).

When the modulation wavelength exactly fits a number of parent unit cells (that is, is

[3] Take care to note that a *modulated* structure is *not* the same as a *modular* structure.

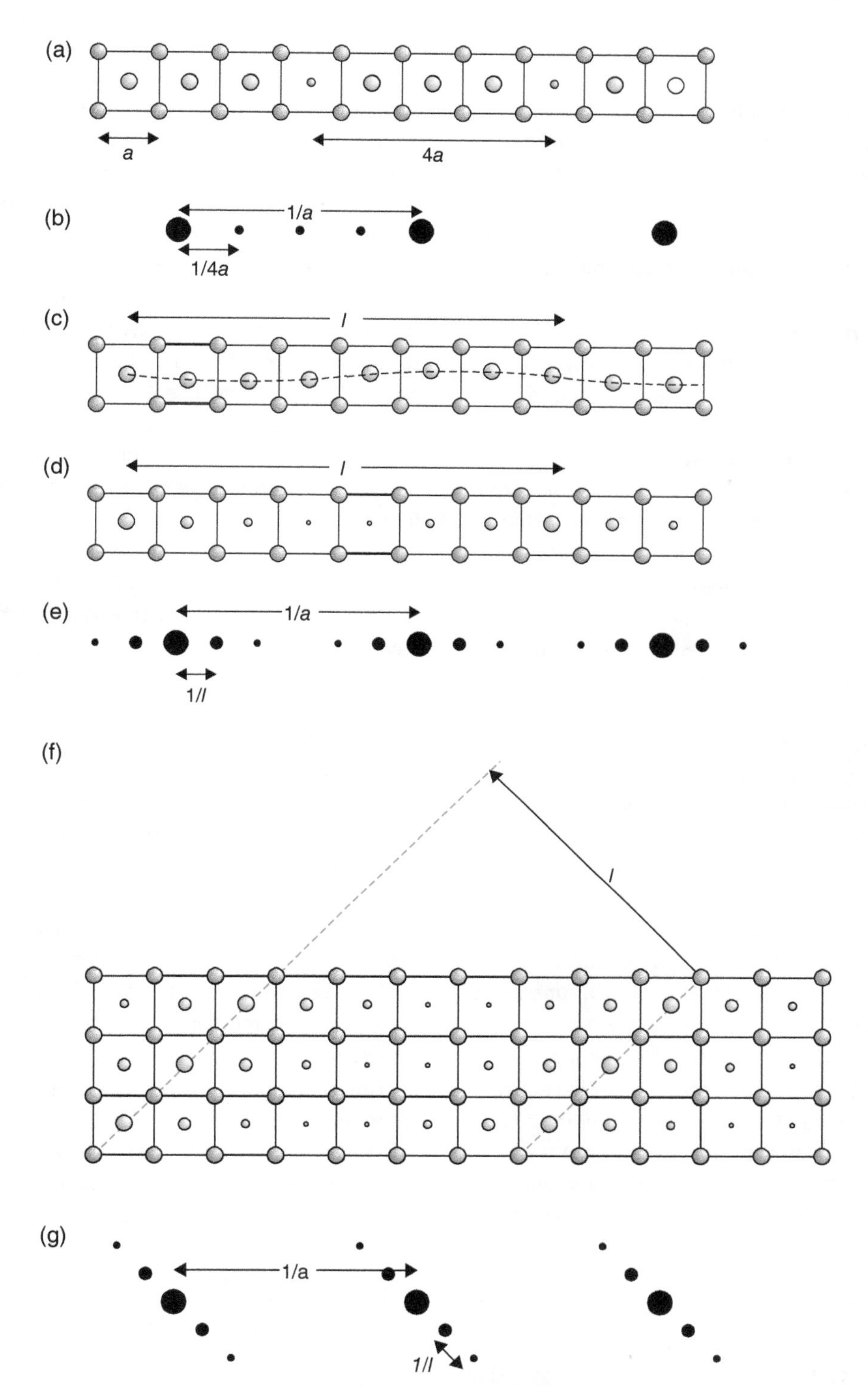

Figure 9.21 Schematic representations of normal and modulated crystal structures and diffraction patterns: (a) a normal superlattice, formed by the repetition of an anion substitution; (b) part of the diffraction pattern of (a); (c) a crystal showing a displacive modulation of the anion positions; (d) a crystal showing a compositional modulation of the anion conditions (the change in the average chemical nature of the anion is represented by differing circle diameters); (e) part of the diffraction pattern from (b) or (c); (f) a modulation wave at an angle to the unmodulated component; (g) part of the diffraction pattern from (f).

commensurate with it), it will be possible to index the reflections in terms of a normal superlattice. Recently a number of structures that were once described in this way have been restudied and found to be better described as modulated structures with commensurate modulation waves – **commensurately modulated structures**.

There are a number of ways in which a structure can be modulated and by and large these were characterised little by little, a procedure that gave rise a number of different naming schemes. Thus, amongst these subsets, classes of compounds described as **vernier structures**, **chimney-ladder structures** and **layer misfit structures** can be found. In addition to these principally chemical modulations, magnetic spins, electric dipoles and other physical attributes of atoms can be modulated in a commensurate or incommensurate fashion.

The crystal structures of modulated materials can be illustrated by reference to the barium iron sulfides of formula $Ba_xFe_2S_4$, and the similar strontium titanium sulfides, Sr_xTiS_3. These phases fall into the subset of modulated structures, which have two interpenetrating chemical substructures. One of these expands or contracts as a smooth function of composition, whilst the other remains relatively unchanged. Over much of the composition range, these components are not in register, so that a series of incommensurately modulated structures form, often with enormous unit cell dimensions. However, the two components come into register for some compositions, and in these cases, a commensurately modulated structure forms with normal unit cell dimensions.

The barium iron sulfides of formula $Ba_xFe_2S_4$, exist over a composition range corresponding to x values from 1.0 to approximately 1.25. The more or less rigid component of the structure is composed of chains of edge-shared FeS_4 tetrahedra. The sharing of edges results in an iron:sulfur ratio of FeS_2 within each chain. These chains run parallel to the **c**-axis, and are arranged so as to give a tetragonal unit cell, with cell parameter a_{FeS} (= b_{FeS}) (Figure 9.22a–c). The Ba atoms form the second component. These atoms lie between the chains of tetrahedra and also form rows parallel to the **c**-axis. The Ba array can also be assigned to a tetragonal unit cell, with cell parameter a_{Ba} (= b_{Ba}). The cell edges a_{FeS} and a_{Ba} are identical for the two subsystems (Figure 9.22d).

The c-parameters of the FeS_2 and Ba components are different. The c-parameter of the FeS_2 chains, c_{FeS}, is constant, but the c-parameter of the Ba component, c_{Ba}, changes smoothly as the Ba content varies. This modulation falls in step with the FeS_2 component at regular intervals governed by the modulation wavelength. Broadly speaking, the spacing of the Ba atoms is equal to the spacing of two FeS_2 units, so that the overall formula of the phases is close to $Ba(Fe_2S_4)$. As an example, the c-axis projection of the phase $Ba_{1.1250}Fe_2S_4$, is shown in Figure 9.22f. The coincidence of the repeat is 9 Ba atoms to 16 FeS_2 units, so that the formula can be written $Ba_9(Fe_2S_4)_8$ and represents a commensurately modulated structure.

The general compositions of these structures can be written as $Ba_p(Fe_2S_4)_q$. The c-parameter of the complete structure, c_S, is given by the period at which the two subcells fit, that is, by the least common multiple of the subcell dimensions, so that:

$$c_S = q\, c_{FeS} = p\, c_{Ba}$$

where c_{FeS} is the average dimension of a pair of (Fe_2S_4) tetrahedra and c_{Ba} is the average

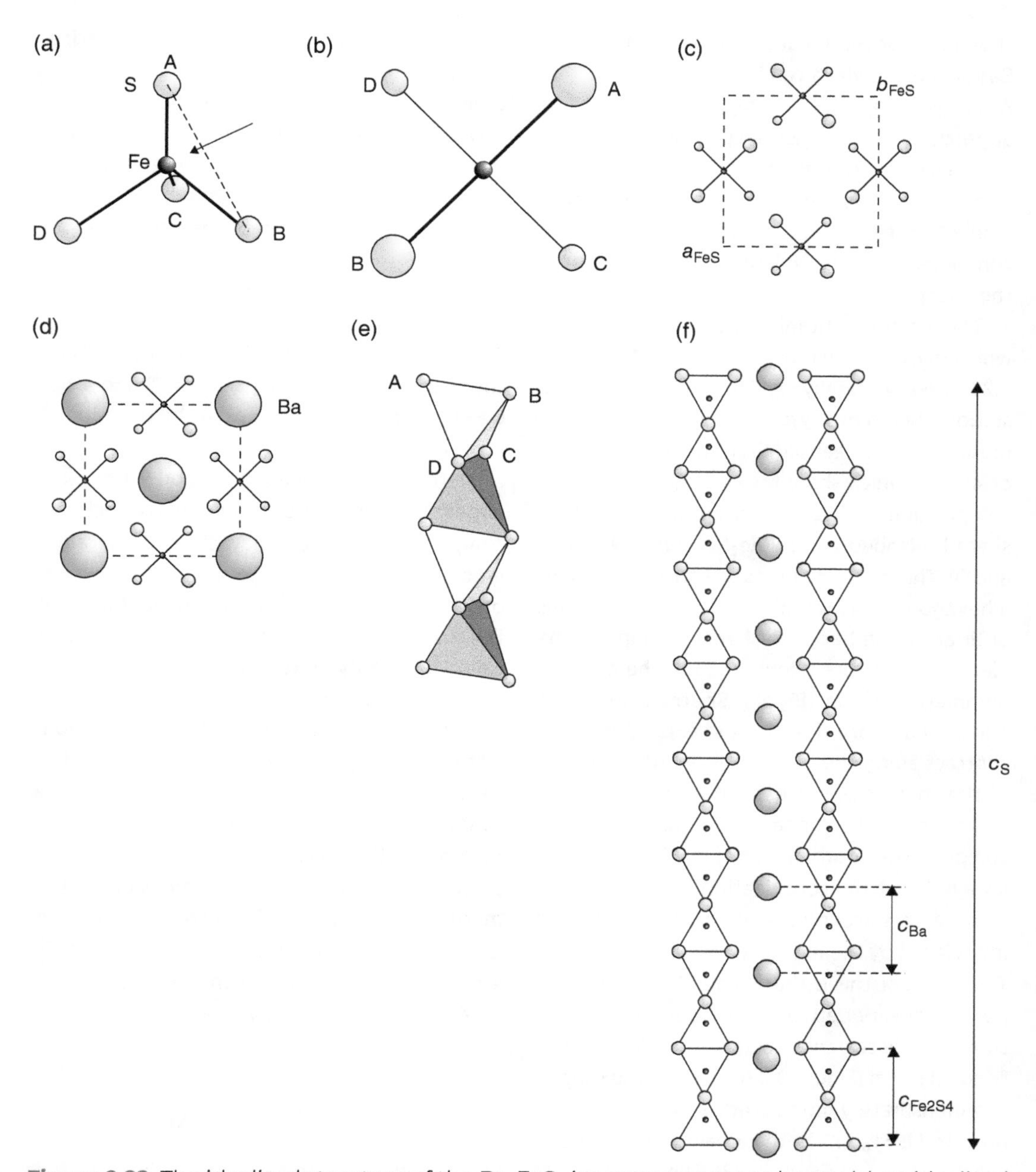

Figure 9.22 The idealised structure of the Ba_xFeS_2 incommensurate phases: (a) an idealised FeS_4 tetrahedron; (b) the same tetrahedron as (a), viewed perpendicular to the A–B edge; (c) the tetragonal unit cell of the edge-shared FeS_4 tetrahedral chains viewed along [001]; (d) the tetragonal unit cell including Ba atoms; (e) chains of edge-shared FeS_4 tetrahedra; (f) a projection of the idealised structure of $Ba_9(Fe_2S_4)_8$, along [110] (the tetrahedra are viewed in a direction C–D with respect to (a)). Only two of the four FeS_4 tetrahedral chains and one set of Ba atoms are shown.

separation of the Ba atoms. For the phase $Ba_{1.1250}Fe_2S_4$, which can be written as $Ba_9(Fe_2S_4)_8$, $c_S = 8c_{FeS} = 9c_{Ba}$. It can be appreciated that if the Ba content is changes infinitesimally, the two subcells will only come into register after an enormous distance. This is true, of course, for most compositions, and the commensurately modulated compositions are the exceptions.

The strontium titanium sulfides, Sr_xTiS_3, which exist between the x values of 1.05 and 1.22, are structurally similar to the barium iron sulfides, but in this system, both subsystems are modulated. In idealised Sr_xTiS_3 columns of TiS_6 ocahedra, which share faces, to give a repeat composition of $TiS_{6/2}$ or TiS_3, replace the edge-shared tetrahedra in $Ba_xFe_2S_4$ (Figure 9.23a and b). These TiS_3 columns are arranged to give a hexagonal unit cell (Figure 9.23c), and chains of Sr atoms lie between them to complete the idealised structure (Figure 9.23d). The a-parameters of the TiS_3 and Sr arrays are equal. The Sr chains are flexible, and expand or contract along the **c**-axis as a smooth function of the composition x in Sr_xTiS_3. The real structures of these phases are much more complex. The coordination of the Ti atoms is always 6, but the coordination polyhedron of sulfur atoms around the metal atoms is in turn modulated by the modulations of the Sr chains. The result of this is that some of the TiS_6 polyhedra vary between octahedra and a form some way between an octahedron and a trigonal prism (Figure 9.23e). One of the simpler commensurately modulated structures reported is $Sr_8(TiS_3)_7$, the idealised structure of which is drawn in Figure 9.23f. The vast majority of compositions give incommensurately modulated structures with enormous unit cells.

As in the case of the barium iron sulfides, the compositions of the (real or idealised) materials can be written as $Sr_p(TiS_3)_q$. The c-lattice parameter of the complete structure, c_S, is given by the least common multiple of the c-lattice parameters of the (more or less fixed) TiS_3 chains, c_{TiS}, and the flexible Sr chains, c_{Sr}. However, these phases differ from the barium iron sulfides in that the wavelength of the Sr chain contains 2 Sr atoms, rather than 1, so the correct expression for c_S becomes:

$$c_S = q\ c_{TiS} = (p/2)\ c_{Sr}$$

Thus, with reference to Figure 9.23f, $Sr_8(TiS_3)_7$, there are 7 c_{TiS} wavelengths, ($q = 7$) and 4 c_{Sr} wavelengths ($p = 8$, $p/2 = 4$) in the overall unit cell.

Defects as described in Section 9.1 do not appear to form in these compounds. Each composition, no matter how close, chemically, it is to any other, appears to generate a unique ordered structure, often with an giant unit cell. Because of this, such structures are sometimes called **'infinitely adaptive structures'**.

These complex phases cannot be described in terms of the classical ideas of crystallography. Instead, the crystal reciprocal lattice must be viewed mathematically in a higher dimensional space. The incommensurate spacings observed on a diffraction pattern are then regarded as projections from this higher dimension. The mathematical aspects of the crystallography of these interesting materials are beyond the scope of this book, but further information is given in the Bibliography.

9.9 QUASICRYSTALS

Experimentally, a perfect crystal gives a diffraction pattern consisting of sharp reflections or spots. This is the direct result of the translational order that characterises the crystalline state. To recapitulate, a crystal can only be built from a unit cell consistent with the seven crystal systems and the 14 Bravais

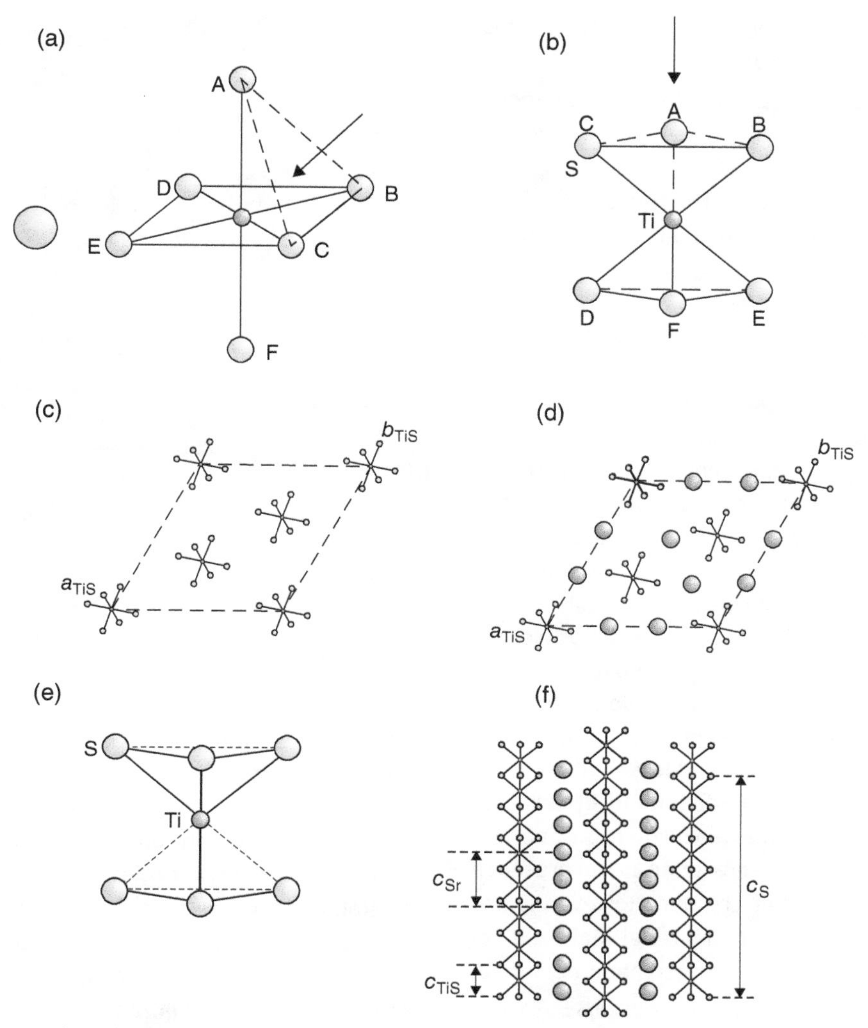

Figure 9.23 The idealised structure of the Sr_xTiO_3 incommensurate phases: (a) an idealised TiS_6 octahedron; (b) the same octahedron as in (a), viewed in the direction of the arrow, approximately normal to a triangular face (bonds projecting above the plane of the paper are shown as full lines, and those below the plane of the paper as dotted lines). The chains of face-shared octahedra in Sr_xTiS_3 are linked by faces arrowed in (b); (c) the idealised hexagonal unit cell, formed by columns of edge-sharing TiS_6 octahedra; (d) the idealised unit cell of Sr_xTiS_3 showing the location of the Sr atoms between the TiS_3 chains; (e) a trigonal prism; (f) the idealised structure of Sr_xTiS_3 projected onto (110).

lattices. The unit cell can be translated to build up the crystal, but no other transformation, such as reflection or rotation, is allowed. This totally rules out a unit cell with overall fivefold, or greater than sixfold, rotation symmetry, just as a floor cannot be tiled completely with regular pentagons. Additionally, the symmetry of the structure plays an important part in the modifying the intensity of diffracted beams, one consequence of which is that the intensities of a pair of reflections *hkl* and $\overline{h}\,\overline{k}\,\overline{l}$ are equal in magnitude. This will cause the diffraction pattern from a crystal to appear centrosymmetric even for crystals that lack a centre of symmetry, and the point symmetry of any sharp diffraction pattern will belong to one of the 11 Laue classes.

In 1984 this changed with the discovery of a metallic alloy of composition approximately $Al_{88}Mn_{12}$ which gave a sharp electron diffraction pattern that clearly showed the presence of an apparently 10-fold symmetry axis. The crystallographic problem is that the sharp diffraction pattern indicated long-range translational order. However, this was incompatible with the 10-fold rotation axis, which indicated that the unit cell possessed a forbidden rotation axis. Moreover, the overall diffraction pattern did *not* belong to one of the 11 Laue classes, seeming to suggest that the unit cell showed icosahedral symmetry (Figure 4.5d). Initially effort was put into trying to explain the rotational symmetry as an artefact due to the presence of defects in 'normal' crystals. However, it was soon proven that the material really did have 10-fold rotational symmetry, and belonged to the icosahedral point group $m\overline{3}\,\overline{5}$. This totally violated the laws of classical crystallography.

Since then, many other alloys that give rise to sharp diffraction patterns and which show 5-, 8-, 10- and 12-fold rotation symmetry have been discovered. The resulting materials became known as **quasiperiodic crystals**, or more compactly as **quasicrystals**. Figure 9.24 shows the electron diffraction pattern from a quasicrystalline alloy of composition $Al_{72}Ni_{20}Co_8$, clearly showing 10-fold rotation symmetry.

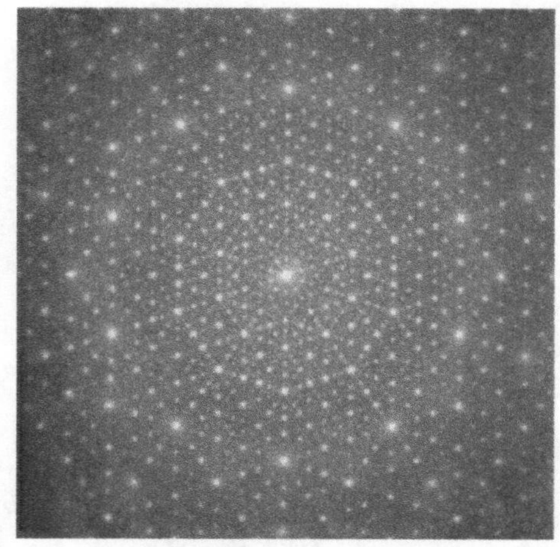

Figure 9.24 Electron diffraction pattern from the quasicrystalline alloy $Al_{72}Ni_{20}Co_8$, showing 10-fold rotation symmetry. (*Source*: Photograph courtesy of Dr Koh Saitoh, see Saitoh, K., Tanaka, M., Tsai, A.P. et al. (2004). ALCHEMI studies on quasicrystals. *JEOL News* **39**(1): 20, reproduced with permission.

There are a number of ways that the contradiction between classical crystallography and the structures of quasicrystals can be resolved, all of which require a slight relaxation in the strict rules of conventional crystallography. The simplest is to consider a quasicrystal as an imperfectly crystallised liquid. The liquid state of many metal alloys contains metal icosahedra that form in a transitory fashion, because this polyhedral arrangement represents a very efficient way of packing a small number of spheres together (Figure 9.25a). Thus, a liquid metal can often be regarded as being composed of icosahedral clusters of metal atoms that are continually forming and breaking apart. On cooling a liquid metal, the icosahedra freeze. If the cooling rate is slow enough, the icosahedra order onto a lattice, say cubic or tetragonal, which then extinguishes the overall icosahedral symmetry of the structure. A large number of metal alloy

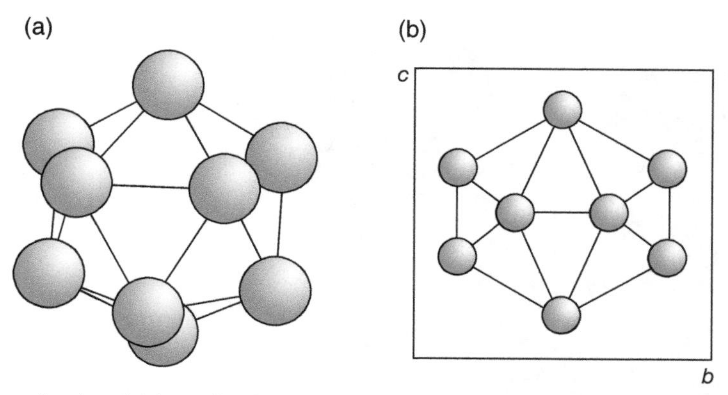

Figure 9.25 Icosahedra: (a) icosahedral geometry; (b) part of the As_3Co skutterudite structure, projected down [100], showing an icosahedron of As atoms. (Note that only a fraction of the As atoms and none of the Co atoms in the As_3Co unit cell are shown. The atoms in both (a) and (b) touch. They are drawn smaller here to show the icosahedral geometry.)

structures are known that contain ordered icosahedra, including $CoAs_3$ with the cubic skutterudite structure, in which As atoms take on this geometry (Figure 9.25b).

If the cooling rate of a liquid metal is very rapid the icosahedra freeze at random, to give an **icosahedral glass**. Over a range of intermediate temperatures, the icosahedra all freeze in the same orientation, but they do not have the energy or time to order on a lattice. A quasicrystal can then be regarded as made up of icosahedral clusters of metal atoms, all oriented in the same way and separated by variable amounts of disordered material (Figure 9.26). That is, the materials show orientational order but not translational order.

Quasicrystals can also be considered as three-dimensional analogues of Penrose tilings. Penrose tilings are aperiodic: they cannot be described as having unit cells and do not show translational order (Section 3.8). However, a Penrose tiling has a sort of translational order, in the sense that parts of the pattern, such as all of the decagons, are oriented identically, but they are not spaced in such a way as to generate a unit cell (Figure 9.27a). Moreover, if the nodes in a Penrose tiling are replaced by 'atoms', there are well-defined regularly spaced atom planes running through the tilings that would suffice to give sharp diffraction spots, following Bragg's law (Figure 9.27b). In fact, the diffraction patterns computed from an array of atoms placed at the nodes of Penrose tilings show sharp spots and 5- and 10-fold rotation symmetries.

The Penrose model of quasicrystals consists of a three-dimensional Penrose tiling using effectively two 'unit cells', corresponding to the dart and kite units of the plane patterns. These can be joined to give an aperiodic structure which has the same sort of order as the two-dimensional tiling. That is, all of the icosahedra are in the same orientation, but not arranged on a lattice, and once again the model structures show orientational order but not translational order. As with the planar tilings, these three-dimensional analogues give diffraction patterns that show both sharp spots and forbidden rotation symmetries.

The Penrose tiling model of quasicrystals has been reinforced by the preparation of quasicrystal arrays from nanoparticles

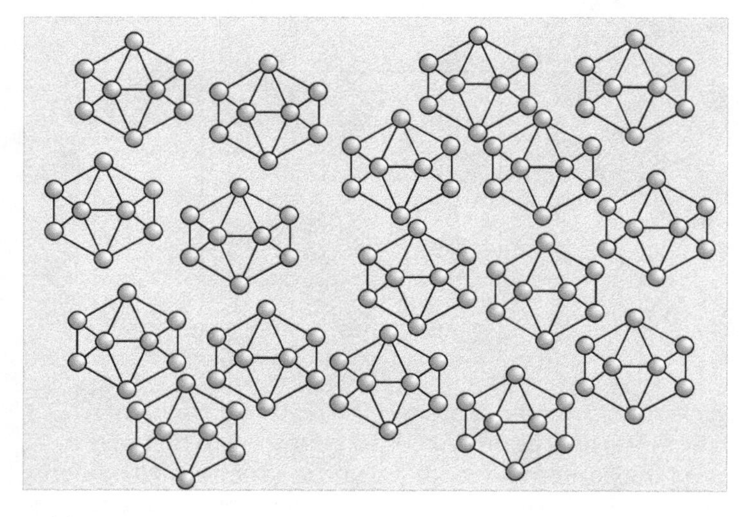

Figure 9.26 Icosahedra, arranged with the same orientation, but not on a lattice, can be taken as one model of a quasicrystal alloy.

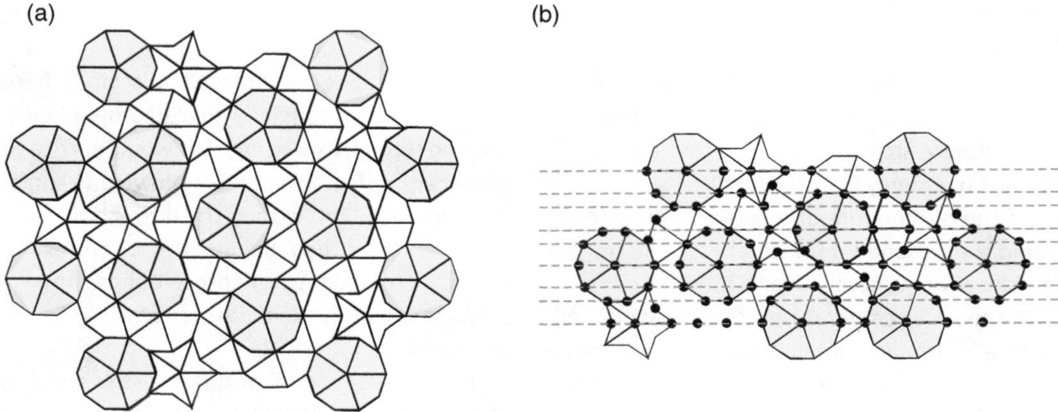

Figure 9.27 A Penrose tiling: (a) the decagons are all oriented in the same way but do not lie upon a lattice: (b) 'atoms' placed on the nodes of the tiling fall onto a fairly well-defined set of planes.

consisting of a tetrahedrally shaped core of CdSe covered with a coating of CdS. These tetrahedral core–shell nanoparticles were aggregated into a decagon made up of five pairs of tetrahedral units surrounding a central pair of tetrahedra, so that, in projection, they appear like the dodecahedra in Figure 9.27a. These decagons are flexible, due to the suppleness of the chemical bonds between the tetrahedra, and this allows them to pack together in a variety of non-crystallographic Penrose tilings, to form quasicrystals which, in projection, resemble the sort of arrangement shown in Figure 9.27a.

ANSWERS TO INTRODUCTORY QUESTIONS

What are modular structures?

Modular structures are those that can be considered to be built from slabs of one or more parent structures. Slabs can be sections from just one parent phase, as in many *perovskite*-related structures, *CS* phases and polytypes, or they can come from two or more parent structures, as in the mica–pyroxene intergrowths. Some of these crystals possess enormous unit cells, of some hundreds of nm in length. In many materials the slab thicknesses may vary widely, in which case the slab boundaries will not fall on a regular lattice, and so form planar defects.

What are incommensurately modulated structures?
In general, incommensurately modulated structures have two fairly distinct parts. One part of the crystal structure is conventional, and behaves like a normal crystal. An additional, more or less independent part exists that is modulated in one, two, or three dimensions. For example, the fixed part of the structure might be the metal atom array, whilst the modulated part might be the anion array. The modulation might be in the position of the atoms, called a displacive modulation, or the modulation might be in the occupancy of a site, for example, the gradual replacement of O by F in a compound $M(O,F)_2$, called compositional modulation. In some more complex crystals, modulation in one part of the structure induces a corresponding modulation in the 'fixed' part.

In cases where the wavelength of the modulation fits exactly with the dimensions of the underlying structure, a commensurately modulated crystallographic phase forms. In cases where the dimensions of the modulation are incommensurate (that is, do not fit) with the underlying structure, the phase is an incommensurately modulated phase.
The modulation produces sets of extra reflections on the diffraction pattern that may (commensurate) or may not (incommensurate) fit with the reflections from the parent non-modulated component.

What are quasicrystals?
Quasicrystals or quasiperiodic crystals are metallic alloys that yield sharp diffraction patterns displaying 8-, 10-, or 12-fold rotational symmetry axes – forbidden by the rules of classical crystallography. The first quasicrystals discovered, and most of those that have been investigated, have icosahedral symmetry. Two main models of quasicrystals have been suggested. In the first, a quasicrystal can be regarded as made up of icosahedral clusters of metal atoms, all oriented in the same way, and separated by variable amounts of disordered material. Alternatively, quasicrystals can be considered to be three-dimensional analogues of Penrose tilings. In either case, the material does not possess a crystallographic unit cell in the conventional sense.

PROBLEMS AND EXERCISES

Quick Quiz

1. The oxides Al_2O_3 and Cr_2O_3 form a complete solid solution, $Al_xCr_{2-x}O_3$. The site occupancy factor for the cations in the crystal $AlCrO_3$ is:
   a. 1.0
   b. 0.5
   c. 0.3
2. The *fluorite* structure non-stoichiometric oxide $Ca^{2+}{}_{0.1}Zr^{4+}{}_{0.9}O_{1.9}$ has a site occupancy factor for O equal to:
   a. 1.9
   b. 0.95
   c. 0.1
3. A solid containing interstitial atom point defects will have a theoretical density:
   a. Higher than the parent crystal
   b. Lower that the parent phase
   c. The same as the parent phase

4. A solid is prepared by heating a 1 : 1 mixture of $SrTiO_3$ and $Sr_4Nb_4O_{14}$ to give a member of the $Sr_n(Nb,Ti)_nO_{3n+2}$ series of phases. The value of n of the new material is:
   a. 2.5
   b. 5
   c. 10
5. Carborundum is a form of:
   a. Al_2O_3
   b. ZnS
   c. SiC
6. The Ramsdell symbol 2H applies to polytypes with the same structure as:
   a. Wurzite
   b. Zincblende
   c. Carborundum I
7. A family of phases with compositions represented by a general formula such as W_nO_{3n-2} is called:
   a. A homogeneous series
   b. A heterogeneous series
   c. A homologous series
8. Ruddlesden–Popper phases, Aurivillius phases and high temperature superconductors all contain slabs of structure similar to that of:
   a. Rutile
   b. Perovskite
   c. Wurtzite
9. If the positions of a set of atoms in a structure follow a wave-like pattern, the modulation is described as:
   a. Compositional
   b. Displacive
   c. Incommensurate
10. Quasicrystals are alloys that are characterised by fivefold or higher symmetry and:
   a. Orientational but not translational order
   b. Glass-like disorder
   c. Modulated order

Calculations and Questions

9.1. Both TiO_2 and SnO_2 adopt the rutile structure (see Chapter 1) and crystals with intermediate compositions $Ti_xSn_{1-x}O_2$ are often found in nature. A crystal of composition $Ti_{0.7}Sn_{0.3}O_2$ has lattice parameters $a = 0.4637$ nm, $c = 0.3027$ nm. The scattering factors appropriate to the (200) reflection from this unit cell are $f_{Ti} = 15.573$, $f_{Sn} = 39.405$ and $f_O = 5.359$. (a) Calculate the metal atom site scattering factor, the structure factor for the 200 reflection and the intensity of the reflection (see Section 7.8). Repeat the calculations for (b) pure TiO_2 and (c) SnO_2 samples, assuming that the lattice parameter of these materials is identical to that of the solid solution.

9.2. The compound $Y_3Ga_5O_{12}$ crystallises with the garnet structure.[4] A solid solution, $(Y_xTm_{1-x})_3Ga_5O_{12}$, forms when Y is replaced by Tm. The lattice parameters of several members of this solid solution series are given in the table below. (a) Plot these against composition to determine how well Vegard's law is followed. (b) Estimate the lattice parameter of $Tm_3Ga_5O_{12}$. (c). What is the composition of the phase with a lattice parameter of 1.22400 nm?

[4] The garnet structure is cubic, with a lattice parameter of approximately 1.2 nm. The atoms are situated in three different coordination sites, and the formula is conveniently represented as $\{A^{3+}\}_3[B^{3+}]_2(C^{3+})_3O_{12}$, where {X} represents cations in cubic sites, [Y] represents cations in octahedral sites and (Z) represents cations in tetrahedral sites. In the garnets $Y_3Ga_5O_{12}$ and $Tm_3Ga_5O_{12}$, the cations Y or Tm are in cubic sites and the Ga ions occupy both the octahedral and tetrahedral sites.

Composition, x^a	Lattice parameter / nm
0	—
0.20	1.22316
0.35	1.22405
0.50	1.22501
0.65	1.22563
0.80	1.22638
1.00	1.22734

[a]Adapted from data given by Liu, F.S., Sun, B.J. Liang, J.K., et al. (2005). *Journal of Solid State Chemistry* **178**: 1064–1070.

9.3. The compound $LaMnO_{3.165}$ adopts a distorted perovskite structure, in which the unit cell has rhombohedral symmetry, with hexagonal lattice parameters $a = 0.55068$ nm, $c = 1.33326$ nm, $Z = 6$ $LaMnO_{3.165}$. The oxygen excess can arise either as oxygen interstitials or as metal atom vacancies. (a) Write the formula of each alternative (assume that metal vacancies occur in equal numbers on both the La and Mn positions). (b) Determine the theoretical density of each. (Data adopted from Barnabé, A., Gaudon, M., Bernard, C., et al. (2004). *Materials Research Bulletin* **39** : 725–735.

9.4. (a) The packing symbol of a 15R ZnS polytype is (cchch). Sketch the zig-zag structure (projected onto the hexagonal $(11\bar{2}0)$ plane) and determine the Zhadanov symbol for the phase. (b) The Zhadanov symbol for a 24H ZnS polytype is (7557). Sketch the zig-zag structure (projected onto the hexagonal $(11\bar{2}0)$ plane) and determine the packing sequence. (c) The zig-zag structure (projected onto the hexagonal $(11\bar{2}0)$ plane) of a ZnS polytype is shown in the figure below. Determine the Zhadanov symbol and the stacking sequence.

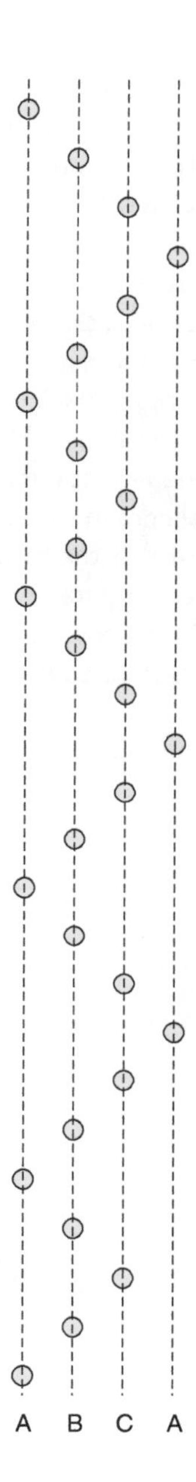

9.5. (a) *CS* planes that lie between {102} and {103} are known in the reduced tungsten oxides. What plane would the *CS* plane composed of alternating blocks of 4 and 6 edge-sharing octahedra lie along, and what is the series formula of materials containing ordered arrays of these *CS*

planes? (b) What is the general formula of a Ruddlesden–Popper structure oxide composed of the slab sequence ...344... of the Ruddlesden–Popper phases $n = 3$ and $n = 4$? (c) What is the general formula of an Aurivillius structure oxide composed of the slab sequence ...23... of the Aurivillius phases $n = 2$ and $n = 3$?

9.6. (a) The (average) spacing of Ba atoms along the *c*-axis in a sample of $Ba_xFe_2S_4$ is 0.5 nm. The average repeat dimension of a pair of FeS_4 tetrahedra in the same direction is 0.56 nm. What is the formula of the phase and its *c*-lattice parameter? (Data adapted from Grey, I.E. (1975). *Acta Crystallographica* **B31**: 45.) (b) Assuming the same dimensions for barium atom separation and iron sulfur tetrahedra hold, express the formula of the phase $Ba_{1.077}(Fe_2S_4)$ in terms of integers p and q, and estimate the *c*-lattice parameter. (c) The composition of a strontium titanium sulfide is $Sr_{1.145}TiS_3$. The repeat distance of the TiS_3 chains, c_{TiS}, was found to be 0.2991 nm and that of the Sr chains, c_{Sr}, was found to be 0.5226 nm. Express the formula in terms of (approximate) indices $Sr_p(TiS_3)_q$, and determine the approximate *c*-parameter of the phase. (Data adopted from Onoda, M., Saeki, M., Yamamoto, A., et al. (1993). *Acta Crystallographica* **B49**: 929–936.)

Appendix A Vector Addition and Subtraction

Vectors are used to specify quantities that have a direction and a magnitude. Unit cell edges are specified by vectors **a**, **b** and **c**, which have a direction, and a magnitude, which is a scalar (an ordinary number), *a*, *b* and *c*.

A vector is often represented by an arrow (Figure A.1a).

A vector **a** multiplied by a scalar $+a$ is a vector with the same direction and a times as long (Figure A.1b).

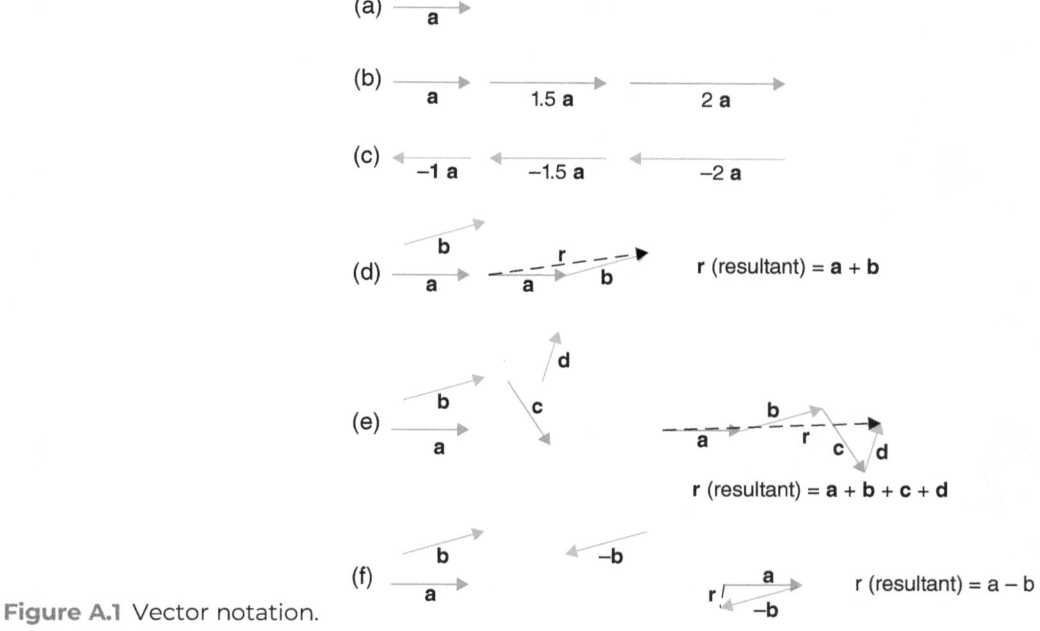

Figure A.1 Vector notation.

Crystals and Crystal Structures, Second Edition. Richard J. D. Tilley.

A vector **a** multiplied by a scalar $-a$ is a vector pointing in the opposite direction to **a** and a times as long (Figure A.1c).

Two vectors, **a** and **b**, can be **added** to give **a** + **b** by linking them so that the tail of the second vector, **b**, is placed in contact with the head of the first one, **a**. The vector joining the tail of **a** to the head of **b** is the **vector sum**, or **resultant**, **r**, which is also a vector (Figure A.1d). The resultant, **r**, of a large number of vector additions is found by successive application of this process (Figure A.1e).

Two vectors, **a** and **b**, can be **subtracted** to give **a** – **b** by 'adding' the negative vector, –**b** to **a** (Figure A.1f). The resultant, **r**, of a large number of vector subtractions is found by successive application of this process.

Appendix B
Crystallographic Data for Some Inorganic Crystal Structures

Copper (A1)

Structure: cubic; $a = 0.3610$ nm; $Z = 4$; space group, $Fm\bar{3}m$ (No. 225)

Atom positions: Cu: 4*a* 0, 0, 0; ½, ½, 0; 0, ½, ½; ½, 0, ½

Tungsten (A2)

Structure: cubic; $a = 0.3160$ nm; $Z = 2$; space group, $Im\bar{3}m$ (No. 229)

Atom positions: W: 2*a* 0, 0, 0; ½, ½, ½

Magnesium (A3)

Structure: hexagonal; $a = 0.3210$ nm, $c = 0.5210$ nm; $Z = 2$; space group, $P6_3/mmc$, (No. 194)

Atom positions: Mg: 2*d* ⅔, ⅓, ¼; ⅓, ⅔, ¾

Diamond (A4)

Structure: cubic; $a = 0.3560$ nm; $Z = 8$; space group, $Fd\bar{3}m$ (No. 227)

Atom positions: C: 8*a* 0, 0, 0; ½, ½, 0; 0, ½, ½; ½, 0, ½; ¼, ¼, ¼; ¾, ¾, ¼; ¾, ¼, ¾; ¼, ¾, ¾

Graphite

Structure: hexagonal, $a = 0.2460$ nm, $c = 0.6701$ nm; $Z = 4$; space group, $P6_3mc$ (No. 186)

Atom positions: C1: 2*a* 0, 0, 0; 0, 0, ½
C2: 2*b* ⅓, ⅔, 0; ⅔, ⅔, ½

Halite, NaCl (B1)

Structure: cubic; $a = 0.5630$ nm; $Z = 4$; space group, $Fm\bar{3}m$ (No. 225)

Atom positions: Na: 4*a* 0, 0, 0; ½, ½, 0; ½, 0, ½; 0, ½, ½
Cl: 4*b* ½, 0, 0; 0, 0, ½; 0, ½, 0; ½, ½, ½

(or vice versa)

Caesium chloride, CsCl (B2)

Structure: cubic $a = 0.4110$ nm; $Z = 2$; space group, $Pm\bar{3}m$ (No. 221)

Atom positions: Cs: 1*a* 0,0,0
Cl: 1*b* ½,½,½

(or vice versa)

Crystals and Crystal Structures, Second Edition. Richard J. D. Tilley.

Zinc blende (sphalerite), ZnS (B3)

Structure: cubic; $a = 0.5420$ nm, $Z = 4$; space group, $F\bar{4}3m$ (No. 216)

Atom positions: Zn: 4*a* 0, 0, 0; ½, ½, 0; ½, 0, ½; 0, ½, ½
S: 4*c* ¼, ¼, ¼; ¾, ¾, ¼; ¾, ¼, ¾; ¼, ¾, ¾

Wurtzite, ZnS (B4)

Structure: hexagonal; $a = 0.3810$ nm, $c = 0.6230$ nm; $Z = 2$; space group, $P6_3mc$ (No. 186)

Atom positions: Zn: 2*b* ⅓, ⅔, ½; ⅔, ⅓, 0
S: 2*b* ⅓, ⅔, ⅜; ⅔, ⅓, ⅞

Nickel arsenide, NiAs

Structure: hexagonal; $a = 0.3610$ nm, $c = 0.5030$ nm; $Z = 2$; space group, $P6_3/mmc$ (No. 194)

Atom positions: Ni: 2*b* 0, 0, ¼; 0, 0, ¾
As: 2*c* ⅓, ⅔, ¼; ⅔, ⅓, ¾

Boron nitride, BN

Structure: hexagonal; $a = 0.2500$ nm, $c = 0.6660$ nm; $Z = 2$; space group, $P6_3/mmc$, (No. 194)

Atom positions: B: 2*c* ⅓, ⅔, ¼; ⅔, ⅓, ¾
N: 2*d* ⅓, ⅔, ¾; ⅔, ⅓, ¼

Corundum, Al_2O_3

Structure: trigonal, hexagonal axes; $a = 0.4763$ nm, $c = 1.3009$ nm; $Z = 6$; space group, $R\bar{3}c$ (No. 167)

Atom positions: each of (0, 0, 0); (⅔, ⅓, ⅓); (⅓, ⅔, ⅔); plus

Al: 12*c*: 0, 0, z; 0, 0, $\bar{z}$ + ½; 0, 0, $\bar{z}$; 0, 0, z + ½;
O: 18*e*: x, 0, ¼; 0, x, ¼; $\bar{x}$, $\bar{x}$, ¼; $\bar{x}$, 0, ¾; 0, $\bar{x}$, ¾; x, x, ¾

The x coordinate (O) and the z coordinate (Al) can be approximated to ⅓. In general, these can be written $x = ⅓ + u$ and $z = ⅓ + w$, where u and w are both small. Taking typical values of $x = 0.306$, $z = 0.352$, the positions are:

Al: 12*c*: 0, 0, 0.352; 0, 0, 0.148; 0, 0, 0.648; 0, 0, 852
O: 18*e*: 0.306, 0, ¼; 0, 0.306, ¼; 0.694, 0.694, ¼; 0.694, 0, ¾; 0, 0.694, ¾; 0.306, 0.306, ¾

Fluorite, CaF_2 (C1)

Structure: cubic; $a = 0.5463$ nm, $Z = 4$; space group, $Fm\bar{3}m$ (No. 225)

Atom positions: Ca: 4*a* 0, 0, 0; ½, ½, 0; ½, 0, ½; 0, ½, ½
F: 8*c* ¼, ¼, ¼; ¼, ¾, ¾; ¾, ¼, ¾; ¾, ¾, ¼; ¼, ¼, ¾; ¼, ¾, ¼; ¾, ¼, ¼; ¾, ¾, ¾

Pyrites, FeS_2 (C2)

Structure: cubic; $a = 0.5440$ nm, $Z = 4$; space group, $Pa\bar{3}$ (No. 205)

Atom positions: Fe: 4*a* 0,0,0; ½,½,0; ½,0,½; 0,½,½
S: 8*c* x,x,x; $\bar{x}$ + ½, $\bar{x}$, x + ½; $\bar{x}$, x + ½, $\bar{x}$ + ½;
x + ½, $\bar{x}$ + ½, $\bar{x}$; $\bar{x},\bar{x},\bar{x}$; x + ½, x, $\bar{x}$ + ½; x, $\bar{x}$ + ½, x + ½; $\bar{x}$ + ½, x + ½, x

The x coordinate for O is close to ⅓. Taking a typical value of $x = 0.375$, the positions are:

S: 8*c* 0.378, 0.378, 0.378; 0.122, 0.622, 0.878; 0.622, 0.878, 0.122; 0.578, 0.122, 0.622; 0.622, 0.622, 0.622; 0.878, 0.378, 0.122; 0.378, 0.122, 0.878; 0.122, 0.878, 0.378

Rutile, TiO_2

Structure: tetragonal; $a = 0.4594$ nm, $c = 0.2959$ nm, $Z = 2$; space group, $P\,4_2/m\,n\,m$ (No. 136)

Atom positions: Ti: 2*a* 0,0,0; ½,½,½

O: 4*f* $x,x,0$; $\overline{x},\overline{x},0$; $\overline{x}$ + ½, x + ½, ½; x + ½, $\overline{x}$ + ½, ½

The x coordinate for O is close to ⅓. Taking a typical value of $x = 0.305$, the positions are:

O: 4*f* 0.305, 0.305, 0; 0.695, 0.695, 0; 0.195, 0.805, ½; 0.805, 0.195, ½

Rhenium trioxide, ReO_3

Structure: cubic; $a = 0.3750$ nm, $Z = 1$; space group, $P\,m\,\overline{3}\,m$ (No. 221)

Atom positions: Re: 1*a* 0,0,0

O: 3*d* ½,0,0; 0,½,0; 0,0,½

Strontium titanate (ideal perovskite), $SrTiO_3$

Structure: cubic; $a = 0.3905$ nm, $Z = 1$; space group, $P\,m\,\overline{3}\,m$ (No. 221)

Atom positions: Ti: 1*a* 0,0,0

Sr: 1*b* ½,½,½

O: 3*c* 0,½,½; ½,0,½; ½,½,0

Spinel, $MgAl_2O_4$

Structure: cubic; $a = 0.8090$ nm, $Z = 8$; Space group, F d $\overline{3}$ m (No. 227)

Atom positions: each of (0,0,0); (0,½,½); (½,0,½); (½,½,0); plus

Mg: 8*a* 0,0,0; ¾,¼,¾

Al: 16*d* ⅝, ⅝, ⅝; ⅜, ⅞, ⅛; ⅞, ⅛, ⅜; ⅛, ⅜, ⅞

O: 32*e* x,x,x; $\overline{x}$, $\overline{x}$ + ½, x + ½; $\overline{x}$ + ½, x + ½, $\overline{x}$; x + ½, $\overline{x}$, $\overline{x}$ + ½; x + ¾, x + ¼, $\overline{x}$ + ¾; $\overline{x}$ + ¼, $\overline{x}$ + ¼, $\overline{x}$ + ¼; x + ¼, $\overline{x}$ + ¾, x + ¾; $\overline{x}$ + ¾, x + ¾, x + ¼

The x coordinate for O is approximately equal to ⅜, and is often given in the form ⅜ + u, where u is of the order of 0.01. Taking a typical value of $x = 0.388$, in which $u = 0.013$, the positions are:

O: 32*e* 0.388, 0.388, 0.388; 0.612, 0.112, 0.888; 0.112, 0.888, 0.612; 0.888, 0.612, 0.112; 0.138, 0.638, 0.862; 0.862, 0.862, 0.862; 0.368, 0.362, 0.138; 0.362, 0.138, 0.638

The structural details given above are representative and taken from sources listed in the Bibliography. *Strukturbericht* symbols are given for some compounds.

Appendix C
Schoenflies Symbols

Schoenflies symbols are widely used to describe molecular symmetry, the symmetry of atomic orbitals, and in chemical group theory. The terminology of the important symmetry operators and symmetry elements used in this notation are given in Table C.1.

The symbol E represents the identity operation, that is, the combination of symmetry elements that transforms the object (molecule) into a copy identical in every way to the original. There is one important feature to note. The improper rotation axis defined here is **not** the same as the improper rotation axis defined via Hermann-Mauguin symbols, but is a rotoreflection axis (see Chapter 4 for details).

An object such as a molecule can be assigned a collection of symmetry elements that characterise the **point group** of the shape. The main part of the Schoeflies symbol describing the point group is a letter symbol describing the principle rotation symmetry, as set out in Table C.2.

The symbol *C* represents a (proper) rotation axis. The symbol *D* represents a (primary)

Table C.2 Schoenflies symbols for point groups

Rotation group	Symbol
Cyclic	*C*
Dihedral	*D*
Tetrahedral	*T*
Octahedral	*O*

Table C.1 Symmetry operations and symmetry elements

Symmetry element	Symmetry operation	Symbol
Whole object (molecule)	Identity	E
n-fold axis of rotation	Rotation by $2\pi/n$	C_n
Mirror plane	Reflection	σ
Centre of symmetry	Inversion	i
n-fold improper axis	Rotation by $2\pi/n$ plus *reflection*	S_n

Crystals and Crystal Structures, Second Edition. Richard J. D. Tilley.

Table C.3 Combinations of symmetry elements and group symbols.

Symmetry elements	Group designation
E only	C_1
σ only	C_s
i only	C_i
C_n only	C_n
S_n only, n even	S_n ($S_2 \equiv C_i$)
S_n only, n odd	C_{nh} (C_n axis plus horizontal mirror)
C_n + perpendicular 2-fold axes	D_n
$C_n + \sigma_h$	C_{nh} (C_n is taken as vertical)
$C_n + \sigma_v$	C_{nv}
C_n + perpendicular σ_h	D_{nh}
C_n + perpendicular 2-fold axes + σ_d	D_{nd}
Linear molecule with no symmetry plane $\perp$ molecule axis	$C_{\infty v}$
Linear molecule with symmetry plane $\perp$ molecule axis	$D_{\infty h}$
Three mutually $\perp$ 2-fold axes	T
Three mutually $\perp$ 4-fold axes	O
C_5 axes + i	I_h

rotation axis together with another (supplementary) rotation axis normal to it. The symbol T represents tetrahedral symmetry, essentially the presence of four threefold axes and three twofold axes. The symbol O represents octahedral symmetry, essentially four threefold axes and three fourfold axes. These main symbols are followed by one or two subscripts giving further information on the order of the rotation and position. For example, the symbol C_n represents a (proper) rotation axis with a rotation of $2\pi/n$ around the axis. The symbol D_n represents a rotation axis with a rotation of $2\pi/n$ around the axis together with another rotation axis normal to it. Subscripts are also added to other symbols in the same way. Thus, a mirror plane perpendicular to the main rotation axis, which is regarded as vertical, is written as σ_h. A mirror plane that includes the vertical axis can be of two types. If all are identical, they are labelled σ_v, and if both types are present they are labelled σ_v and σ_d, where the d stands for dihedral. In general σ_v planes include the horizontal two-fold axes, while σ_d planes lie between the horizontal two-fold axes. The nomenclature in brief is set out in Table C.3.

The correspondence between the Schoenflies and Hermann-Mauguin notation for the 32 crystallographic point groups is given in Table C.4.

Table C.4 Crystallographic point group symbols

Hermann-Mauguin full symbol	Hermann-Mauguin short symbol[a]	Schoenflies symbol
1		C_1
$\overline{1}$		C_i
2		C_2
M		C_s
$2/m$		C_{2h}
222		D_2
$mm2$		C_{2v}
$2/m\ 2/m\ 2/m$	mmm	D_{2h}
4		C_4
$\overline{4}$		S_4
$4/m$		C_{4h}
422		D_4
$4mm$		C_{4v}
$\overline{4}2m$ or $\overline{4}m2$		D_{2d}
$4/m\ 2/m\ 2/m$	$4/mmm$	D_{4h}
3		C_3
$\overline{3}$		C_{3i}
32 or 321 or 312		D_3
$3m$ or $3m1$ or $31m$		C_{3v}
$\overline{3}\ 2/m$ or $\overline{3}\ 2/m\ 1$ or $\overline{3}1\ 2/m$	$\overline{3}m$ or $\overline{3}m1$ or $\overline{3}1m$	D_{3d}
6		C_6
$\overline{6}$		C_{3h}
$6/m$		C_{6h}
622		D_6
$6mm$		C_{6v}
$\overline{6}m2$ or $\overline{6}2m$		D_{3h}
$6/m\ 2/m\ 2/m$	$6/mmm$	D_{6h}
23		T
$2/m\ \overline{3}$	$m\overline{3}$	T_h
432		O
$\overline{4}3m$		T_d
$4/m\ \overline{3}\ 2/m$	$m\overline{3}m$	O_h

[a]Only given if different from the full symbol.

Appendix D
The 230 Space Groups

Point group[a]	Space group[b]				
			Triclinic		
1, C_1	1 $P1$				
	C_1^1				
$\bar{1}$, C_i	2 $P\bar{1}$				
	C_i^1				
			Monoclinic		
2, C_2	3 $P\,1\,2\,1$	4 $P\,1\,2_1\,1$	5 $C\,1\,2\,1$		
	$P\,2$, C_2^1	$P\,2_1$, C_2^2	$C\,2$, C_2^3		
m, C_s	6 $P\,1\,m\,1$	7 $P\,1\,c\,1$	8 $C\,1\,m\,1$	9 $C\,1\,c\,1$	
	$P\,m$, C_s^1	$P\,c$, C_s^2	$C\,m$, C_s^3	$C\,c$, C_s^4	
$2/m$, C_{2h}	10 $P\,1\,2/m\,1$	11 $P\,1\,2_1/m\,1$	12 $C\,1\,2/m\,1$	13 $P\,1\,2/c\,1$	14 $P\,1\,2_1/c\,1$
	$P\,2/m$, C_{2h}^1	$P\,2_1/m$, C_{2h}^2	$C\,2/m$, C_{2h}^3	$P\,2/c$, C_{2h}^4	$P\,2_1/c$, C_{2h}^5
	15 $C\,1\,2/c\,1$				
	$C\,2/c$, C_{2h}^6				
			Orthorhombic		
2 2 2, D_2	16 $P\,2\,2\,2$	17 $P\,2\,2\,2_1$	18 $P\,2_1\,2_1\,2$	19 $P\,2_1\,2_1\,2_1$	20 $C\,2\,2\,2_1$
	D_2^1	D_2^2	D_2^3	D_2^4	D_2^5
	21 $C\,2\,2\,2$	22 $F\,2\,2\,2$	23 $I\,2\,2\,2$	24 $I\,2_1\,2_1\,2_1$	
	D_2^6	D_2^7	D_2^8	D_2^9	
$m\,m\,2$, C_{2v}	25 $P\,m\,m\,2$	26 $P\,m\,c\,2_1$	27 $P\,c\,c\,2$	28 $P\,m\,a\,2$	29 $P\,c\,a\,2_1$
	C_{2v}^1	C_{2v}^2	C_{2v}^3	C_{2v}^4	C_{2v}^5
	30 $P\,n\,c\,2$	31 $P\,m\,n\,2_1$	32 $P\,b\,a\,2$	33 $P\,n\,a\,2_1$	34 $P\,n\,n\,2$
	C_{2v}^6	C_{2v}^7	C_{2v}^8	C_{2v}^9	C_{2v}^{10}
	35 $C\,m\,m\,2$	36 $C\,m\,c\,2_1$	37 $C\,c\,c\,2$	38 $A\,m\,m\,2$	39 $A\,b\,m\,2$
	C_{2v}^{11}	C_{2v}^{12}	C_{2v}^{13}	C_{2v}^{14}	C_{2v}^{15}

Crystals and Crystal Structures, Second Edition. Richard J. D. Tilley.

	40 A m a 2 C_{2v}^{16}	41 A b a 2 C_{2v}^{17}	42 F m m 2 C_{2v}^{18}	43 F d d 2 C_{2v}^{19}	44 I m m 2 C_{2v}^{20}
	45 I b a 2 C_{2v}^{21}	46 I m a 2 C_{2v}^{22}			
m m m, D_{2h}	47 P 2/m 2/m 2/m P m m m, D_{2h}^{1}	48 P 2/n 2/n 2/n P n n n, D_{2h}^{2}	49 P 2/c 2/c 2/m P c c m, D_{2h}^{3}	50 P 2/b 2/a 2/n P b a n, D_{2h}^{4}	51 P 2_1/m 2/m 2/a P m m a, D_{2h}^{5}
	52 P 2/n 2_1/n 2/a P n n a, D_{2h}^{6}	53 P 2/m 2/n 2_1/a P m n a, D_{2h}^{7}	54 P 2_1/c 2/c 2/a P c c a, D_{2h}^{8}	55 P 2_1/b 2_1/a 2/m P b a m, D_{2h}^{9}	56 P 2_1/c 2_1/c 2/n P c c n, D_{2h}^{10}
	57 P 2/b 2_1/c 2_1/m P b c m, D_{2h}^{11}	58 P 2_1/n 2_1/n 2/m P n n m, D_{2h}^{12}	59 P 2_1/m 2_1/m 2/n P m m n, D_{2h}^{13}	60 P 2_1/b 2/c 2_1/n P b c n, D_{2h}^{14}	61 P 2_1/b 2_1/c 2_1/a P b c a, D_{2h}^{15}
	62 P 2_1/n 2_1/m 2_1/a P n m a, D_{2h}^{16}	63 C 2/m 2/c 2_1/m C m c m, D_{2h}^{17}	64 C 2/m 2/c 2_1/a C m c a, D_{2h}^{18}	65 C 2/m 2/m 2/m C m m m, D_{2h}^{19}	66 C 2/c 2/c 2/m C c c m, D_{2h}^{20}
	67 C 2/m 2/m 2/a C m m a, D_{2h}^{21}	68 C 2/c 2/c 2/a C c c a, D_{2h}^{22}	69 F 2/m 2/m 2/m F m m m, D_{2h}^{23}	70 F 2/d 2/d 2/d F d d d, D_{2h}^{24}	71 I 2/m 2/m 2/m I m m m, D_{2h}^{25}
	72 I 2/b 2/a 2/m I b a m, D_{2h}^{26}	73 I 2_1/b 2_1/c 2_1/a I b c a, D_{2h}^{27}	74 I 2_1/m 2_1/m 2_1/a I m m a, D_{2h}^{28}		

Tetragonal

4, C_4	75 P 4 C_4^1	76 P 4_1 C_4^2	77 P 4_2 C_4^3	78 P 4_3 C_4^4	79 I 4 C_4^5
	80 I 4_1 C_4^6				
$\bar{4}$, S_4	81 P $\bar{4}$ S_4^1	82 I $\bar{4}$ S_4^2			
4/m, C_{4h}	83 P 4/m C_{4h}^1	84 P 4_2/m C_{4h}^2	85 P 4/n C_{4h}^3	86 P 4_2/n C_{4h}^4	87 I 4/m C_{4h}^5
	88 I 4_1/a C_{4h}^6				
4 2 2, D_4	89 P 4 2 2 D_4^1	90 P 4 2_1 2 D_4^2	91 P 4_1 2 2 D_4^3	92 P 4_1 2_1 2 D_4^4	93 P 4_2 2 2 D_4^5
	94 P 4_2 2_1 2 D_4^6	95 P 4_3 2 2 D_4^7	96 P 4_3 2_1 2 D_4^8	97 I 4 2 2 D_4^9	98 I 4_1 2 2 D_4^{10}
4 m m, C_{4v}	99 P 4 m m C_{4v}^1	100 P 4 b m C_{4v}^2	101 P 4_2 c m C_{4v}^3	102 P 4_2 n m C_{4v}^4	103 P 4 c c C_{4v}^5

	104 $P\,4\,c\,n$	105 $P\,4_2\,m\,c$ C_{4v}^{7}	106 $P\,4_2\,b\,c$ C_{4v}^{8}	107 $I\,4\,m\,m$ C_{4v}^{9}	108 $I\,4\,c\,m$ C_{4v}^{10}
	109 $I\,4_1\,m\,d$ C_{4v}^{11}	110 $I\,4_1\,c\,d$ C_{4v}^{12}			
$\bar{4}\,2\,m$, $\bar{4}\,m\,2$, D_{2d}	111 $P\,\bar{4}\,2\,m$ D_{2d}^{1}	112 $P\,\bar{4}\,2\,c$ D_{2d}^{2}	113 $P\,\bar{4}\,2_1\,m$ D_{2d}^{3}	114 $P\,\bar{4}\,2_1\,c$ D_{2d}^{4}	115 $P\,\bar{4}\,m\,2$ D_{2d}^{5}
	116 $P\,\bar{4}\,c\,2$ D_{2d}^{6}	117 $P\,\bar{4}\,b\,2$ D_{2d}^{7}	118 $P\,\bar{4}\,n\,2$ D_{2d}^{8}	119 $I\,\bar{4}\,m\,2$ D_{2d}^{9}	120 $I\,\bar{4}\,c\,2$ D_{2d}^{10}
	121 $I\,\bar{4}\,2\,m$ D_{2d}^{11}	122 $I\,\bar{4}\,2\,d$ D_{2d}^{12}			
$4/m\,m\,m$, D_{4h}	123 $P\,4/m\,2/m\,2/m$ $P\,4/m\,m\,m$, D_{4h}^{1}	124 $P\,4/m\,2/c\,2/c$ $P\,4/m\,c\,c$, D_{4h}^{2}	125 $P\,4/n\,2/b\,2/m$ $P\,4/n\,b\,m$, D_{4h}^{3}	126 $P\,4/n\,2/n\,2/c$ $P\,4/n\,n\,c$, D_{4h}^{4}	127 $P\,4/m\,2_1/b\,2/m$ $P\,4/m\,b\,m$, D_{4h}^{5}
	128 $P\,4/m\,2_1/n\,2/c$ $P\,4/m\,n\,c$, D_{4h}^{6}	129 $P\,4/n\,2_1/m\,2/m$ $P\,4/n\,m\,m$, D_{4h}^{7}	130 $P\,4/n\,2_1/c\,2/c$ $P\,4/n\,c\,c$, D_{4h}^{8}	131 $P\,4_2/m\,2/m\,2/c$ $P\,4_2/m\,m\,c$, D_{4h}^{9}	132 $P\,4_2/m\,2/c\,2/m$ $P\,4_2/m\,c\,m$, D_{4h}^{10}
	133 $P\,4_2/n\,2/b\,2/c$ $P\,4_2/n\,b\,c$, D_{4h}^{11}	134 $P\,4_2/n\,2/n\,2/m$ $P\,4_2/n\,n\,m$, D_{4h}^{12}	135 $P\,4_2/m\,2_1/b\,2/c$ $P\,4_2/m\,b\,c$, D_{4h}^{13}	136 $P\,4_2/m\,2_1/n\,2/m$ $P\,4_2/m\,n\,m$, D_{4h}^{14}	137 $P\,4_2/n\,2_1/m\,2/c$ $P\,4_2/n\,m\,c$, D_{4h}^{15}
	138 $P\,4_2/n\,2_1/c\,2/m$ $P\,4_2/n\,c\,m$, D_{4h}^{16}	139 $I\,4/m\,2/m\,2/m$ $I\,4/m\,m\,m$, D_{4h}^{17}	140 $I\,4/m\,2/c\,2/m$ $I\,4/m\,c\,m$, D_{4h}^{18}	141 $I\,4_1/a\,2/m\,2/d$ $I\,4_1/a\,m\,d$, D_{4h}^{19}	142 $I\,4_1/a\,2/c\,2/d$ $I\,4_1/a\,c\,d$, D_{4h}^{20}

Trigonal

3, C_3	143 $P\,3$ C_3^{1}	144 $P\,3_1$ C_3^{2}	145 $P\,3_2$ C_3^{3}	146 $R\,3$ C_3^{4}	
$\bar{3}$, C_{3i}	147 $P\,\bar{3}$ C_{3i}^{1}	148 $R\,\bar{3}$ C_{3i}^{2}			
$3\,2$, $3\,1\,2$, $3\,2\,1$, D_3	149 $P\,3\,1\,2$ D_3^{1}	150 $P\,3\,2\,1$ D_3^{2}	151 $P\,3_1\,1\,2$ D_3^{3}	152 $P\,3_1\,2\,1$ D_3^{4}	153 $P\,3_2\,1\,2$ D_3^{5}
	154 $P\,3_2\,2\,1$ D_3^{6}	155 $R\,3\,2$ D_3^{7}			
$3\,m$, $3\,m\,1$, $3\,1\,m$, C_{3v}	156 $P\,3\,m\,1$ C_{3v}^{1}	157 $P\,3\,1\,m$ C_{3v}^{2}	158 $P\,3\,c\,1$ C_{3v}^{3}	159 $P\,3\,1\,c$ C_{3v}^{4}	160 $R\,3\,m$ C_{3v}^{5}
	161 $R\,3\,c$ C_{3v}^{6}				
$\bar{3}m$, $\bar{3}\,1\,m$, $\bar{3}m\,1$, D_{3d}	162 $P\,\bar{3}\,1\,2/m$ $P\,\bar{3}\,1\,m$, D_{3d}^{1}	163 $P\,\bar{3}\,1\,2/c$ $P\,\bar{3}\,1\,c$, D_{3d}^{2}	164 $P\,\bar{3}\,2/m\,1$ $P\,\bar{3}\,m\,1$, D_{3d}^{3}	165 $P\,\bar{3}\,2/c\,1$ $P\,\bar{3}c1$, D_{3d}^{4}	166 $R\,\bar{3}\,2/m$ $R\,\bar{3}\,m$, D_{3d}^{5}
	167 $R\,\bar{3}\,2/c$ $R\,\bar{3}\,c$, D_{3d}^{6}				

			Hexagonal		
6, C_6	168 *P* 6 C_6^1	169 *P* 6_1 C_6^2	170 *P* 6_5 C_6^3	171 *P* 6_2 C_6^4	172 *P* 6_4 C_6^5
	173 *P* 6_3 C_6^6				
$\bar{6}$, C_{3h}	174 *P* $\bar{6}$ C_{3h}^1				
6/*m*, C_{6h}	175 *P* 6/*m* C_{6h}^1	176 *P* 6_3/*m* C_{6h}^2			
6 2 2, D_6	177 *P* 6 2 2 D_6^1	178 *P* 6_1 2 2 D_6^2	179 *P* 6_5 2 2 D_6^3	180 *P* 6_2 2 2 D_6^4	181 *P* 6_4 2 2 D_6^5
	182 *P* 6_3 2 2 D_6^6				
6 *m m*, C_{6v}	183 *P* 6 *m m* C_{6v}^1	184 *P* 6 *c c* C_{6v}^2	185 *P* 6_3 *c m* C_{6v}^3	186 *P* 6_3 *m c* C_{6v}^4	
$\bar{6}$*m* 2, $\bar{6}$ 2 *m*, D_{3h}	187 *P* $\bar{6}$ *m* 2 D_{3h}^1	188 *P* $\bar{6}$ *c* 2 D_{3h}^2	189 *P* $\bar{6}$ 2 *m* D_{3h}^3	190 *P* $\bar{6}$ 2 *c* D_{3h}^4	
6/*m m m*, D_{6h}	191 *P* 6/*m* 2/*m* 2/*m* *P* 6/*m m m*, D_{6h}^1	192 *P*6/*m* 2/*c* 2/*c* *P* 6/*m c c*, D_{6h}^2	193 *P* 6_3/*m* 2/*c* 2/*m* *P* 6_3/*m c m*	194 *P* 6_3/*m* 2/*m* 2/*c* *P* 6_3/*m m c*, D_{6h}^4	
			Cubic		
2 3, *T*	195 *P* 2 3 T^1	196 *F* 2 3 T^2	197 *I* 2 3 T^3	198 *P* 2_1 3 T^4	199 *I* 2_1 3 T^5
m$\bar{3}$, T_h	200 *P* 2/*m* $\bar{3}$ *P m* $\bar{3}$, T_h^1	201 *P* 2/*n* $\bar{3}$ *P n* $\bar{3}$, T_h^2	202 *F* 2/*m* $\bar{3}$ *F m* $\bar{3}$, T_h^3	203 *F* 2/*d* $\bar{3}$ *F d* $\bar{3}$, T_h^4	204 *I* 2/*m* $\bar{3}$ *I m* $\bar{3}$, T_h^5
	205 *P* 2_1/*a* $\bar{3}$ *P a* $\bar{3}$, T_h^6	206 *I* 2_1/*a* $\bar{3}$ *I a* $\bar{3}$, T_h^7			
432, *O*	207 *P* 4 3 2 O^1	208 *P* 4_2 3 2 O^2	209 *F* 4 3 2 O^3	210 *F* 4_1 3 2 O^4	211 *I* 4 3 2 O^5
	212 *P* 4_3 3 2 O^6	213 *P* 4_1 3 2 O^7	214 *I* 4_1 3 2 O^8		
$\bar{4}$3*m*, T_d	215 *P* $\bar{4}$ 3 *m* T_d^1	216 *F* $\bar{4}$ 3 *m* T_d^2	217 *I* $\bar{4}$ 3 *m* T_d^3	218 *P* $\bar{4}$ 3 *n* T_d^4	219 *F* $\bar{4}$ 3 *c* T_d^5
	220 *I* $\bar{4}$ 3 *d* T_d^6				
m$\bar{3}$*m*, O_h	221 *P* 4/*m* $\bar{3}$ 2/*m* *P m* $\bar{3}$ *m*, O_h^1	222 *P* 4/*n* $\bar{3}$ 2/*n* *P n* $\bar{3}$ *n*, O_h^2	223 *P* 4_2/*m* $\bar{3}$ 2/*n* *P m* $\bar{3}$ *n*, O_h^3	224 *P* 4_2/*n* $\bar{3}$ 2/*m* *P n* $\bar{3}$ *m*, O_h^4	225 *F* 4/*m* $\bar{3}$ 2/*m* *F m* $\bar{3}$ *m*, O_h^5
	226 *F* 4/*m* $\bar{3}$ 2/*c* *F m* $\bar{3}$ *c*, O_h^6	227 *F* 4_1/*d* $\bar{3}$ 2/*m* *F d* $\bar{3}$ *m*, O_h^7	228 *F* 4_1/*d* $\bar{3}$ 2/*c* *F d* $\bar{3}$ *c*, O_h^8	229 *I* 4/*m* $\bar{3}$ 2/*m* *I m* $\bar{3}$ *m*, O_h^9	230 *I* 4_1/*a* $\bar{3}$ 2/*d* *I a* $\bar{3}$ *d*, O_h^{10}

[a]The first entry is the Hermann-Mauguin short symbol. The full symbol is given in Table 4.4 and in Appendix C. The second entry is the Schoenflies symbol.

[b]The first entry is the space group number, the second is the full Hermann-Mauguin symbol, the third is the short Hermann-Mauguin symbol when it differs from the full symbol, or else the Schoenflies symbol, and the fourth entry, when it appears, is the Schoenflies symbol.

Appendix E Complex Numbers

A complex number is a number written

$$z = a + \mathrm{i}b$$

where i is defined as $\mathrm{i} = \sqrt{-1}$, a is called the **real part** of the number and b the **imaginary part**. The **modulus** of a complex number $z = a + \mathrm{i}b$ is written $|z|$ and is given by

$$|z| = \sqrt{a^2 + b^2}$$

An Argand diagram is a graphical method of representing a complex number, z, written as $a + \mathrm{i}\,b$. The real part of the complex number, a, is plotted along the horizontal axis and the imaginary part, $\mathrm{i}\,b$, along the vertical axis. The complex number z is then represented by a point whose Cartesian coordinates are (a, b) (Figure E.1). (These diagrams are also referred to as *representations in the Gaussian plane* or *representations in the Argand plane.*)

The complex number

$$z = a + \mathrm{i}b$$

can also be written in polar form, with **z** taken to represent a radial vector of length r, where

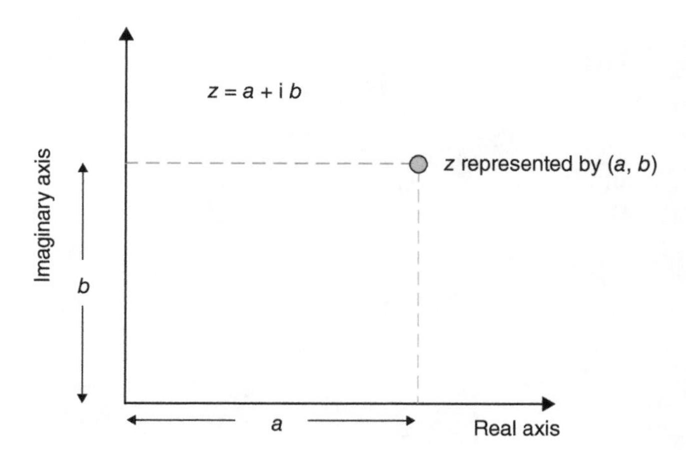

Figure E.1 Representation of a complex number by an Argand diagram.

Crystals and Crystal Structures, Second Edition. Richard J. D. Tilley.

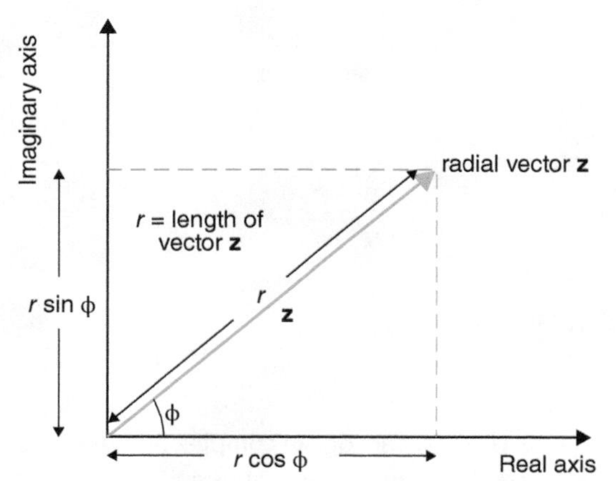

Figure E.2 Representation of a vector as a complex number.

$$r = |\mathbf{z}| = \sqrt{a^2 + b^2}$$

r is a scalar, that is, a positive number with no directional properties, given by the modulus of vector **z**, written |**z**| (Figure E.2). The angle between the vector **z** and the horizontal axis, ϕ, measured in an anticlockwise direction from the axis, is called its **argument**. The real and imaginary parts of **z** are given by the projections of **z** on the horizontal and vertical axes, i.e.

$$\text{Real part} = a = r \cos \phi$$
$$\text{Imaginary part} = b = r \sin \phi$$

and $\mathbf{z} = r \cos \phi + \mathrm{i}\, r \sin \phi = r(\cos \phi + \mathrm{i} \sin \phi)$

The value of ϕ is given by

$$\phi = \arctan\,(r \sin \phi / r \cos \phi) = \arctan\,(b/a)$$

The **complex conjugate number** to z, written z^*, is obtained by replacing i with −i:

$$z = a + \mathrm{i}b$$
$$z^* = a - \mathrm{i}b$$

The product of a complex number with its complex conjugate is always a real number:

$$z\,z^* = (a + \mathrm{i}b)(a - \mathrm{i}b) = a^2 - \mathrm{i}^2 b^2 = a^2 + b^2$$

Appendix F
Complex Amplitudes

The addition of vectors that represent the scattering of a beam of radiation by several objects can be achieved algebraically by writing them in the form:

$$\mathbf{a} = a\,e^{i\phi} \quad \text{or} \quad a\,e^{-i\phi}$$

where a is the scalar magnitude of the vector and ϕ is the phase. The vector **a** is called a **complex amplitude**. The value of i is taken as positive when the phase is ahead of a 'standard phase' and negative when the phase is behind the standard phase. (The 'standard phase' in X-ray diffraction is set by the atom at the origin of the unit cell, (0,0,0).)

This terminology allows the scattering of X-rays by atoms in a unit cell to be added algebraically, by writing the scattering by an atom in a unit cell as a complex amplitude **f**:

$$\mathbf{f} = f\,e^{i\phi}$$

where f is the scalar magnitude of the scattering and ϕ is the phase of the scattered wave. Using Euler's formula:

$$e^{i\phi} = \cos\phi + i\sin\phi$$

the scattering can be written

$$\mathbf{f} = f\{\cos\phi + i\sin\phi\}$$

This complex number can be represented in polar form as a vector **f** plotted on an Argand diagram.

$$\mathbf{f} = f\{\cos\phi + i\sin\phi\}$$

where f is the length of the scattering vector and ϕ is the argument (i.e. the phase angle) associated with **f** (Figure F.1).

The advantage of using this representation is that algebraic addition is equivalent to vector addition. For example, suppose that it is necessary to add two vectors $\mathbf{f}_1$ with magnitude f_1 and phase ϕ_1 and $\mathbf{f}_2$ with magnitude f_2 and

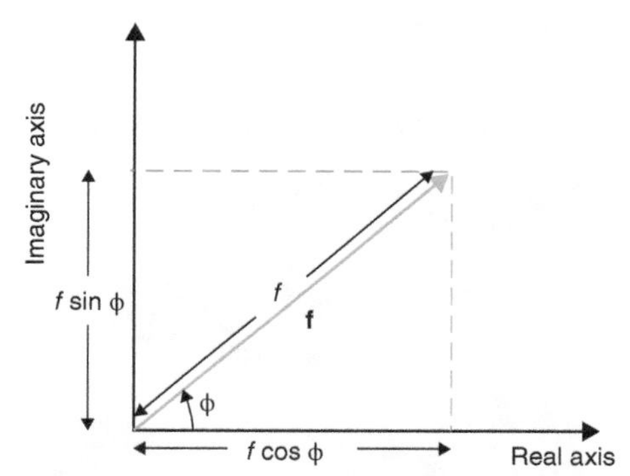

Figure F.1 Representation of an atomic scattering factor as a complex amplitude vector.

Crystals and Crystal Structures, Second Edition. Richard J. D. Tilley.

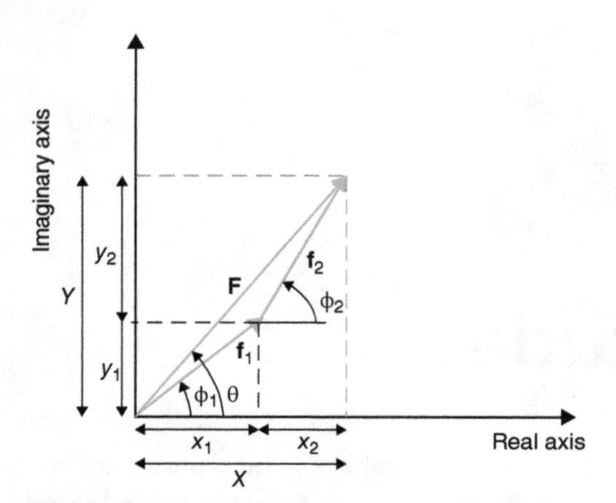

Figure F.2 The addition of two complex amplitude vectors.

phase ϕ_2 to obtain a resultant vector **F**, with magnitude F and phase θ, that is:

$$\mathbf{F} = F\,e^{i\phi} = f_1 e^{i\phi_1} + f_2 e^{i\phi_2}$$

Drawing these on a diagram in the Argand plane (Figure F.2) shows that the total displacement along the real axis, X, is:

$$X = x_1 + x_2 = f_1 \cos\phi_1 + f_2 \cos\phi_2$$

and along the imaginary axis, Y, is:

$$Y = y_1 + y_2 = f_1 \sin\phi_1 + f_2 \sin\phi_2$$

$$\mathbf{F} = (X + iY)$$

The length of **F** is then F:

$$F = \sqrt{X^2 + Y^2}$$

and the phase angle is given by:

$$\tan\theta = Y/X$$

Clearly this can be repeated with any number of terms to obtain the algebraic sums:

$$X = x_1 + x_2 + x_3 \ldots = f_1 \cos\phi_1 + f_2 \cos\phi_2 + f_3 \cos\phi_3 \ldots = \sum_{n=1}^{N} x_n = \sum_{n=1}^{N} f_n \cos\phi_n$$

$$Y = y_1 + y_2 + y_3 \ldots = f_1 \sin\phi_1 + f_2 \sin\phi_2 + f_3 \sin\phi_3 \ldots = \sum_{n=1}^{N} y_n = \sum_{n=1}^{N} f_n \sin\phi_n$$

with $F = \sqrt{X^2 + Y^2}$ and $\tan\theta = Y/X$.

The intensity of a scattered beam, I, is given by multiplying **F** by its complex conjugate, $\mathbf{F}^*$(see Appendix E) to give a real number:

$$\mathbf{F} \times \mathbf{F}^* = (X + iY)(X - iY) = X^2 + Y^2 = F^2$$

or

$$\mathbf{F} \times \mathbf{F}^* = F\,e^{i\phi} \times F\,e^{-i\phi} = F^2 = I$$

Answers to Problems and Exercises

CHAPTER 1

Quick Quiz

1c, 2b, 3b, 4a, 5b, 6c, 7b, 8c, 9c, 10a.

Calculations and Questions

1.1

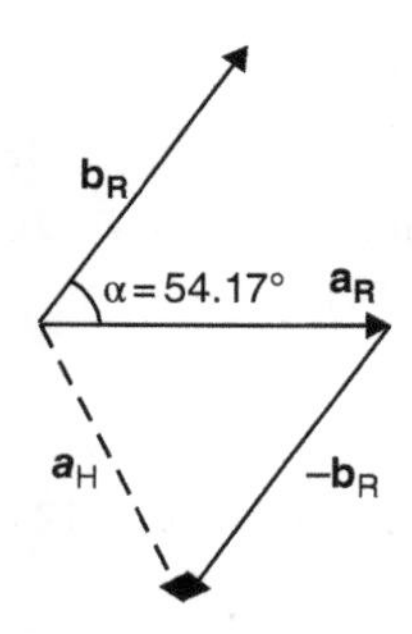

Graphical, $a_H \approx 0.372$ nm; arithmetical, $a_H = 0.375$ nm, $c_H = 1.051$ nm.

1.2

Sn atoms at: 000; 0.2369 nm, 0.2369 nm, 0.1594 nm;

O atoms at: 0.1421 nm, 0.1421 nm, 0; 0.3790 nm, 0.09476 nm, 0.1594 nm; 0.3317 nm, 0.3317 nm, 0; 0.09476 nm, 0.3790 nm, 0.1594 nm;

Volume = 0.0716 nm^3.

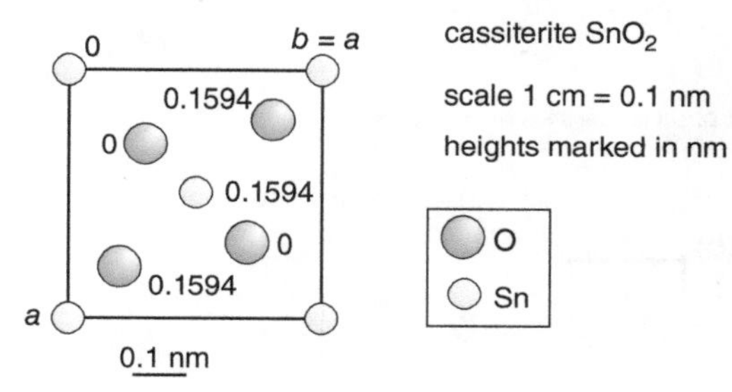

1.3 There are six equal bonds of length 0.1952 nm. Regular octahedron.

1.4 (a) 4270 kg m^{-3}; (b) 4886 kg m^{-3}.

1.5 0.7014 nm.

1.6 95.9 g mol^{-1}.

1.7 23.8 g mol^{-1}, (Mg).

1.8 2

1.9 0.1366 nm

1.10 0.126 nm

Crystals and Crystal Structures, Second Edition. Richard J. D. Tilley.

CHAPTER 2

Quick Quiz

1b, 2c, 3a, 4a, 5c, 6a, 7c, 8c, 9b, 10b.

Calculations and Questions

2.1 (a) Not a lattice; (b) rectangular; (c) hexagonal; (d) centred rectangular; (e) not a lattice; (f) oblique.

2.2

(a)

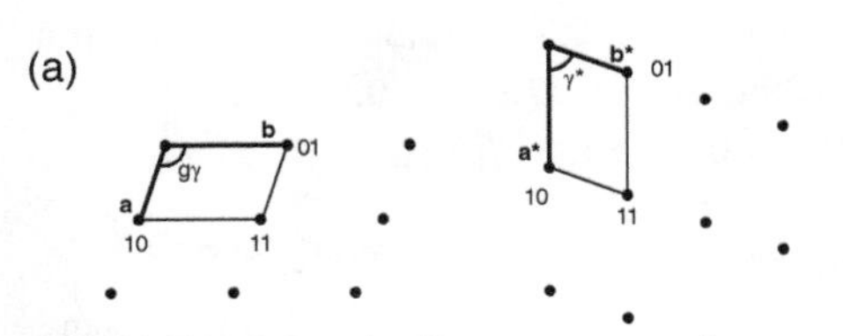

Direct lattice: a = 8 nm, b = 12 nm, γ = 110°.

Reciprocal lattice: a^* = 0.125 nm^{-1}, b^* = 0.083 nm^{-1}, γ^* = 70°.

(b)

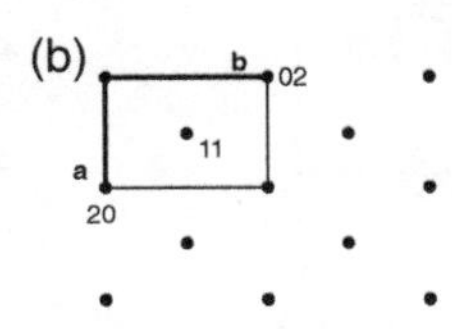

Direct lattice: a = 10 nm, b = 14 nm.

Reciprocal lattice: a^* = 0.10 nm^{-1}, b^* = 0.071 nm^{-1}.

(c)

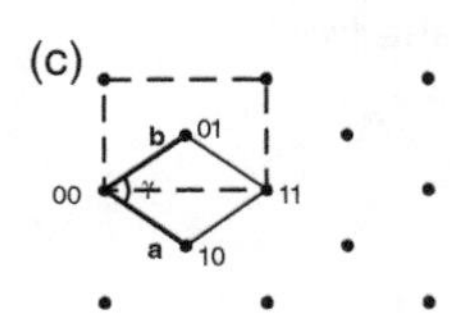

Direct lattice, primitive: $a = b$ = 17.2 nm, γ = 71°.

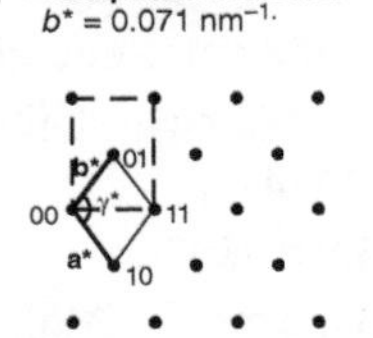

Reciprocal lattice: $a^* = b^*$ = 0.058 nm^{-1}, γ^* = 109°.

2.3

(a)

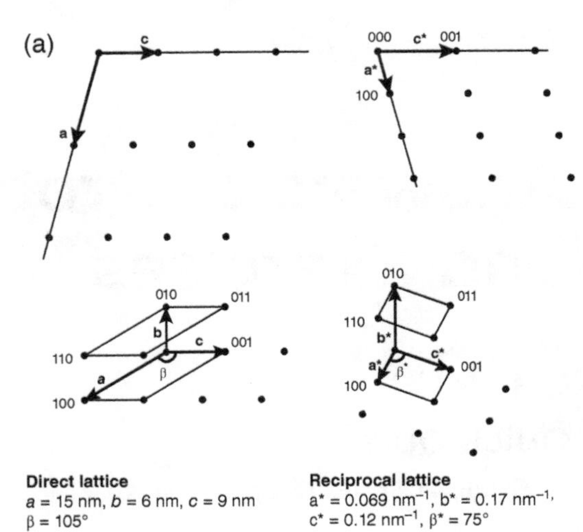

Direct lattice
a = 15 nm, b = 6 nm, c = 9 nm
β = 105°

Reciprocal lattice
a* = 0.069 nm^{-1}, b* = 0.17 nm^{-1},
c* = 0.12 nm^{-1}, β^* = 75°

2.4 (a) (230); (b) (120); (c) $(1\bar{1}0)$ or $(\bar{1}10)$; (d) (310); (e) $(5\bar{3}0)$ or $(\bar{5}30)$.

2.5 (a) (110), $(11\bar{2}0)$; (b) $(1\bar{1}0)$, $(1\bar{1}00)$; (c) $(3\bar{2}0)$, $(3\bar{2}\bar{1}0)$; (d) (410), $(41\bar{5}0)$; (e) $(3\bar{1}0)$, $(3\bar{1}\bar{2}0)$.

2.6 (a) [110]; (b) $[\bar{1}20]$; (c) $[\bar{3}\bar{1}0]$; (d) [010]; (e) $[\bar{1}\bar{3}0]$.

2.7 (a) [001]; (b) $[12\bar{2}]$; (c) $[01\bar{1}]$; (d) [101].

2.8 (a) 0.2107 nm; (b) 0.2124 nm; (c) 0.2812 nm; (d) 0.1359 nm; (e) 0.3549 nm.

2.9 (No answer required.)

2.10 (110)/(120) 18.4°; (111)/$(11\bar{1})$ 70.5°; (112)/(103) 25.35°.

2.11 (110)/(211) 29.8°; (111)/$(11\bar{1})$ 69.9°; (111)/(121) 19.5°.

2.12 (110)/(112): 45.0°; (111)/(11$\bar{1}$): 53.2°; (111)/(112) 18.4°; (111)/(211): 19.3°.

2.13 (211)/(112): 18.1°; (111)/(1$\bar{1}$1): 86.1°; (101)/(211): 22.1°; (110)/(121): 13.3°.

CHAPTER 3

Quick Quiz

1c, 2b, 3b, 4c, 5b, 6a, 7c, 8a, 9b, 10c.

Calculations and Questions

3.1

a. 5, *m* (through each apex); 5*m*.
b. 2, *m* (horizontal), *m* (vertical), centre of symmetry; 2*mm*.
c. *m* (vertical); *m*.
d. 4, *m* (vertical), *m* (diagonal), centre of symmetry; 4*mm*.
e. 2, *m* (horizontal), *m* (vertical), centre of symmetry; 2*mm*.

3.2

a. Fourfold rotation axis at the centre of the pattern.
b. Twofold rotation axis at the centre of the pattern plus a vertical mirror through the centre of the pattern.

3.3

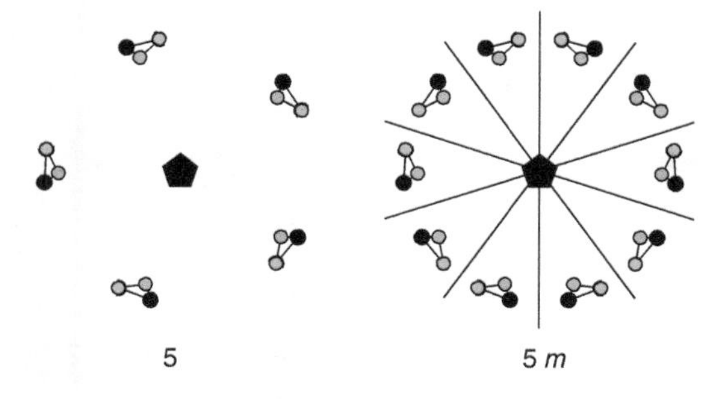

3.4 (a) *p*2; (b) *pg*; (c) p2*gg*.

3.5 (a) *p*4; (b) *p*4*mm*; (c) *p*31*m*.

3.6 (0.125, 0.475); (0.875, 0.525); (0.875, 0.475); (0.125, 0.525); (0.625, 0.975); (0.375, 0.025); (0.375, 0.975); (0.625, 0.025).

3.7 (0.210, 0.395); (0.790, 0.605); (0.605, 0.210); (0.395, 0.790); (0.290, 0.895); (0.710, 0.105); (0.895, 0.710); (0.105, 0.290).

CHAPTER 4

Quick Quiz

1b, 2a, 3c, 4b, 5c, 6a, 7c, 8a, 9b, 10b

Calculations and Questions

4.1

a. 2, 3, m, $\bar{4}$; $\bar{4}3m$; T_d
b. 3, 2, *m* (vertical), *m* (horizontal); $\bar{6}m2$ or $\bar{6}2m$; D_{3h}
c. 3, *m* (vertical); 3*m* (3*m*1 or 31*m*); C_{3v}
d. 6, 2, $\bar{1}$, *m* (vertical), *m* (horizontal); 6/*m mm*; D_{6h}
e. 4, $\bar{4}$, 3, $\bar{3}$, 2, $\bar{1}$, *m* (vertical), *m* (horizontal); $m\bar{3}m$; O_h

4.2 (a) 2/m, (b) *m*, (c) 2.

4.3 (a) *mmm*; (b) *mm*2; (c) 222.

4.4 (a) $\frac{2}{m}\frac{2}{m}\frac{2}{m}$; (b) $\frac{4}{m}\frac{2}{m}\frac{2}{m}$; (c) $\bar{3}\ \frac{2}{m}$; (d) $\frac{6}{m}\frac{2}{m}\frac{2}{m}$; (e) $\frac{4}{m}\ \bar{3}\ \frac{2}{m}$.

4.5

a. See figure; three 4-fold axes through the centre of each square face; four $\bar{3}$ axes along each body diagonal; six 2-fold axes along each diagonal; mirrors perpendicular to each 4 and 2 axis.
b. See figure; to go from $4/m\,\bar{3}\,2/m$ to $2/m\,\bar{3}$ it is necessary to remove *all* the tetrad (4) axes (otherwise the first symbol would remain at 4 and the system would become tetragonal), and replace them with diad (2) axes.

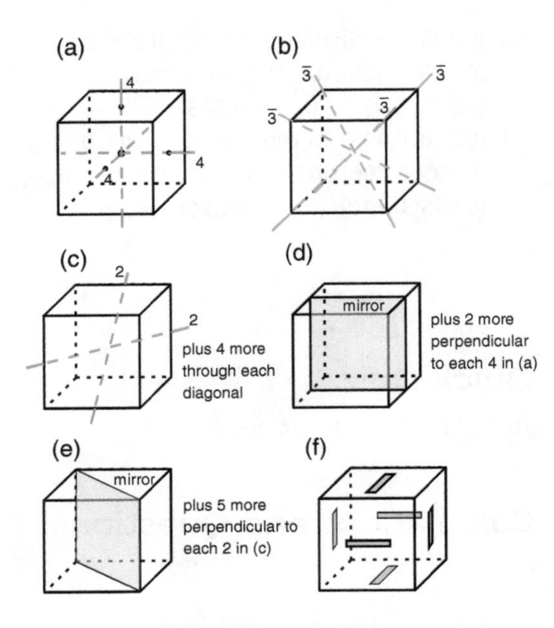

4.6 (a) Cu, $m\bar{3}m$; (b) Fe, $m\bar{3}m$; (c) Mg, $6/mmm$.
4.7 (a) NaCl, $m\bar{3}m$; (b) CaF_2, $m\bar{3}m$; (c) $SrTiO_3$, $m\bar{3}m$.

CHAPTER 5

Quick Quiz

1b, 2b, 3a, 4c, 5c, 6b, 7a, 8c, 9c, 10c.

Calculations and Questions

5.1 $Sn_{0.5}Pb_{0.5}O_2$.
5.2 (a)

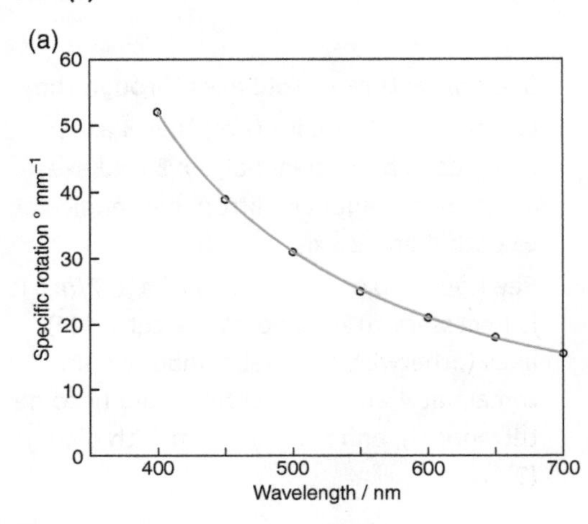

(b) Red light, 5.36 mm; violet light, 1.99 mm.
5.3 $p = 0.064 \times 10^{-28}$ Cm; $P_s = 0.1$ Cm^{-2}.
5.4 (a) 0.0375 Cm^{-2}; (b) along the **c**- (tetrad) axis.
5.5 (a) 7×10^{-7} Cm^{-1}; (b) 5.54×10^{-7} Cm^{-1}.
5.6 2.9×10^{-7} Cm^{-1} along the axes, 5×10^{-7} Cm^{-1} along [111].
5.7 (a) cubic; (b) cubic; (c) hexagonal.

CHAPTER 6

Quick Quiz

1b, 2a, 3b, 4c, 5c, 6a, 7c, 8b, 9b, 10c.

Calculations and Questions

6.1

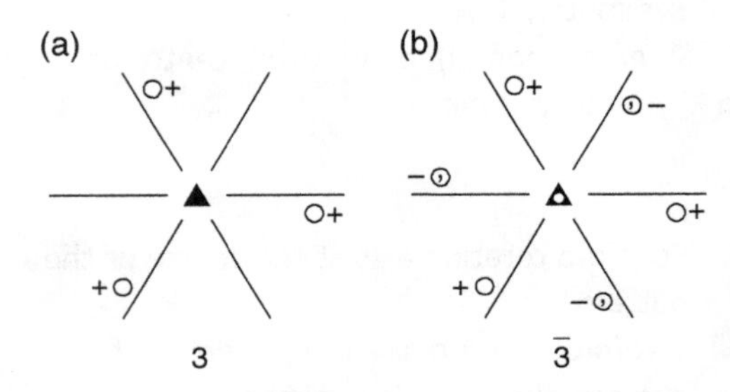

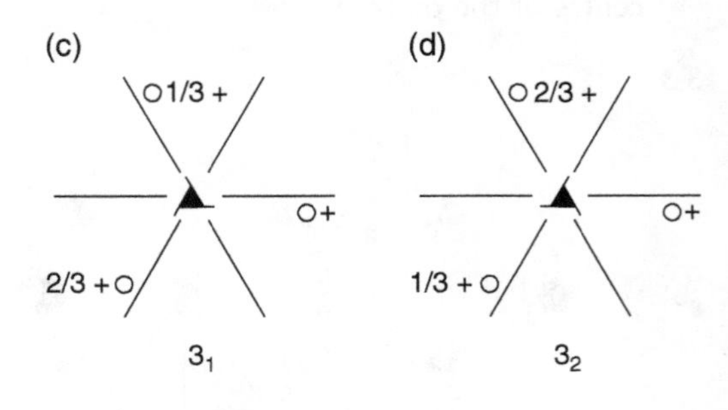

6.2 (a) $P1c1$; (b) $P2/m\,2/m\,2/m$; (c) $P4_2/n\,2/b\,2/c$; (d) $P\,6_3/m\,2/c\,2/m$; (e) $F\,4/m\,\bar{3}\,2/m$.
6.3 (a) tetrad axis; (c) twofold screw axis, 2_1; fourfold screw axis, 4_2.

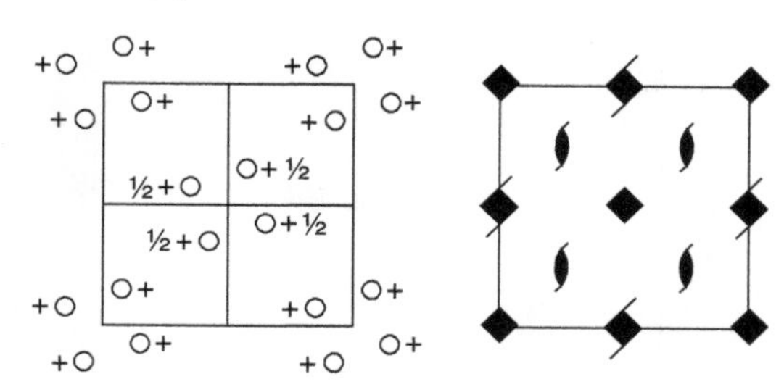

6.4 000 + each of: $x, y, z; \overline{x}, \overline{y}, z; \overline{y}, x, z; y, \overline{x}, z$;
½ ½ ½ + each of: $x, y, z; \overline{x}, \overline{y}, z; \overline{y}, x, z$;
$y, \overline{x}, z$.

6.5

a. Tetrad axis parallel to the **c**-axis; mirrors with normals along [100], [010]; mirrors with normals along [110], [1$\overline{1}$0].
b. Diad axes, glide planes.
c. 2 Pu.
d. Pu(1): 0,0,0.
   Pu(2): ½, ½, 0.4640.
e. 4S.
f. S(1): 0, 0, 0.3670
   S(2): ½, ½, 0.0970
   S(3): ½, 0, 0.7320
   S(4): 0, ½, 0.7320.

6.6 *Pnma* 4 *a* 0,0,0 ½,0,½ 0,½,0 ½,½,½
Pbnm 4 *a* 0,0,0 ½,½,0 0,0,½ ½,½,½
Pnma 4 *b* 0,0,½ ½,0,0 0,½,½ ½,½,0
Pbnm 4 *b* ½,0,0 0,½,0 ½,0,½ 0,½,½
Pnma 4 *c* *x*,¼,*z* −*x* + ½, ¾, *z* + ½
−*x*,¾,−*z* *x* + ½, ¼, −*z* + ½
Pbnm 4 *c* *x*,*y*,¼ *x* + ½, −*y* + ½, ¾
−*x*,−*y*,¾ −*x* + ½, *y* + ½, ¼
Pnma 8*d* *x*,*y*,*z* −*x* + ½, −*y*, *z* + ½
−*x*, *y* + ½, −*z* *x* + ½, −*y* + ½, −*z* + ½
−*x*,−*y*,−*z* *x* + ½, *y*, −*z* + ½
x, −*y* + ½,*z* −*x* + ½, *y* + ½, *z* + ½
Pbnm 8 *d* *x*,*y*,*z* *x* + ½, −*y* + ½, −*z*
−*x*, −*y*, *z* + ½ −*x* + ½, *y* + ½, −*z* + ½
−*x*,−*y*,−*z* −*x* + ½, *y* + ½, *z*
x, *y*, −*z* + ½ *x* + ½, −*y* + ½, *z* + ½

6.7 (a)

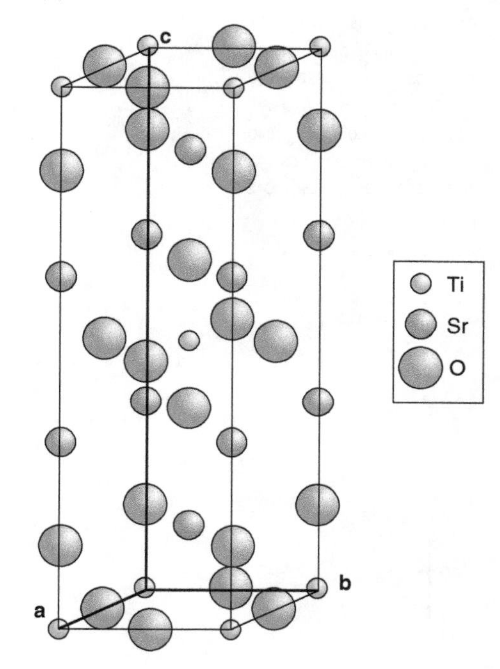

(b) Sr_2TiO_4 is composed of stacked perovskite layers, one TiO_6 octahedron thick (see Figure 9.17).

CHAPTER 7

Quick Quiz

1b, 2c, 3a, 4c, 5a, 6b, 7a, 8c, 9b, 10c.

Calculations and Questions

7.1 (111), 9.54°; (200), 11.04°; (220), 15.71°; (311), 18.51°; (222), 19.36°; (400), 34.68°.

7.2 (a) 0.2291 nm; (b) 10.2 nm.

7.3 Note: the value of λl will depend upon the scale of reproduction. If the distance from 000 to 020 in (a) is, say, 20 mm, λl will be 3.30 mm nm.

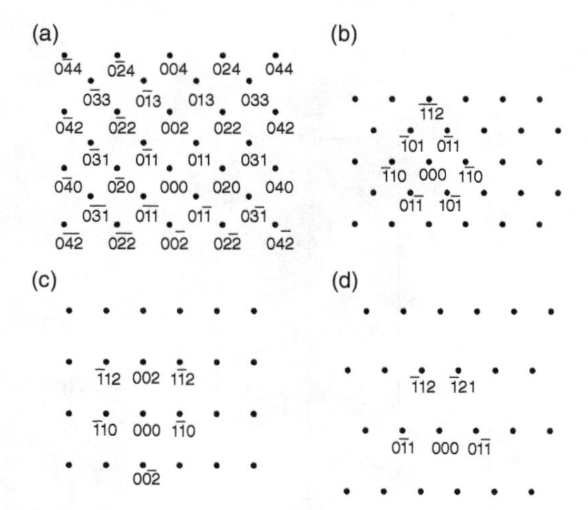

7.4

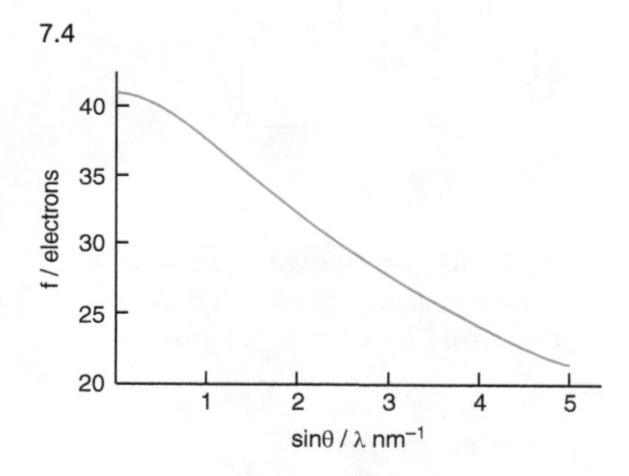

7.5

(a)

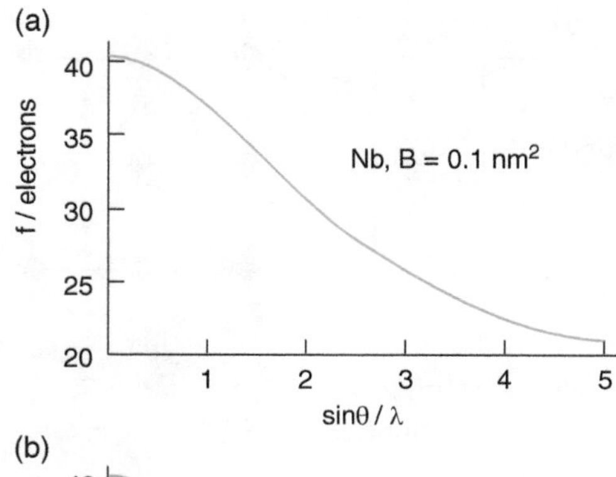

(b)

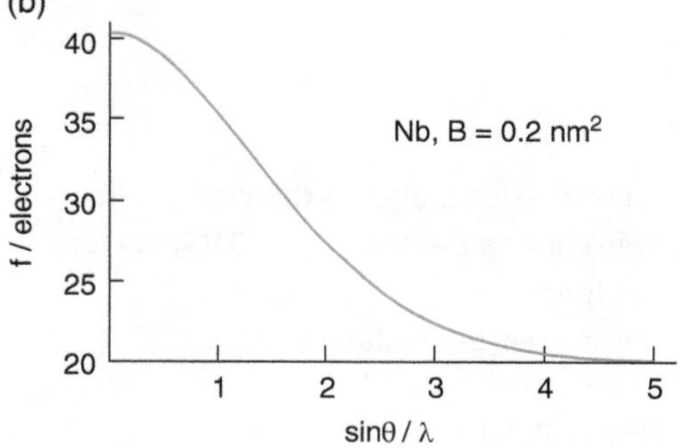

(a) 0.036 nm; (b) 0.050 nm.

7.6 $F_{110} = 37.46$, phase = 0; the vector diagram gives a value of 37.5 for F.

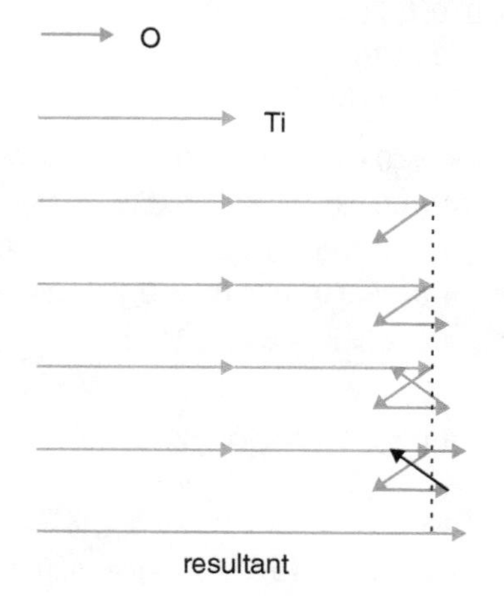

7.7 (a) 4447; (b) 144 187.

7.8

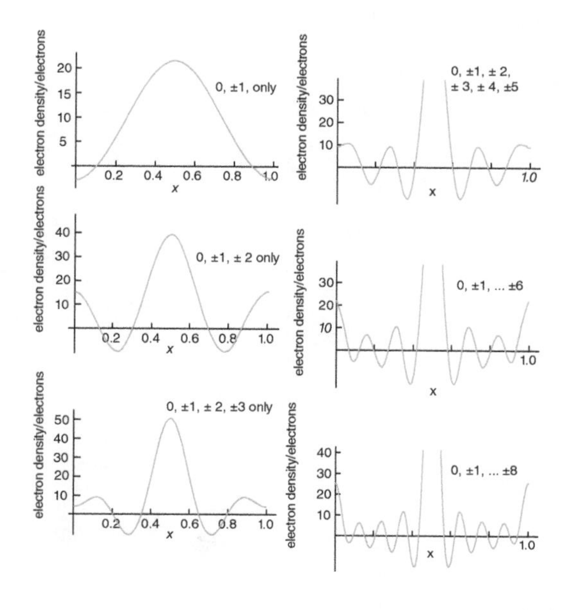

CHAPTER 8

Quick Quiz

1c, 2b, 3b, 4a, 5c, 6a, 7a, 8b, 9c, 10b.

Calculations and Questions

8.1

a. Room temperature, $a = 0.2838$ nm; high temperature, $a = 0.3603$ nm.

b. Room temperature, $a = 0.2924$ nm, $c = 0.4775$ nm; high temperature, a = 0.3286 nm.

c. Room temperature, $a = 0.5583$ nm; high temperature, a = 0.4432 nm.

8.2 r(Cl), 0.181 nm; r(Li), 0.075 nm; r(Na), 0.101 nm; r(K), 0.134 nm; r(Rb), 0.148 nm; r(Cs), 0.170 nm.

8.3 The normal distribution fits best.

8.4 Site 1 containing Nb^{5+} and Site 2 containing Zr^{4+} gives the best fit.

8.5 (a) Li_2O; (b) $FeTiO_3$; (c) Ga_2S_3; (d) $CdCl_2$; (e) Fe_2SiO_4; (f) Cr_5O_{12}.

8.6 Clearly ionic radii give the poorest description. Metallic radii are surprisingly similar to van der Waals radii, and so reasonable.

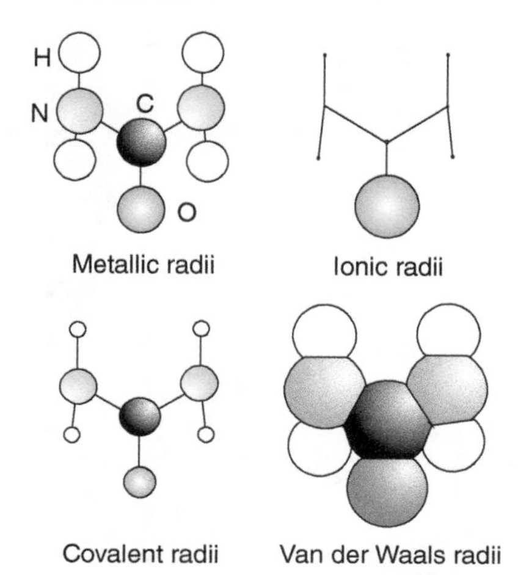

CHAPTER 9

Quick Quiz

1b, 2b, 3a, 4b, 5c, 6a, 7c, 8b, 9b, 10a.

Calculations and Questions

9.1 (a), f (mixed sites) = 22.72, $F_{200} = 28.50$; intensity = 789.9; (b) F_{200} (pure TiO_2) = 13.80, intensity (pure TiO_2) = 190.6; (c) F_{200} (pure SnO_2) = 61.5, intensity (pure SnO_2) = 3778.6.

9.2 (a) Vegard's law is obeyed; (b) ≈ 1.2223 nm; (c) $x \approx 0.27$, i.e. $(Y_{0.27}Tm_{0.73})_3Ga_5O_{12}$.

9.3 (a) Vacancies, $La_{0.948}Mn_{0.948}O_3$, interstitials, $LaMnO_{3.165}$; (b) vacancies, 6594.2 kg m^{-3}, interstitials, 6956.9 kg m^{-3}.

9.4 (a)

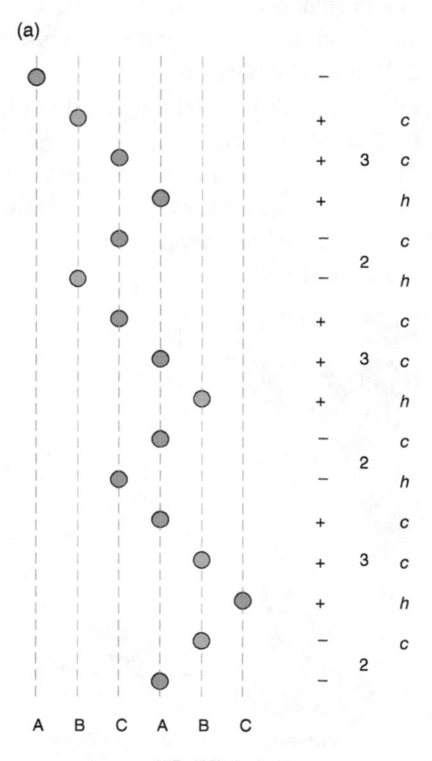

15R, (32), (cchch)

(b)

24H, (7557), $(hc_4)_2(hc_6)_2$

(c)

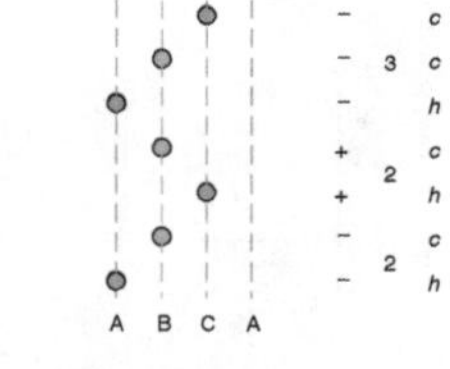

16H, (332233), $(hcc)_4(hc)_2$

9.5 (a) The *CS* plane lies midway between (102) and (103), $W_nO_{3n-1.5}$; (b) $A_{4.666}B_{3.666}O_{12}$, $n = 3.666$ ($3\frac{2}{3}$); (c) $Bi_2A_{1.5}B_{2.5}O_{10.5}$, $n = 2.5$ ($2\frac{1}{2}$).

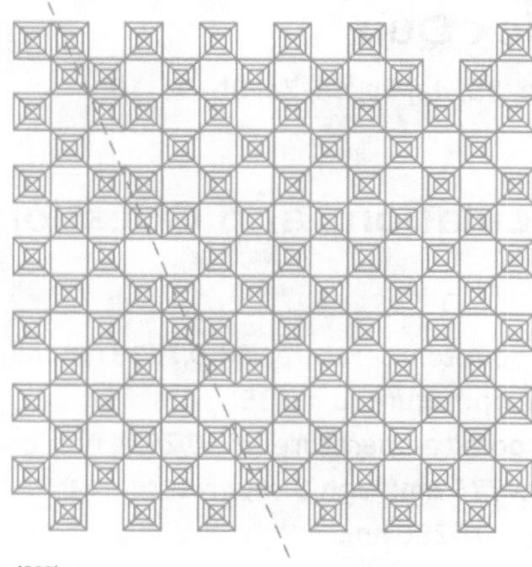

9.6 (a) $Ba_{1.111}Fe_2S_4$, $c = 5.02$ nm (average value); (b) $Ba_{14}(Fe_2S_4)_{13}$, $c = 7.14$ nm (average value); (c) $Sr_8(TiS_3)_7$, $c = 2.092$ nm (average).

Bibliography

GENERAL

Molčanov, K. and Stilinovic, V. (2014). Chemical crystallography before X-ray diffraction. *Angew. Chem. Int. Ed.* **53**: 638–652.

Krivovichev, S.K. (2014). Inorganic crystal structures. *Angew. Chem. Int. Ed.* **53**: 654–661.

Various Authors (2014). Crystallography at 100. *Science* **243** (Special Issue): 1092–1116.

Various Authors (2014). The crystal century. *Nature* **505** (Special Issue): 601–606.

ATOMIC RADII AND SIZE

Alcock, N.W. (1990). *Structure and Bonding.* Chichester: Ellis Horwood.

A large amount of information on atomic size and structure is to be found in many of the chapters in: O'Keeffe, M. and Navrotsky, A. (eds.) (1981). *Structure and Bonding in Crystals*, vol. **I and II.** New York: Academic Press especially:.

O'Keefe, M. and Hyde, B.G. Chapter 10, *The role of nonbonded forces in crystals.*

Baur, W.H. Chapter 15, *Interatomic distance predictions for computer simulation of crystal structures.*

Shannon, R.D. Chapter 16, *Bond distances in sulfides and a preliminary table of sulfide crystal radii.*

BOND-VALENCE

Brown, I.D. (2002). *The Chemical Bond in Inorganic Chemistry*, International Union of Crystallography Monographs on Crystallography No. 12. Oxford: Oxford University Press.

Brese, N. and O'Keeffe, M. (1991). Bond-valence parameters for solids. *Acta Crystallogr.* **B47**: 192–197.

Müller, P., Köpke, S., and Sheldrick, G.M. (2003). Is the bond-valence method able to identify metal atoms in protein structures? *Acta Crystallogr.* **D59**: 32–37.

CRYSTALLOGRAPHY, CRYSTAL CHEMISTRY AND PHYSICS

Bragg, W.L. (1992). *The Development of X-ray Analysis.* New York: Dover.

Giacovazzo, C., Monaco, H.L., Artioli, G. et al. (2002). *Fundamentals of Crystallography*, 2e. Oxford: Oxford University Press.

Hammond, C. (1990). *Introduction to Crystallography*, Royal Microscopical Society Microscopy Handbooks 19. Oxford: Oxford University Press.

Crystals and Crystal Structures, Second Edition. Richard J. D. Tilley.

Hyde, B.G. and Andersson, S. (1989). *Inorganic Crystal Structures.* New York: Wiley-Interscience.

O'Keeffe, M. and Hyde, B.G. (1996). *Crystal Structures, I. Patterns and Symmetry.* Washington, DC: Mineralogical Society of America.

Pearson, W.B. (1972). *The Crystal Chemistry and Physics of Metals and Alloys.* New York: Wiley- Interscience.

Smith, J.V. (1982). *Geometrical and Structural Crystallography.* New York: Wiley.

Wells, A.F. (1984). *Structural Inorganic Chemistry,* 5e. Oxford: Oxford University Press.

Klein, C. and Hurlbut Jr, C.S. (1999). *Manual of Mineralogy* (after J.D. Dana), 21e, revised. New York: John Wiley.

The determination of crystal structures is described by:

Clegg, W. (1998). *Crystal Structure Determination,* Oxford Chemistry Primers, No. 60. Oxford: Oxford University Press.

Massa, W. (ed.) (trans. R.O. Gould) (2001). *Crystal Structure Determination.* New York: Springer.

Viterbo, D. (2002). Chapter 6, *Solution and refinement of crystal structures.* In: *Fundamentals of Crystallography,* 2e (eds. C. Giacovazzo, H.L. Monaco, G. Artioli, et al.), 413–501. Oxford: Oxford University Press.

The point groups of molecules are introduced in:

Shriver, D.F., Atkins, P.W., and Langford, C.H. (1994). *Inorganic Chemistry,* 2e. Oxford: Oxford University Press.

Atkins, P.W. (1994). *Physical Chemistry,* 5e. Oxford: Oxford University Press.

Symmetry is detailed in:

International Union of Crystallography (1987). *International Tables for Crystallography, Vol. A: Space-Group Symmetry,* 2e (ed. T. Hahn). Dordrecht: International Union of Crystallography/D. Reidel.

Magnetic point groups are illustrated in the report:

de Graef, M. (2009). *Teaching Crystallographic and Magnetic Point Group Symmetry Using Three-dimensional Rendered Visualisations.* International Union of Crystallography.

Symmetry in nature is discussed by:

Stewart, I. (2001). *What Shape Is a Snowflake?* London: Weidenfeld and Nicolson.

A survey of Pauling's rules is in:

Bloss, F.D. (1971). *Crystallography and Crystal Chemistry.* New York: Holt, Rinehart and Winston.

Crystal structures in terms of anion-centred polyhedra are discussed by:

O'Keeffe, M. and Hyde, B.G. (1985). An alternative approach to non-molecular crystal structures. *Struct. Bond.* **61**: 77–144.

The physical properties of crystals are discussed in:

Bloss, F.D. (1971). *Crystallography and Crystal Chemistry.* New York: Holt, Rinehart and Winston.

Catti, M. (2002). Chapter 10, *Physical properties of crystals: phenomenology and modelling.* In: *Fundamentals of Crystallography,* 2e (eds. C. Giacovazzo, H.L. Monaco, G. Artioli, et al.), 759–813. Oxford: Oxford University Press.

Nye, J.F. (1985). *Physical Properties of Crystals: Their Representation by Tensors and Matrices.* Oxford University Press.

Electron diffraction and microscopy:

Hirsch, P.B., Howie, A., Nicholson, R.B. et al. (1965). *Electron Microscopy of Thin Crystals.* London: Butterworths.

Williams, D.B. and Carter, C.B. (1996). *Transmission Electron Microscopy.* Kluwer Academic/Plenum.

SFX AND XFELS – FEMTOSECOND CRYSTALLOGRAPHY

Ice, G.E., Budai, J.D., and Pang, J.W.L. (2011). The race to microbeam and nanobeam science. *Science* **224**: 1234–1239.

Mitchell Waldrop, M. (2014). The big guns. *Nature* **505**: 604–606.

Fromme, P. (2015). XFELs open a new era in structural chemical biology. *Nat. Chem. Biol.* **11**: 895–899.

CRYO-EM

Callaway, E. (2015). The revolution will not be crystallized. *Nature* **525**: 172–174.

Raunser, S. (2017). Cryo-EM revolutionizes the structure determination of big molecules. *Angew. Chem. Int. Ed.* **56**: 16450–16452.

Cheng, Y. (2018). Single-particle cryo-EM – how did it get here and where will it go. *Science* **361**: 876–880.

Nobel lectures with respect to cryo-EM:

Henderson, R. (2018). From electron crystallography to single-particle cryo-EM. *Angew. Chem. Int. Ed.* **57**: 10804–10825.

Frank, J. (2018). Single-particle reconstruction of biological molecules – story in a sample. *Angew. Chem. Int. Ed.* **57**: 10836–10841.

Dubochet, J. (2018). On the development of electron cryo-microscopy. *Angew. Chem. Int. Ed.* **57**: 10842–10846.

Incommensurately modulated structures:

Giacovazzo, C. (2002). Chapter 4, *Beyond ideal crystals*. In: *Fundamentals of Crystallography*, 2e (eds. C. Giacovazzo, H.L. Monaco, G. Artioli, et al.), 227–294. Oxford: Oxford University Press.

Janssen, T. and Janner, A. (1987). Incommensurabilty in crystals. *Adv. Phys.* **36**: 519–624.

Makovicky, E. and Hyde, B.G. (1992). *Incommensurate, two-layer structures with complex crystal chemistry: minerals and related synthetics. Mater. Sci. Forum* **100 & 101**: 1–100.

van Smaalen, S. (1995). *Incommenurate crystal structures. Crystallogr. Rev.* **4**: 79–202.

Wiegers, G.A. (1996). Misfit layer compounds: structures and physical properties. *Prog. Solid State Chem.* **24**: 1–139.

Withers, R.L., Schmid, S., and Thompson, J.G. (1998). *Compositionally and/or displacively flexible systems and their underlying crystal chemistry. Prog. Solid State Chem.* **26**: 1–96.

Information concerning incommensurate diffraction patterns is given in:

International Tables for Crystallography, Vol. B (2001). *Reciprocal space*, 2e (ed. U. Schmueli). Dordrecht: Kluwer.

REFERENCES TO $Ba_xFe_2S_4$

Grey, I.E. (1975). *The structure of $Ba_5Fe_9S_{18}$. Acta Crystallogr.* **B31**: 45–48.

Onoda, M. and Kato, K. (1991). Refinement of structures of the composite crystals $Ba_xFe_2S_4$ (x = 10/9 and 9/8) in a four-dimensional formalism. *Acta Crystallogr.* **B47**: 630–634.

and references therein.

REFERENCES TO Sr_xTiS_3

Saeki, M., Ohta, M., Kurashima, K., and Onoda, M. (2002). Composite crystal $Sr_{8/7}TiS_y$ with y = 2.84 – 2.97. *Mater. Res. Bull.* **37**: 1519–1529.

Gourdon, O., Petricek, V., and Evain, M. (2000). A new structure type in the hexagonal perovskite family; structure determination of the modulated misfit compound $Sr_{9/8}TiS_3$. *Acta Crystallogr.* **B56**: 409–418.

and references therein.

MODULAR STRUCTURES:

Baronet, A. (1992). *Polytypism and stacking disorder*, Chapter 7. In: *Reviews in Mineralogy*, vol. **27** (ed. P.R. Busek). Mineralogical Soc. America.

Ferraris, G. (2002). Chapter 7, *Mineral and inorganic crystals*. In: *Fundamentals of Crystallography*, 2e (eds. C. Giacovazzo, H.L. Monaco, G. Artioli, et al.), 503–584. Oxford: Oxford University Press.

Ferraris, G., Mackovicky, E., and Merlino, S. (2004). *Crystallography of Modular Materials*, International Union of Crystallography Monographs on Crystallography No. 15. Oxford: Oxford University Press.

Tilley, R.J.D. (2008). *Defects in Solids*, Chapter 4. Hoboken: Wiley.

Tilley, R.J.D. (2016). *Perovskites*, Chapter 4. Chichester: Wiley.

Veblen, D.R. (1992). *Electron microscopy applied to nonstoichiometry, polysomatism and replacement reactions in minerals*, Chapter 6. In: *Reviews in Mineralogy*, vol. **27** (ed. P.R. Busek). Mineralogical Society America.

Veblen, D.R. (1991). Polysomatism and polysomatic series: a review and applications. *Am. Mineral.* **76**: 801–826.

NETS:

O'Keeffe, M. and Hyde, B.G. (1980). Plane nets in crystal chemistry. *Phil. Trans. R. Soc. Lond.* **295**: 553–623.

Delgardo-Friedrichs, O., Foster, M.D., O'Keeffe, M. et al. (2005). *What do we know about three-periodic nets? J. Solid State Chem.* **178**: 2533–2544.

Delgardo-Friedrichs, O. and O'Keeffe, M. (2005). Crystal nets as graphs: terminology and definitions. *J. Solid State Chem.* **178**: 2480–2485.

PROTEINS

The first reported protein structure was myoglobin:

Kendrew, J.C., Bodo, G., Dintzis, H.M. et al. (1958). A three-dimensional model of the myoglobin molecule obtained by X-ray analysis. *Nature* **181**: 662–666.

Whitford, D. (2005). *Proteins, Structure and Function*. Chichester: Wiley.

Viterbo, D. (2002). Chapter 6, *Solution and refinement of crystal structures*. In: *Fundamentals of Crystallography*, 2e (eds. C. Giacovazzo, H.L. Monaco, G. Artioli, et al.). Oxford: Oxford University Press.

Zanotti, G. (2002). Chapter 9, *Protein crystallography*. In: *Fundamentals of Crystallography*, 2e (eds. C. Giacovazzo, H.L. Monaco, G. Artioli, et al.). Oxford: Oxford University Press.

QUASICRYSTALS:

The first report of a quasicrystal was:

Schechtman, D., Blech, I., Gratias, D., and Cahn, J. W. (1984). Metallic phase with long-range orientational order and no translational symmetry. *Phys. Rev. Lett.* **53**: 1951–1953.

Giacovazzo, C. (2002). Chapter 4, *Beyond ideal crystals*. In: *Fundamentals of Crystallography*, 2e (eds. C. Giacovazzo, H.L. Monaco, G. Artioli, et al.), 227–294. Oxford: Oxford University Press.

Nelson, D.R. (1986). *Quasicrystals. Sci. Am.* **255** (2): p32.

Stephens, P.W. and Goldman, A.I. (1991). *The structure of quasicrystals. Sci. Am.* **264** (4): p24.

Yamamoto, A. (1996). Crystallography of quasiperiodic crystals. *Acta Crystallogr.* **A52**: 509–560.

Senechal, M. (1995). *Quasicrystals and Geometry.* Cambridge: Cambridge University Press.

The linking of core-shell nanoparticles to form quasicrystals is described by:

Wu, S. and Sun, Y. (2018). Tesselating tiny tetrahedrons. *Science* **362**: 1354.

Nagaoka, Y., Zhu, H., Eggerts, D., and Chen, O. (2018). Single-component quasicrystalline nanocrystal superlattices through flexible polygon tiling rule. *Science* **362**: 1396.

TILING:

Gardiner, M. (1975). *Mathematical Games. Sci. Am.* **233** [1] July: p112. [This was used to construct Figure 3.18.].

Gardner, M. (1975). *Mathematical Games. Sci. Am.* **233** [2] August: p112.

Gardner, M. (1977). *Mathematical Games. Sci. Am.* **236**[1] January: p110. [This was used to construct Figures 3.17 and 3.18.].

Giacovazzo, C. (2002). Chapter 4, *Beyond ideal crystals.* In: *Fundamentals of Crystallography*, 2e (eds. C. Giacovazzo, H.L. Monaco, G. Artioli, et al.), 227–294. Oxford: Oxford University Press.

Grünbaum, B. and Shephard, G.C. (1987). *Tilings and Patterns.* New York: W.H. Freeman.

Index

Crystals and Crystal Structures, Second Edition. Richard J. D. Tilley.

D

E

F

Q

R

S

T

U

V

W

Z

www.ingramcontent.com/pod-product-compliance
Lightning Source LLC
Chambersburg PA
CBHW080551270125
20834CB00018B/236

* 9 7 8 1 1 1 9 5 4 8 3 8 6 *